REVIEW AND EVALUATION OF ALTERNATIVE CHEMICAL DISPOSAL TECHNOLOGIES

Panel on Review and Evaluation
of Alternative Chemical Disposal Technologies
Board on Army Science and Technology
Commission on Engineering and Technical Systems
National Research Council

NATIONAL ACADEMY PRESS
WASHINGTON, D.C. 1996

NOTICE: The project that is the subject of this report was approved by the Governing Board of the National Research Council, whose members are drawn from the councils of the National Academy of Sciences, the National Academy of Engineering, and the Institute of Medicine. The members of the committee responsible for the report were chosen for their special competencies and with regard for appropriate balance.

This report has been reviewed by a group other than the authors according to procedures approved by a Report Review Committee consisting of members of the National Academy of Sciences, the National Academy of Engineering, and the Institute of Medicine.

The National Academy of Sciences is a private, nonprofit, self-perpetuating society of distinguished scholars engaged in scientific and engineering research, dedicated to the furtherance of science and technology and to their use for the general welfare. Upon the authority of the charter granted to it by the Congress in 1863, the Academy has a mandate that requires it to advise the federal government on scientific and technical matters. Dr. Bruce Alberts is president of the National Academy of Sciences.

The National Academy of Engineering was established in 1964, under the charter of the National Academy of Sciences, as a parallel organization of outstanding engineers. It is autonomous in its administration and in the selection of its members, sharing with the National Academy of Sciences the responsibility for advising the federal government. The National Academy of Engineering also sponsors engineering programs aimed at meeting national needs, encourages education and research, and recognizes the superior achievements of engineers. Dr. William A. Wulf is interim president of the National Academy of Engineering.

The Institute of Medicine was established in 1970 by the National Academy of Sciences to secure the services of eminent members of appropriate professions in the examination of policy matters pertaining to the health of the public. The Institute acts under the responsibility given to the National Academy of Sciences by its congressional charter to be an adviser to the federal government and, upon its own initiative, to identify issues of medical care, research, and education. Dr. Kenneth I. Shine is president of the Institute of Medicine.

The National Research Council was organized by the National Academy of Sciences in 1916 to associate the broad community of science and technology with the Academy's purposes of furthering knowledge and advising the federal government. Functioning in accordance with general policies determined by the Academy, the Council has become the principal operating agency of both the National Academy of Sciences and the National Academy of Engineering in providing services to the government, the public, and the scientific and engineering communities. The council is administered jointly by both Academies and the Institute of Medicine. Dr. Bruce M. Alberts and Dr. William A. Wulf are chairman and interim vice chairman, respectively, of the National Research Council.

This is a report of work supported by Contract DAAH04-95-C-0049 between the U.S. Army and the National Academy of Sciences.

Library of Congress Catalog Card Number 96-61747
International Standard Book Number 0-309-05525-3

Copies available from the:
National Academy Press
2101 Constitution Avenue, N.W.
Box 285
Washington, DC 20418
800-624-6242, 202-334-3313 (in the Washington Metropolitan Area)

Printed in the United States of America.

PANEL ON REVIEW AND EVALUATION OF ALTERNATIVE CHEMICAL DISPOSAL TECHNOLOGIES

RICHARD S. MAGEE, *Chair*, New Jersey Institute of Technology, Newark
JOAN B. BERKOWITZ, Farkas Berkowitz & Company, Washington D.C.
GENE H. DYER, Consultant, San Rafael, California
FREDERICK T. HARPER, Sandia National Laboratories, Albuquerque, New Mexico
JOSEPH A. HEINTZ, Consultant, Schererville, Indiana
DAVID A. HOECKE, Enercon Systems Inc., Elyria, Ohio
DAVID S. KOSSON, Rutgers, the State University of New Jersey, Piscataway
WALTER G. MAY, University of Illinois, Urbana
ALVIN H. MUSHKATEL, Arizona State University, Tempe
LAURANCE ODEN, U.S. Bureau of Mines (retired), Albany, Oregon
GEORGE W. PARSHALL, Dupont Company (retired), Wilmington, Delaware
L. DAVID PYE, Alfred University, Alfred, New York
ROGER W. STAEHLE, Consultant, North Oaks, Minnesota
WILLIAM TUMAS, Los Alamos National Laboratory, Los Alamos, New Mexico

Board on Army Science and Technology Liaison

ROBERT A. BEAUDET, University of Southern California, Los Angeles

Staff

BRUCE A. BRAUN, Director, Division of Military Science and Technology
MICHAEL A. CLARKE, Study Director
ROBERT J. KATT, Technical Writer/Consultant
MARGO L. FRANCESCO, Administrative Associate
DEBORAH B. RANDALL, Senior Secretary/Project Assistant

Preface

In 1985, Public Law 99-145 mandated an "expedited" effort to dispose of M55 rockets containing unitary chemical warfare agents because of the potential for self-ignition of these particularly hazardous munitions during storage. This program soon expanded into the Army Chemical Stockpile Disposal Program (CSDP), whose mission was to eliminate the entire stockpile of unitary chemical weapons. The CSDP developed the current baseline incineration system. In 1992, after setting several intermediate goals and dates, Congress enacted Public Law 102-484, which directed the Army to dispose of the entire stockpile of unitary chemical warfare agents and munitions by December 31, 2004. Since 1987, the Committee on Review and Evaluation of the Army Chemical Stockpile Disposal Program (the Stockpile Committee) of the National Research Council (NRC) has overseen the Army's disposal program and has endorsed the baseline incineration process as an adequate technology for destroying the stockpile.

Growing public concerns about and opposition to incineration, coupled with the rising cost of the CSDP, have raised interest in alternatives. The Stockpile Committee, which has been following the state of alternative technologies, reviewed a NRC study of alternative technologies by a separate NRC committee and in 1994 recommended that the Army continue research on neutralization.

In the summer of 1995, the assistant secretary of the Army for research, development and acquisition informally explored the issue of examining alternative chemical disposal technologies with the Stockpile Committee. Following numerous discussions between the Army and the NRC, a decision was made to conduct a new NRC study to reexamine the status of a limited number of maturing alternative chemical disposal technologies (including the two neutralization-based processes on which the Army was currently conducting research) for possible implementation at the two bulk-storage sites at Aberdeen Proving Ground, Maryland, and the Newport Chemical Activity, Indiana.

The NRC established the Panel on Review and Evaluation of Alternative Disposal Technologies (the AltTech Panel) to conduct the new study. The panel includes six members of the Stockpile Committee, who have accumulated experience in dealing with the complex issues involved in monitoring the destruction of the unitary chemical agent stockpile, and eight new members who possess specific expertise for thoroughly evaluating the alternative technologies.

The panel received detailed briefings from the Army and the three companies that had proposed alternative technologies for the Army's consideration (hereafter, the technology proponent companies, or TPCs). Before the briefings on individual technologies, the panel compiled a questionnaire to elicit information needed to evaluate the technologies on a range of factors. The questionnaire was sent to the TPCs and to the Army team for neutralization-based technologies. The responses to the questionnaires and subsequent follow-up conversations were supplemented with site visits by teams of panel members to inspect each TPC's technology.

In addition to gathering technical information on the alternative technologies, the AltTech Panel met with members of the public from the communities near the Aberdeen and Newport sites. These meetings included public forums, which were open to all, and meetings with the Citizens Advisory Commissions for Maryland and Indiana. (These commissions are formal groups established as a channel of communication with communities near stockpile sites.) The panel also met with regulators from the state agencies responsible for review and approval of permits required by agent-destruction facilities and for implementing other relevant regulations and state laws.

Parallel with the AltTech Panel activities and under Army supervision, the TPCs conducted small-scale tests of their technologies on actual chemical agent. The Army also contracted with MitreTek Systems, Inc., to perform a preliminary accident hazard assessment for each technology. The test results and the contractor's report were provided to the panel for consideration.

The activities described above formed the basis for the findings and recommendations in this report.

To the members of the Stockpile Committee who agreed to perform double duty by serving on the AltTech Panel, I owe a great deal of gratitude. To the new members, I want to express my appreciation for the fresh insights they provided. Without their help, the evaluations would have suffered. I thank all these volunteers for the time and energy they contributed at the expense of other responsibilities. The travel and inconvenience of conducting a fast-track study were considerable; each member spent a great deal of time analyzing information, arriving at consensus evaluations and judgments, and capturing the results in writing. On behalf of the National Research Council, I thank each of them.

The AltTech panel recognizes and appreciates the substantial support provided by the Army staff and the program office for chemical demilitarization. The panel also recognizes the efforts of the TPCs. You were all cordial, responsive, forthcoming, and generous with your time. Thank you.

The panel greatly appreciates the support of panel activities and the timely production of the report by NRC staff members Michael Clarke, Margo Francesco, and Deborah Randall as well as the services of the reports officer of the Commission on Engineering and Technical Systems, Carol Arenberg, the consulting technical writer, Robert Katt, the electronic composition by Mary Beth Mason and Sally Naas and the graphics by consultant James Butler.

Richard S. Magee, Chair
Panel on Review and Evaluation of
Alternative Chemical Disposal Technologies

Contents

APPENDICES

Tables and Figures

TABLES

FIGURES

Abbreviations and Acronyms

ACAMS	automatic continuous air monitoring system
APG	Aberdeen Proving Ground
ASME	American Society of Mechanical Engineers
ALTTECH	Panel on Review and Evaluation of Alternative Chemical Disposal Technologies
BDAT	best demonstrated available technology
CAC	Citizens Advisory Commission
CAIN	Citizen Against Incineration at Newport
CEP™	catalytic extraction processing
CFR	Code of Federal Regulations
CPU	catalytic processing unit
CSDP	Army Chemical Stockpile Disposal Program
CSEPP	Chemical Stockpile Emergency Preparedness Program
CWC	Chemical Weapons Convention
CWWG	Chemical Weapons Working Group
DAAMS	depot area agent monitoring system
DC	direct current
DCS	distributed control system
DDT	dichlorodiphenyltrichloroethane
DRE	destruction removal efficiency
EMPA	ethylmethylphosphonic acid
EPA	Environmental Protection Agency
FMEA	failure modes and effects analysis
FOTW	federally owned treatment works
GB	sarin (a nerve agent, o-isopropylmethylphosphonofluoridate)
GC/MS	gas chromatography followed by mass spectrometry
HD	distilled mustard agent, bis(2-chloroethyl sulfide)
HLE	high level exposure (a statutory standard for exposure to an airborne hazardous substance)
HRT	hydraulic residence time
HVAC	heating, ventilation, and air conditioning
IDLH	immediately dangerous to life and health (a statutory standard for exposure to an airborne hazardous substance
JACADS	Johnston Atoll Chemical Agent Disposal System
LD50	lethal dose to 50 percent of a test population
MEA	monolethanolamine
MLSS	mixed liquors suspended solids
MPA	methylphosphonic acid
MPL	maximum permissible limit (a statutory standard for exposure to an airborne hazardous substance)

NPDES	national pollutant discharge elimination system
NRC	National Research Council
OPMAT&A	Office of the Product Manager for Alternative Technologies and Approaches
PCB	polychlorinated biphenyl
PMCD	program manager for chemical demilitarization
POTW	publicly owned treatment works
PPB	parts per billion
PPE	personal protective equipment
RCRA	Resource Conservation and Recovery Act
RPU	radioactive processing unit
SBR	sequencing batch reactor
SBV	sequential batch vaporizer
SCADA	supervisory control and data acquisition
SRT	solid residence time
TAGA	trace atmospheric Gas Analyzer
TPC	technology proponent company
TSDF	treatment, storage, and disposal facility
TWA	time weighted average
TOCDF	Tooele Chemical Agent Disposal Facility
VOC	volatile organic compound
VX	a nerve agent (O-ethyl-S[2-diisopropyl amino)ethyl] methylphosphonothiolate

Executive Summary

Congress has assigned the U.S. Army the responsibility for destroying the stockpile of aging unitary chemical warfare agents. Of the eight sites in the contiguous United States where chemical weapons are stockpiled, two sites contain only one type of agent each, which is stored only in bulk containers called "ton containers." These two sites are Aberdeen Proving Ground, Maryland, and the Newport Chemical Activity, Indiana. These two sites contain about 9.5 percent of the total stockpile. The remainder of the stockpile contains a complex mix of agents and explosive-configured agent-containing weapons. To destroy all types of agent-containing munitions at all the stockpile sites, as well as the ton containers of agents, the Army has developed a complete processing system, called the baseline system, which uses incineration technology in four separate process streams to destroy chemical agents, energetics (explosives and propellants), and dunnage (e.g., packaging materials) and to decontaminate metal containers and parts.

In August 1995, the Army advertised for information on technologies not resembling incineration that were sufficiently developed to be considered as options for destruction of the stockpiles at Aberdeen and Newport. In November 1995, a contractor hired by the Army selected three technologies that best met the Army's advertised selection criteria. The Army asked the National Research Council to conduct a technical review of these three alternative technologies and two alternatives the Army had been pursuing on its own. The Army intends to use this technical review as one factor in deciding whether to proceed with pilot-testing of one or more alternative technologies at Aberdeen and Newport. The Army plans to present its recommendations to the Department of Defense in October 1996. The National Research Council was not asked to compare the alternative technologies with the baseline system. Nor was it asked to consider the application of the alternatives to other stockpile sites.

The three technologies selected from the submitted information were (1) a process that uses a high-temperature, molten-metal bath to break complex compounds (such as chemical warfare agents) into simple substances; (2) electrochemical oxidation mediated by ionic silver in aqueous solution; and (3) gas-phase chemical reduction with high-temperature hydrogen and steam. The two technologies from the Army program were (1) stand-alone neutralization, which is a chemical hydrolysis that breaks agent molecules into two fragments that are far less toxic than the agent and (2) neutralization followed by biodegradation. (Biodegradation here refers to using microorganisms to break down the fragments from chemical hydrolysis into simpler compounds that are not hazardous to humans or the environment.)

THE ALTTECH PANEL

To conduct the review requested by the Army, the National Research Council formed the Panel on Review and Evaluation of Alternative Chemical Disposal Technologies (AltTech Panel). This report contains the panel's findings and recommendations. It also details the factual data, the information supplied by the proponent for each technology, and the analyses and arguments that support the findings and recommendations. Chapter 1 describes the context for the panel's work, including the history of the Chemical Stockpile Disposal Program, the role of the National Research Council and its committees in reviewing and advising that program, the nature of the agent stockpiles at Aberdeen and Newport, and the Army Alternative Technology Program. Chapter 2 is a discussion of the broad set of evaluation factors that the panel assembled for organizing information about the five alternatives with respect to (1) the technical requirements of agent-destruction processes; (2) safety, health, and environmental considerations; and (3) the implications of these requirements and considerations for the time required to implement each technology as a fully operational, yet fully tested and proven, facility to destroy chemical agents at Newport or Aberdeen.

EVALUATING THE ALTERNATIVE TECHNOLOGIES

The panel had to do much more than evaluate the conceptual design packages submitted by companies that advocated alternative technologies or by the Army (in the case of the two neutralization alternatives). To acquire as much information as possible that would be relevant to the evaluation, the panel sent a lengthy questionnaire to each technology proponent company (TPC) and to the Army, to which they responded in writing. Chapter 3 describes the development of the TPC questionnaire as a framework for gathering information. Teams of panel members followed up the questionnaire with visits to the facilities or demonstration sites of the TPCs. These teams conducted probing interactions with the TPCs, consisting of a series of written or verbal questions, requests for further information, and face-to-face inquiries during site visits.

Chapters 4, 5, and 6 summarize what the panel learned about the three technologies selected for review. The panel decided that the alternatives proposed by the Army for neutralization and neutralization followed by biodegradation should be evaluated with respect to specific chemical agents. Therefore, Chapter 7 discusses neutralization and biodegradation options for the blister agent called mustard or HD, which is the only agent stockpiled at Aberdeen. Chapter 8 does the same for the nerve agent VX, which is the only agent stored at Newport. These five technical chapters are similar in format; after a short introduction to the technology, each chapter presents the scientific principles underlying the agent-destruction process, the developmental status of the technology, operational requirements and other detailed process considerations, instrumentation and control, stability and reliability of the process, materials of construction, utility and scale-up requirements, safety issues, and an estimate of the time required to completely destroy the stockpiles at the two sites.

NEIGHBORING COMMUNITIES AND STATE REGULATORS

The most significant impetus for seeking alternative technologies to destroy chemical agents has been opposition to incineration—and support for an alternative—by members of the communities around the stockpile sites. Fully aware of the importance placed on community involvement in previous stockpile- related reports by National Research Council committees and others, the AltTech Panel decided that the views and values of these communities were important to consider in the panel's criteria for comparing technologies. Chapter 9 describes the open forums conducted by the panel in the communities near the Aberdeen and Newport sites and the meetings with citizen commissions set up in each state as part of the Army's public participation efforts. The chapter explains how the panel interpreted the opinions it heard and how they relate to the evaluation criteria. Also summarized are meetings with Indiana and Maryland regulators who will be evaluating the permit applications required for any agent-destruction facility to be pilot-tested or operated at full-scale in their states.

FINDINGS AND RECOMMENDATIONS

After six months of intensive information-gathering from the TPCs and the affected communities, the panel honed the broad set of evaluation factors to a tighter set of evaluation criteria. These criteria focus on characteristics that differentiate among the candidate technologies with respect to process performance and engineering; concerns about safety, health, and the environment; and the implications of the preceding factors for the time required to destroy the stockpiles. Chapter 10 explains the criteria and presents summary evaluations of each candidate technology. These crosscutting evaluations are the basis for the panel's findings and recommendations, which are listed in abbreviated form below. Chapter 11 contains the full statement of the findings and recommendations, together with supporting narrative.

General Findings

General Finding 1. Since the 1993 National Research Council report, *Alternative Technologies for the Destruction of Chemical Agents and Munitions,* there has been sufficient development to warrant re-evaluation of alternative technologies for chemical agent destruction. Because the developmental status of the technologies varies widely, the time required to complete pilot demonstrations will also vary.

General Finding 2. All the technologies selected for the panel to review have successfully demonstrated the ability to destroy agent at laboratory scale.

General Finding 3. Members of the communities near the Aberdeen and Newport sites want an alternative to incineration that has the following characteristics: operation at low temperature and low pressure; simplicity; the capability of testing all process residuals prior to release; and minimal potential for detrimental effects, short term or long term, on public health and the environment. Although the communities do not want treaty or legislative schedules to drive decisions on technology options, they want the stockpiles at the two sites to be destroyed as quickly as possible.

General Finding 4. Based on the panel's discussions with state regulators, all the technologies appear to be permittable under the Resource Conservation and Recovery Act and associated state regulations within one to two years of submitting the applications. The actual time will depend on the complexity of the technology and the regulators' familiarity with it.

General Finding 5. As complete processing systems for chemical agent, all the technologies reviewed are of moderate to high complexity. Although components of each process are standard and proven, no alternative is an off-the-shelf solution as an agent-destruction process. Any one of them will require extensive design review, hazard and operability studies, materials selection, and related work as it moves through the piloting stage to full-scale demonstration and operation. During this necessary preparation for implementing an agent-destruction system, everyone involved should bear in mind that most failures in complex, engineered systems occur not during steady-state, normal operations but during transient conditions such as startup, shutdown, or operator responses to deviations from design conditions.

Specific Findings and Recommendations

Specific Finding 1. The Army required each TPC to demonstrate the capacity of its processes to destroy agents in a government-approved laboratory. Each TPC supplied test results to the panel indicating it had successfully destroyed both blister (HD) and nerve (VX) agents. Due to time constraints, the panel was not able to review and analyze in depth the data from these important tests. However, two key issues stand out.

First, the tests were conducted under conditions of varying similarity to conditions in a pilot-scale or fully operational facility.[1] It is therefore inappropriate to expect that the particular destruction removal efficiencies (DREs) attained in the tests would be the same as DREs attained in an operating facility.[1] It is also inappropriate to compare technologies only on the basis of DRE results. Given the lack of comparability between the test conditions and scaled-up facility for an individual technology and the differences in test conditions for different technologies, the panel has used the test results only to address, in yes-or-no fashion, whether a technology can destroy agent.

Second, the by-products of any agent-destruction process are of significant concern to the panel, the neighboring communities, and the regulators. A DRE value gives no information on the composition and concentration of by-products that may be hazardous to human health or the environment. An in-depth, independent analysis of these test data will be necessary to support future Department of Defense decisions about proceeding with pilot-testing. This analysis may show that further independent testing is needed.

Recommendation 1. For any technology that is to be pilot-tested, the Army should support an in-depth analysis of the agent-destruction test results by a competent, independent third party not associated with the Army or any of the TPCs.

Specific Finding 2. Current Army prohibitions on the off-site treatment and disposal of process residuals unduly restrict the options for stockpile destruction. No toxicologic, or risk, basis for the proposed Army release standards has been developed. In addition, there appears to be an inconsistency among the limits for airborne exposure and residual concentrations in liquid and solid materials that are to be released from toxic handling facilities to off-site facilities for subsequent treatment and disposal.

Recommendation 2a. Standards for releasing wastes should be evaluated on a clearly defined regulatory and risk basis that takes existing practices into account. Standards should be revised or established as necessary.

---

[1]DRE is calculated as the percentage of agent destroyed or removed. A DRE of 99.99 percent is often referred to as "four 9's," a DRE of 99.9999 percent as "six 9's," and so on.

Recommendation 2b. The Army should review and revise current restrictions on off-site treatment and disposal of process liquid and solid residual streams to allow treatment and disposal of the process effluents from agent destruction at permitted off-site treatment, storage, and disposal facilities and at permitted federally owned treatment works for wastewater.

Specific Finding 3. The panel determined that the development status of the technologies assessed and the lack of long-term experience with their use for the destruction of chemical agent necessitate a comprehensive design review of any selected technology prior to the construction of a pilot plant. Reliability of the facility, as affected by system design, control, operation, maintenance, monitoring, and material selection, must be thoroughly evaluated.

Recommendation 3. A detailed, comprehensive design review of any selected technology or technologies should be performed prior to starting pilot-plant construction. This review should examine reliability as affected by system design, controls, operation, maintenance, monitoring, and materials selection.

Specific Finding 4. The panel has found that, no matter which technology is selected for potential use at either site, the affected communities insist that they be included in a meaningful way in the process leading up to key decisions, including the decision to proceed to pilot demonstration.

Recommendation 4. The Army should take immediate steps, if it has not already done so, to involve the communities around the Aberdeen and Newport sites in a meaningful way in the process leading up to the Army recommendation to the Defense Acquisition Board on whether to pilot-test one or more alternative technologies.

Specific Finding 5. The results of independent risk assessments performed on the alternative technologies at the same time as this study were not available to the AltTech Panel until very late in the preparation of this report. The panel assumes that more rigorous, site-specific assessments will be done at an appropriate time before a full-scale facility for agent destruction is built and operations on agent begin. The required assessments include a quantitative risk assessment and a health and environmental risk assessment.

Recommendation 5. Before any technology is implemented at a stockpile site, an independent, site-specific quantitative risk assessment and a health and environmental risk assessment should be completed, evaluated, and used in the Army's risk management program.

HD at Aberdeen

Specific Finding 6. Aqueous neutralization of the chemical agent HD followed by biodegradation of the hydrolysate surpasses the other alternative technologies with respect to the panel's priority criteria (see Chapter 11).

Recommendation 6. The Army should demonstrate the neutralization of HD at Aberdeen on a pilot scale.

- The AltTech Panel recommends biodegradation of hydrolysate from HD at an off-site treatment, storage, and disposal facility as the most attractive neutralization configuration presented for review.

- The second best configuration is neutralization with biodegradation on site, followed by disposal of the aqueous effluent through a federally owned treatment works. If this option is selected, the panel recommends separating the volatile organic compounds prior to biodegradation, followed by off-site treatment and disposal of these compounds.

VX at Newport

Specific Finding 7. Neutralization of chemical agent VX with sodium hydroxide solution destroys agent effectively and substantially lowers the toxicity of the process stream. With respect to the panel's priority criteria (Chapter 11), this technology followed by off-site treatment and disposal of the hydrolysate has the same relative advantages as neutralization of HD. One difference, however, is the uncertainty about the appropriate disposal method for VX hydrolysate. It is possible, although not yet established by adequate testing, that the hydrolysate has sufficiently low toxicity associated with its organic products that complete biodegradation prior to discharge may not be necessary. Furthermore, treatment of VX hydrolysate by existing processes other than biodegradation is likely to be possible. The residual concentrations of agent or agent precursors allowable under the Chemical Weapons Convention are likely to be less stringent than the concentrations required by the environmental permits for the destruction and downstream disposal facilities.

Recommendation 7a. The Army should pilot-test VX neutralization followed by off-site treatment of the hydrolysate at a permitted treatment, storage, and disposal facility, for potential use at the Newport site, but only if the effluent discharged from the off-site facility has been shown to have acceptably low toxicity and to result in minimal environmental burden.

Recommendation 7b. If on-site disposal of VX hydrolysate is preferred to shipping it off site for treatment, existing commercial processes other than biodegradation should be considered. The panel does not recommend on-site biodegradation because of the need for cofeeding a substantial amount of carbon substrate and because of limited success to date in testing on-site biodegradation.

Specific Finding 8. Electrochemical oxidation is the next best alternative for destroying VX at the Newport site. Although the developmental status of this technology is not as advanced as the status of other technologies considered, the panel is confident that the remaining development can lead to a successful pilot demonstration.

Recommendation 8. If successful off-site treatment of VX hydrolysate at an existing treatment, storage, and disposal facility is not confirmed by appropriate treatability studies, and successful on-site treatment of VX hydrolysate with existing commercial processes cannot be demonstrated, then the Army should pilot-test the electrochemical oxidation of VX for potential use at the Newport site.

1

Introduction

THE CALL FOR DISPOSAL

The United States has maintained a stockpile of highly toxic chemical agents and munitions for more than half a century. Chemical agents are extremely hazardous, which is why they have been used in weapons. The manufacture of chemical agents and munitions and their subsequent stockpiling were undertaken in the belief that they had value as deterrents to the use of similar materials against U.S. forces. Today, other deterrents are considered more appropriate. In an attempt to avoid the worldwide risk posed by chemical warfare, the United States is entering into an agreement with many other nations to rid the world of all chemical weapons and munitions. Even apart from this agreement, the United States can no longer justify the continuing risk and expense of storing them. Consequently, there is ample incentive for the United States to dispose of its chemical agents and munitions as soon as this can be done safely.

In 1985, Public Law 99-145 mandated an "expedited" effort to dispose of M55 rockets because these particularly hazardous munitions have the potential for self-ignition during storage. The M55 rockets are loaded with chemical agent, a fuse, an explosive designed to disperse the agent (a burster), and ignition-ready rocket propellant. This mandate soon expanded into the Army Chemical Stockpile Disposal Program (CSDP), whose mission was to eliminate the entire stockpile of unitary[1] chemical weapons. The CSDP developed the current baseline incineration system for this purpose. In 1992, after setting several intermediate goals and dates, Congress enacted Public Law 102-484, which directed the Army to dispose of the entire unitary chemical warfare agent and munitions stockpile by December 31, 2004.

DESCRIPTION OF THE STOCKPILE

Agents

The principal unitary chemical agents in the U.S. stockpile are the two nerve agents (GB and VX)[2] and three related forms of blister, or mustard, agent (H, HD, and HT). These agents are stored and exist largely as liquids: nerve agent VX, a high-boiling point liquid that will adhere to surfaces for days or weeks; nerve agent GB (sarin), a liquid that has a volatility similar to water and therefore evaporates relatively quickly; and a blister agent (mustard) that evaporates slowly. These agents are stored in a variety of munitions and containers. The stockpile consists of 30,600 tons of unitary agents (U.S. Army, 1996h).

Nerve agents are organophosphonate compounds; that is, they contain phosphorus double-bonded to an oxygen atom and single-bonded to a carbon atom. They are highly toxic and lethal in both liquid and vapor forms. They can kill in a matter of minutes by interfering with respiratory and nervous system functions. In pure form, nerve agents are practically colorless and odorless. GB evaporates at about the same rate as water and is relatively nonpersistent in the environment. VX evaporates much more slowly and can persist for a long time under average weather conditions.

Bis(2-chloroethyl)sulfide is the principal active ingredient in blister agents, or mustard.[3] Mustard has a

---

[1]The term unitary distinguishes a single chemical loaded in munitions or stored as a lethal material. More recently, binary munitions have been produced in which two relatively safe chemicals are loaded in separate compartments to be mixed to form a lethal agent after the munition is fired or released. The components of binary munitions are stockpiled in separate states. They are not included in the present CSDP. However, under the Chemical Weapons Convention of 1993, they are included in the munitions that will be destroyed.

[2]GB is O-isopropyl methylphosphonofluoridate. VX is O-ethyl-S[2-(diisopropyl amino) ethyl]-methylphosphonothiolate.

[3]Names such as mustard gas, sulfur mustard, and yperite have also been applied to this agent. The term mustard "gas" is often used, but the chemical is a liquid at ambient temperature.

TABLE 1-1 Physical Properties of Chemical Warfare Agents

Agent Characteristic	VX (Nerve Agent)	HD (Blister Agent)
Chemical formula	$C_{11}H_{26}NO_2PS$	$(ClCH_2CH_2)_2S$
Molecular weight	267.38	159.08
Boiling point, °C	298	217
Freezing point, °C	< –51	14.45
Vapor pressure, mm Hg	0.0007 @ 25°C	0.072 @ 20°C
Volatility, mg/m^3	10.5 @ 25°C	75 @ 0°C (solid) 610 @ 20°C (liquid)
Surface tension, dynes/cm	32.0 @ 20°C	43.2 @ 20°C
Viscosity, cS	12.256 @ 20°C	3.95 @ 20°C
Liquid density g/cm^3 at 20°C	1.0083	1.2685
Solubility, g/100 g of distilled water	5 @ 25°C; best solvents are dilute mineral acids	0.92 @ 22°C; soluble in acetone, CCl_4, CH_3Cl, tetrachloroethane, ethyl benzoate, ether
Heat of combustion Btu/lb (cal/g)	15,000 (8.33)	8,100 (4.5)

Source: NRC, 1993.

garlic-like odor and is hazardous on contact and as a vapor. Because it is practically insoluble in water, mustard is very persistent in the environment. Table 1-1 lists some of the physical properties of VX and HD.

Containers and Munitions

Unitary chemical agents are stored in spray tanks, bulk storage (ton) containers,[4] and a variety of munitions including land mines, M55 rockets, bombs, and artillery and mortar projectiles. Some munitions contain no explosives or propellant, whereas others contain some combination of fuse, booster, burster, and propellant. These components are referred to collectively as energetics. They incorporate a variety of chemical compounds that must also be eliminated as part of the CSDP.

Geographical Distribution

The unitary chemical stockpile is located at eight continental U.S. storage sites (see Figure 1-1) and at Johnston Atoll in the Pacific Ocean about 700 miles southwest of Hawaii. Table 1-2 gives the composition of the stockpile at each continental U.S. site by type of container or munition and by type of agent.

As specified in the study panel's statement of task, only the two sites at Aberdeen Proving Ground, Maryland, (Aberdeen site) and at the Newport Chemical

[4]Although bulk containers are commonly referred to as "ton containers," they actually weigh 635.6 kg (1400 lb.) empty and contain an additional 681 to 726 kg (1500 to 1600 lb.) of agent. The total weight is approximately 1407 kg (3100 lb.) (U.S. Army, 1988).

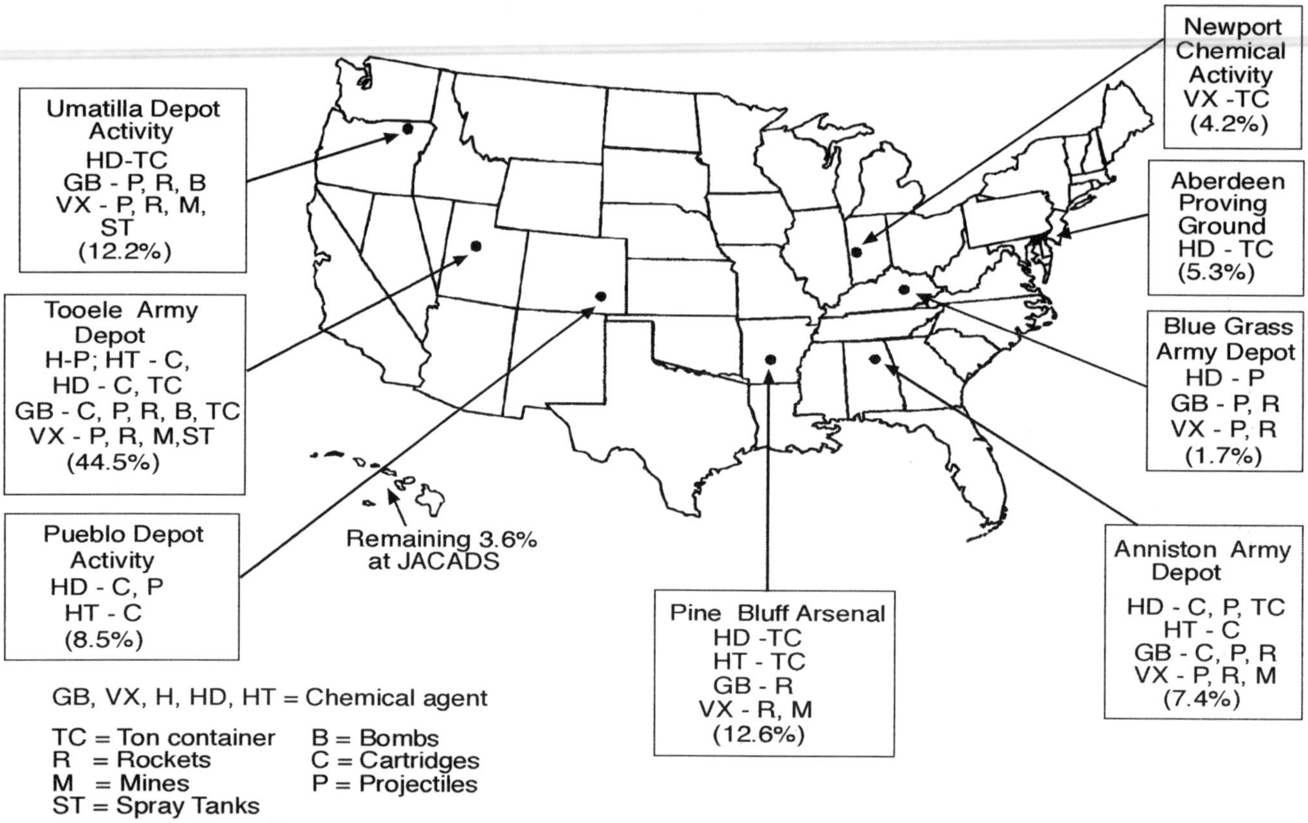

FIGURE 1-1 Types of agent and munitions and percentage of total agent stockpile at each storage site. Derived from OTA, 1992; NRC, 1996.

Activity, Indiana, (Newport site) are considered in this report. The unitary agent stockpile at the Aberdeen site consists entirely of HD (1,625 tons in 1,818 ton containers), and the stockpile at the Newport site consists entirely of VX (1,269 tons in 1,689 ton containers) (U.S. Army, 1996h). Because munitions containing agent and energetics are not present, the process requirements for disposing of only ton containers of agent are less demanding than the processing requirements for the more complex stockpiles at other sites.

The VX nerve agent stored at the Newport site is 90.5 to 94.8 percent pure. It was formulated with 1 to 3 percent diisopropyl carbodiimide as a stabilizer to protect it against decomposition by traces of water. During the 30 to 40 years that the VX has been in storage, some of the stabilizer has hydrolyzed, but most of the nerve agent has not been affected. Traces of a toxic compound, called "pyro,"[5] are present from VX hydrolysis. An impurity, called "bis," which is formed during VX manufacture, hydrolyzes to give EA-2192, which is also highly toxic. In a recent survey conducted by the Army, gas chromatographic analysis of the materials in containers of VX (32 containers were randomly selected and sampled) revealed the presence of the compounds shown in Table 1-3 (U.S. Army, 1996f). Other components, such as bis, have also been detected in some samples by ^{31}PNMR (phosphorus 31 nuclear magnetic resonance spectroscopy).

The HD agent stored at the Aberdeen site was distilled when produced, but it also contains several impurities formed either during manufacture or from decomposition of the HD during storage. The Army estimates that each ton container of HD contains about 14 pounds of "land-banned" chemical impurities (chemicals subject to strict hazardous waste regulations, including limitations on landfill disposal). These strictly regulated impurities include 1,2-dichloroethane, trichloroethylene, tetrachloroethylene, 1,1,2,2-tetrachloroethane, and hexachloroethane. There are also about 30 pounds of dithiane per container and varying amounts of chloroethyl sulfides other than HD. In a

[5]The VX hydrolysis product called "pyro" is $[CH_3P(O)(OC_2H_5)]_2O$. The VX impurity called "bis" is $CH_3P(O)[SCH_2CH_2N(CH(CH_3)_2)_2]_2$. The hydrolysis product of bis called "EA-2192" is $CH_3P(O)(OH)[SCH_2CH_2N(CH(CH_3)_2)_2]$.

TABLE 1-2 Chemical Munitions Stored in the Continental United States

Chemical Munitions (Agent)	APG	ANAD	BGAD	NECA	PBA	PUDA	TEAD[a]	UMDA
Mustard agent (H, HD, or HT)								
105-mm projectile (HD)		X				X		
155-mm projectile (H, HD)		X	X			X	X	
4.2-in. mortar (HD, HT)		X				X	X	
Ton container (HD)	X	X			X		X	X
Ton container (HT)					X			
Agent GB								
105-mm projectile		X						
155-mm projectile		X	X				X	
8-in. projectile		X	X				X	X
M55 rocket		X	X		X			X
500-lb bomb							X	X
750-lb bomb							X	X
Weteye bomb							X	X
Ton container		X[b]					X	
Agent VX								
155-mm projectile		X	X				X	X
8-in. projectile							X	X
M55 rocket		X	X		X		X	X
M23 land mine		X			X		X	X
Spray tank							X	X
Ton container				X			X	
Miscellaneous								
Ton containers (L)							X	
Ton containers (GA)							X	

[a]Small quantities of Lewisite and tabun (GA) are stored in ton containers at TEAD.
[b]Small quantities of agent drained as part of the Drill and Transfer System assessment for the M55 rockets.

NOTE: APG, Aberdeen Proving Ground, Maryland; ANAD, Anniston Army Depot, Alabama; BGAD, Blue Grass Army Depot, Kentucky; NECA, Newport Chemical Activity, Indiana; PBA, Pine Bluff Arsenal, Arkansas; PUDA, Pueblo Depot Activity, Colorado; TEAD, Tooele Army Depot, Utah; and UMDA, Umatilla Depot Activity, Oregon

SOURCE: Adapted from NRC, 1996 and U.S. Army, 1996h.

recent survey conducted by the Army, analysis of the materials in 27 randomly selected and sampled containers of HD reveals the compounds shown in Table 1-4 (U.S. Army, 1996g). In addition to these impurities, which are dissolved in the much larger quantity of HD in the container, all containers tested recently at Aberdeen appear to contain solid or semisolid deposits, called a "heel." The quantities and composition of the heel vary from container to container, but it appears to consist largely of sulfonium and iron salts with adsorbed HD. The heel solids appear to dissolve readily in hot water (U.S. Army, 1996b). The relatively high freezing point of HD (14.45°C) and the outside storage of ton containers at Aberdeen will require facilities to thaw HD during cold weather, prior to processing. This requirement is independent of the destruction technology evaluated and is based on the required processing rates and the maximum amount of agent that can be present in a destruction facility at one time.

ROLE OF THE NATIONAL RESEARCH COUNCIL

The National Research Council (NRC) Committee on Demilitarizing Chemical Munitions and Agents was formed in August 1983 to review the status of the

TABLE 1-3 Composition of VX from Ton Containers Stored at Newport (based on gas chromatography analysis)

Compound	Average (weight percent)
VX	93.71
Dimethyl ketone (acetone)	0.01
Diisopropylamine	0.14
N,N-Diisopropylmethylamine	0.01
Diisopropyl Carbodiimide (stabilizer)	1.74
N,N-Diisopropylethylamine	0.01
O-Ethyl methylphosphonate	0.20
1,3-Diisopropylurea	0.03
Diethyl methylphosphonate	0.06
2-(Diisopropylamino) ethane thiol	0.89
O,O-Diethyl methylphosphonate	0.21
O,S-Diethyl methylphosphonothioate	0.07
2-(Diisopropylamino)ethyl ethyl sulfide	0.13
Diethyl dimethylpyrophosphonate (pyro)	0.99
O,O-Diethyl dimethylpyrophosphonothioate	0.23
O-(2-Diisopropylaminoethyl) O-ethylmethylphosphonate	0.26
1,2-bis(ethyl methylphosphonothiolo)ethane	0.62
Unknowns, plus trace metals	0.69
Total	100.00

Source: U.S. Army, 1996g.

stockpile and technologies for disposal. That committee reviewed a range of technologies and, in its final report in 1984, endorsed incineration as an adequate technology for the safe disposal of chemical agents and munitions (NRC, 1984). The committee also concluded that the stockpile was well maintained and posed no imminent danger but expressed concern about future storage risk due to the potential for an increased rate of stockpile deterioration.

In 1987, at the request of the Undersecretary of the Army, the Committee on Review and Evaluation of the Army Chemical Stockpile Disposal Program (referred to as the Stockpile Committee) was established under the aegis of the NRC Board on Army Science and Technology to provide the Army with technical advice and counsel on specific aspects of the disposal program. Under this charter, the Army has requested and received from the Stockpile Committee 15 reports that evaluated stages of progress and specific aspects of the program.

In March 1991, as a result of growing public concerns about and opposition to the baseline incineration system and the rising cost of the CSDP, the Stockpile Committee suggested, and the Army agreed, that a new study of alternatives to incineration for the destruction of the stockpile should be undertaken.

In January 1992, the NRC, at the request of the office of the Assistant Secretary of the Army for Installations, Logistics and Environment, established the Committee on Alternative Chemical Demilitarization Technologies (Alternatives Committee) to develop a comprehensive list of alternative technologies and to review their capabilities and potential as agent and munitions disposal technologies. In June 1993, this committee published its report, *Alternative Technologies for the Destruction of Chemical Agents and Munitions* (NRC, 1993).

The Stockpile Committee, working with the report of the Alternatives Committee and with its own knowledge of the baseline system and disposal requirements, formulated recommendations regarding the investigation of potential alternatives to incineration. This work was reported in February 1994 in *Recommendations for the Disposal of Chemical Agents and Munitions* (NRC, 1994b). The Stockpile Committee concluded that the baseline system is adequate for disposal of the stockpile and that the storage risk will persist until disposal of all stockpile materials is complete. The report recommended that the CSDP proceed expeditiously and with technology that minimizes total risk to the public at each site.

The Stockpile Committee also found, after examination of all the technologies brought to its attention by the Alternatives Committee and others, that four neutralization-based systems offered the most promise for agent destruction (NRC, 1994b). In view of the increasing total risk associated with delays in the disposal program, and recognizing that public opposition might delay the program for a number of reasons, including opposition to incineration, the committee stated that alternative technologies should be developed promptly. The committee also recommended that the Army continue to

TABLE 1-4 Composition of HD from Ton Containers Stored at Aberdeen

Compound	Average (weight percent)
HD	90.20
2 methyl 1-propene	0.021
thiirane	0.017
2-chlorobutane	0.002
1,2-dichoroethane	0.350
1,4-oxathiane	0.070
1,4-dithiane	1.476
trichloroethylene	0.001
1,2,5-trithiepane	0.086
tetrachloroethylene	0.132
1,1,2,2-tetrachloroethane	0.037
2-chloroethyl 3-chloropropyl sulfide	1.092
bis(2-chloropropyl) sulfide	0.366
$C_6H_{12}C_{12}S$ isomers	0.548
2-chloroethyl 4-chlorobutyl sulfide	1.136
bis(2-chloroethyl disulfide)	0.643
2-chloroethyl (2-chloroethoxy) ethyl sulfide	0.054
Q, 1,2-bis(2-chloroethylthio) ethane	2.639
bis(2-chloroethyl) trisulfide	0.072
hexachloroethane	0.152
Unknown	0.015
Copper as $CuCl_2$	0.003
Iron as $FeCl_2$	0.888
Total	100.00

Source: U.S. Army, 1996g.

monitor other research programs and developments involving potential alternatives.

In April 1994, the Army produced its own report, *U.S. Army's Alternative Demilitarization Technology Report for Congress* (U.S. Army, 1994). The Army accepted the Stockpile Committee's recommendation to pursue neutralization-based technologies but limited the Army's research and development to two alternatives: (1) stand-alone neutralization, and (2) neutralization followed by biodegradation. The Army also agreed to monitor additional developments in alternative disposal technologies.

One aspect of the Army's work on neutralization alternatives was to prepare detailed assessment criteria for decisions on proceeding with the development of neutralization technologies. The Army released its draft report of these criteria in April 1995 as *Assessment Criteria to Aid in the Selection of Alternative Technologies for Chemical Demilitarization* (U.S. Army, 1995a, hereafter cited as the Army *Criteria Report*). The Army also asked the Stockpile Committee to evaluate these draft criteria, which it did in *Evaluation of the Army's Draft Assessment Criteria to Aid in the Selection of Alternative Technologies for Chemical Demilitarization* (NRC, 1995, hereafter cited as the NRC *Criteria Report Evaluation*). Both of these reports were particularly pertinent to the present study.

SCOPE AND ORGANIZATION OF THE STUDY

Since these earlier reports, the Army believes that research developments have sufficiently enhanced the database on the performance of some alternative technologies to warrant reexamination of specific alternatives for use at certain sites. In the summer of 1995, the Assistant Secretary of the Army for Research, Development and Acquisition informally explored with the NRC Stockpile Committee the possibility of examining alternative chemical disposal technologies. Following numerous discussions between the Army and the NRC, a decision was made to conduct a new NRC study to reexamine the status of a limited number of alternative chemical disposal technologies to be selected by the Army (including the two neutralization-based processes on which the Army was currently conducting research) for possible use in the CSDP.

In August 1995, the Army advertised in the *Commerce Business Daily* (Appendix A) for alternative disposal technologies other than the two already being evaluated by the Army. The purpose of this announcement was to determine whether any other technologies were capable, within the CSDP schedule, of meeting chemical demilitarization requirements for the two sites where agent is stored only in bulk (the Aberdeen and Newport sites). The announcement requested information from industry on non-incineration technologies that were sufficiently developed to meet the needs of the CSDP. Following a preliminary 30-day screening review, the Army in November 1995 selected three technologies for review and evaluation by the NRC—gas phase reduction, molten metal catalytic extraction, and electrochemical oxidation—in addition to the two processes, neutralization and neutralization followed by biodegradation, that were already being developed by the Army.

In parallel with the Army selection process, the NRC formed the Panel on Review and Evaluation of Alternative Chemical Disposal Technologies (AltTech Panel). The AltTech Panel held its first meeting prior to the announcement of the Army's selection. Anticipating the broad types of technologies that might be selected by the Army, the panel developed a project plan and preliminary report outline, based on its knowledge of the Stockpile Committee reports and activities. The NRC added three members to the panel after submissions were received for the three technologies to be reviewed. The new members were added to supplement the expertise already on the panel and to provide coverage for the specific technologies to be evaluated.

From November 1995 to June 1996, the panel conducted in-depth reviews and evaluations of the five selected technologies. The entire panel met six times; designated panel teams conducted 14 site visits to study the technologies; and panel members met with regulators, citizens advisory commissions (CACs), and local citizens in Maryland and Indiana. The panel's activities are delineated in the following statement of task.

> At the request of the Assistant Secretary of the Army for Research, Development and Acquisition, the National Research Council will carry out a review of alternative chemical agent disposal technologies. To conduct this review, a Panel on Review and Evaluation of Alternative Chemical Disposal Technologies under the auspices of the Board on Army Science and Technology will examine no more than three alternative technologies (to the baseline incineration system), as well as neutralization and neutralization followed by biodegradation for the disposal of chemical agent at Aberdeen, Maryland (mustard agent) and Newport, Indiana (nerve agent) only. The panel will meet, as appropriate, to:
>
> - establish criteria to assess and evaluate selected alternative technologies;
> - conduct site visits as appropriate to assess firsthand the viability and maturity of technologies being reviewed;
> - conduct site visits to possible locations where alternative technologies may be employed and to hold open meetings there to solicit CAC views on the alternative technologies under consideration;
> - assess technical aspects, strengths and weaknesses, and advantages and disadvantages of each technology;
> - consider the option of shipping treated effluents (agent free) to off-site appropriately permitted disposal facilities; and
> - make recommendations regarding which, if any, of these technologies merit full evaluation and presentation to the Defense Acquisition Board[6] as candidates for pilot-plant demonstration by the Army.

Initially the Army also asked the panel to examine technologies to be used solely for the treatment of neutralization hydrolysate. (Hydrolysate is the aqueous solution of products from the neutralization step.) These technologies were not "stand-alone" technologies (like those selected by the Army for consideration for total on-site agent treatment) but were polishing steps to be taken after neutralization. The panel felt the limited time available would not allow for a complete investigation leading to specific recommendations in this report regarding these technologies. However, consistent with earlier Stockpile Committee report recommendations and based on information provided by the Army, the AltTech Panel is aware that the Army continues to examine technologies for this purpose and supports these efforts.

In conducting this review, the panel recognized that, although it had been charged with evaluating technologies, each of the technologies under evaluation was being developed and submitted for consideration by a specific company. (Hereafter, these companies are referred to as technology proponent companies, or TPCs.) Consequently, the present engineering status of each technology is company-dependent, and the panel's evaluations must, by necessity, depend on the TPCs for information. However, the panel's evaluations apply only to the application of each technology, as submitted for the panel's consideration, to agent destruction at the bulk-storage sites, not to the general capabilities of the TPC or to other applications of the technology.

The panel's interactions with the TPCs during the course of this study clearly showed that technology development had continued after the October submissions responding to the announcement in the *Commerce Business Daily*. The panel realized that these technologies will continue to evolve, but to conduct a review within the time provided, the panel requested that all TPCs submit "final" designs by April 4, 1996. Hence

[6]The Defense Acquisition Board is the entity under the Secretary of Defense that makes major acquisition decisions for Department of Defense programs. The board is scheduled to decide on pilot-testing of alternative technologies for the Aberdeen and Newport sites at its October 1996 meeting.

the technology assessments and evaluations in this report reflect the status of each technology as of that date.

The Army required that the TPCs perform supervised tests to obtain data on how the technology performed in destroying actual chemical agent. The tests were conducted by an Army-approved laboratory at the TPCs' expense. The test data were not available to the NRC panel for review until late June, which did not allow enough time for the panel to conduct an assessment of the reported by-products produced during the tests. Consequently, the tests were used by the panel simply to make a yes-or-no determination as to whether the technology can destroy agent.

In addition, all TPCs were required to give the Army projected cost and implementation schedules by March 17, 1996. The cost data were not provided to the panel and their consideration is outside the scope of this study.

Public Law 102-484 identifies safety as a critical factor in the selection of a technology for the alternative technology program. Process safety risk encompasses risk to the health and safety of workers and the public, as well as risk to the environment. The panel insisted, and the Army agreed, that, consistent with the varying depth and scope of available technical information on the proposed alternative technologies and the need to provide timely support to the Defense Acquisition Board's decision-making process and the NRC panel review, the Army would request preliminary risk assessments of the technologies by an independent contractor (MitreTek Systems, Inc.).

The scope of work for this risk assessment required that the contractor provide a preliminary assessment of the potential process safety risks associated with implementing the baseline incineration system as compared with each of the five alternative disposal technologies at the Aberdeen and Newport sites. Significant discriminators of process safety risk among the baseline system and the alternative technologies were to be identified and evaluated. Discrimination was to be based on safety and health risks to workers, safety and health risks to the public, environmental risks, and storage risks. Risks to plant equipment and operations were not to be considered directly. The contractor was required to present results of the preliminary risk assessment in a draft report by April 15, 1996, and to provide a final report by May 31, 1996.

The contractor's analysis was constrained by two factors. (1) Because the technical information and design maturity of the proposed alternative technologies are at present limited in comparison with the baseline system, assessments of certain aspects of risk were limited and qualitative in nature. (2) The time available to perform the analysis precluded detailed analysis of even the limited information available on the alternative technologies.

The AltTech Panel's risk assessment expert participated in some of the contractor's efforts to gather data, performed an independent risk evaluation of the five technologies, and reviewed the contractor's report. These activities enabled the panel to assess, on a qualitative basis, the process safety risks for each alternative technology. The independent risk evaluation focused on characteristics inherent in each technology that had the potential to lead to accidental release and only briefly addressed accident scenarios caused by combinations of system failures (pipes, pumps, valves, power systems, and cooling systems). The hazards of transporting ton containers from storage to the processing area and of the punch-and-drain operations to remove the agent from containers are common to all the technologies being evaluated. However, the mode of feeding agent into the process may be somewhat more hazardous for some technologies than for others (the differences are discussed in Chapters 4 through 8).

At this point, however, no comprehensive, quantitative risk assessment has been performed on any of the alternative technologies.

The *Commerce Business Daily* announcement and the Army's criteria for selecting potential alternative technologies required that TPCs demonstrate the feasibility of using their technology to conduct all the activities required to process agent on site,[7] consistent with the objectives and capabilities of the baseline incineration system. However, since the time of the announcement, the CSDP has continued to explore ways to increase cost effectiveness. Off-site shipping, for example, is already being used for limited quantities of various process wastes, including empty, cleaned ton containers; used decontamination fluids; and hydrolysate from tests of the neutralization technology. Off-site shipping on a larger scale may significantly improve cost effectiveness. Consequently, the statement of task for the AltTech Panel was amended to direct that the panel examine the option of shipping process wastes off site for final treatment. This option is discussed in Chapter 3.

[7]For the purposes of this report, "on site" means within the boundaries of the federal installation within which the stockpile is located. "Off site" means beyond the boundaries.

REPORT ORGANIZATION

The AltTech Panel divided its evaluation into three phases: organization, data gathering, and report preparation. Because time was limited, the principal data-gathering efforts could not exceed six months and each phase had to be carefully planned. The organization of this report reflects these efforts.

Before site visits were undertaken, the panel extracted relevant evaluation factors from the Army *Criteria Report* and the NRC *Criteria Report Evaluation* and developed its own framework for evaluation. This framework became the basis for a questionnaire sent to the TPCs and the Army well before the panel's site visits. Chapter 2 discusses the evaluation factors, and Chapter 3 describes the framework for gathering information.

Chapters 4 through 8 contain specific technology assessments based on the information gathered by the panel. Chapter 4 assesses the catalytic extraction process (molten metal); Chapter 5, electrochemical oxidation; Chapter 6, gas phase reduction; Chapter 7, neutralization of HD; and Chapter 8, neutralization of VX.

The regulatory process and the opinions of the public and other stakeholders can have a dramatic effect on the implementation schedule. Because delays extend the time of exposure to stockpile storage risk, they can increase overall risk. To assess these effects, the panel held meetings with regulators in Maryland and Indiana and conducted public forums, where concerned citizens were encouraged to voice their opinions of the alternatives under consideration. Chapter 9 discusses this aspect of the study.

Chapter 10 presents the panel's comparison of the alternative technologies based on the criteria developed in Chapter 2. Chapter 11 contains the major findings and recommendations that the panel distilled from the technology assessments and from comparing the technologies.

Because of time constraints on preparing this report and because agent test data were not available until very late in the process, the panel was not able to analyze these test data in depth. Also, the panel had time for only a preliminary review of the MitreTek Systems risk assessment report. Both issues are discussed further in Chapter 2.

2

Evaluation Factors

This chapter discusses the factors that the AltTech Panel considers central to evaluating and comparing the alternative technologies. The factors included here were developed from the panel's review of the Army *Criteria Report* and the NRC *Criteria Report Evaluation*, from the concerns and issues raised in public forums conducted by the panel in communities near the two sites, and from the combined expertise and experience of panel members.

The AltTech Panel has essentially adopted three of the four primary factors identified by the Stockpile Committee in the *Criteria Report Evaluation*: process efficacy, process safety, and schedule (NRC, 1995, pp. 14–19). The fourth factor, cost, will be evaluated independently by the Defense Acquisition Board. In adopting these factors, the AltTech Panel modified the wording of the first two factors (modified portions are shown in italics):

1. *Process Efficacy.* Does the alternative agent-destruction process, when integrated with other necessary destruction system components, effectively *and reliably* meet agent-destruction requirements?
2. *Process Safety.* Is the alternative technology safe *and does it protect public health and the environment*? The criterion of "safe" adopted by the Stockpile Committee is minimization of total risk[1] to the public and to the environment (NRC, 1994b).
3. *Schedule.* What are the impacts of implementation of an alternative technology on the schedule for stockpile destruction?

Each primary factor has several subfactors, which may be interdependent. A negative judgment on a technology for a specific subfactor need not imply a negative overall judgment for the primary factor. The subfactors and their interdependencies are discussed below.

PROCESS EFFICACY

Process efficacy encompasses not only the capability of a technology to destroy the agent of interest but also the status of the technology: its stage of maturation along a spectrum from laboratory scale to pilot-plant development and eventual full-scale operation. Process efficacy also includes whether the process can be controlled, whether it is reliable, and whether it meets applicable regulatory and treaty requirements. The AltTech Panel has defined the following subfactors under process efficacy:

- technology status
- capacity to detoxify agent
- satisfaction of treaty requirements
- satisfaction of environmental and other regulatory requirements
- management of process residuals[2]
- process stability, reliability, and robustness
- process monitoring
- natural resource requirements (e.g., energy)
- scale-up requirements
- applicability for treating other wastes

[1] Total risk is the cumulative adverse consequences from all relevant risks—for example, storage, transport, and processing risks—over the full remaining duration of the stockpile's existence and the stockpile disposal program.

[2] In this report, a process residual is defined as any material remaining at the end of the process. Process residuals include not only all materials in gaseous, liquid, or solid waste streams (emissions, effluents, and wastes) but also materials that may be considered products or by-products because they can be used or have economic value. Process residuals include residual agent or other materials that were in the process feeds (water, chemicals, etc.), as well as materials produced during processing.

Technology Status

By the status of an alternative technology, the panel means the stage to which the technology has progressed toward fully operational practice. In general, chemical-process technologies can be located along a developmental continuum from laboratory-scale, proof-of-concept testing to pilot-plant demonstration and ultimately to full-scale operation.

Many considerations are involved in determining whether a technology is ready to move to the next stage or how close it is to being "successfully demonstrated" at a given stage. For instance, at the laboratory scale, assays and chemical analyses are important in establishing that the desired reactions predominate and that unwanted side-reactions can be controlled. At the pilot scale, precise mass and energy balances become essential, along with quantitative characterizations of how key process variables affect outcomes. The documentation for a pilot design must be complete enough for a preliminary assessment of risks related to the hazard inventory (e.g., agent concentrations at each process step, reactive materials, pressure) and the adequacy of safety features, such as process interlocks and safe means of releasing excess material or energy. Assigning a status to a technology is, therefore, not a simple classification but rather a running checklist of what has been accomplished to date and what remains to be done.

In assessing the status of a technology, the AltTech Panel had to consider the extent of documentation and evidence provided, as well as the capabilities, resources, and commitment of the TPC. These company-specific characteristics are critical to the successful implementation of any technology, both at the demonstration stage and during disposal operations.

Capacity to Detoxify Agent

To detoxify a chemical agent such as VX or HD satisfactorily, the reaction that destroys the agent must proceed until the remaining concentration of agent is below a specific limit. The Army specifies this limit in terms of a "destruction removal efficiency" (DRE), defined as the difference between the amount of agent going into the process and the amount remaining, expressed as a percentage of the amount going in. For a process to be acceptable in destroying agent, it must have a DRE of 99.9999 percent or greater. DRE values are often expressed as the number of 9's in the percentage; this DRE is therefore referred to as "six 9's." A DRE of 99.999999 percent is "eight 9's."

In addition to the required DRE for a destruction process, the Army uses the following limits on allowable concentration of agent to determine whether a material must continue to be controlled as (potentially) agent-contaminated, may be released from an agent-control facility for further treatment, or may be released to the environment or to general, "public" use (i.e., any use other than for further treatment to destroy residual agent).

Gases

The release of gases to the atmosphere is constrained by a health-based General Population Limit at the site boundary. The limit values for HD and VX are, respectively, 0.1 and 0.003 g per cubic meter of air.

Liquids

There is no standard established for unconditional release of liquids containing chemical agents. The standard for release of certain specified liquid wastes from incineration facilities to qualified disposal facilities is 200 ppb for HD and 20 ppb for VX. These same limits apply to release of drinking water to soldiers in the field.

Solids

The Army has three primary classifications for solids that may be contaminated with chemical agent. The first classification is for solid material that is potentially contaminated and has not been subject to further decontamination or testing. This material cannot be released from agent-control areas under Army supervision. The second classification, called "3X," is for solids that have been decontaminated to the point that the agent concentration in the air above the solid does not exceed the health-based 8-Hour Worker Limit. The limit values for HD and VX are, respectively, 3 and 0.01 g per cubic meter of air. A 3X material may be handled on an unrestricted basis by plant workers but is not releasable to the environment or for general reuse (i.e., not releasable "to the public."). In specific cases in which approval has been granted, a 3X material can be shipped to approved hazardous waste treatment facilities for landfill disposal. The third classification, called "5X," is for material that has been subjected to thermal treatment of at least 1000°F for 15 minutes to assure essentially complete destruction of all residual agent. A 5X material is releasable to the public.

For this study, the TPCs conducted laboratory tests under Army supervision to determine if the technologies would, in fact, destroy agent. The panel received results of these tests in late June 1996. Although the overall results demonstrated that all the technologies can destroy agent, quantitative data on process residuals were not available to the panel in time for in-depth review. Careful consideration of process residuals will be required for decisions about pilot testing.

Satisfaction of Treaty Requirements

The 1993 Chemical Weapons Convention (CWC) requires destruction of the primary agent and further reaction or destruction so that none of the end products can be readily converted back to the primary agent. (An appendix to the CWC treaty contains a list of compounds that can be readily converted to the agent; these compounds are called "scheduled precursors.") The CWC objective is to remove the military threat from agents, whereas environmental permits are designed to protect human health and the environment. Therefore, the requirements for residual concentrations of agent allowable under treaty negotiations are likely to be less stringent than the requirements under environmental permits for destruction facilities and downstream disposal facilities.

The CWC requires that the destruction system allow for verification that agent has been destroyed. The convention further requires that the destruction of the unitary chemical weapons stockpile be completed within 10 years after the treaty is ratified (ratification was expected in 1996).[3]

Satisfaction of Environmental and Other Regulatory Requirements

The agent-destruction process that is implemented must comply with state and federal regulatory requirements. Key regulatory requirements include specifications for acceptable process residuals and waste- management practices. Other regulatory compliance issues include workplace safety and health requirements (e.g., those set by the Occupational Safety and Health Administration) and management of nonprocess wastes, such as decontamination fluids and personal protection equipment.

Management of Process Residuals

Disposal of process residuals is a critical aspect of any agent-destruction system. The process residuals from alternative technologies differ in physical state, composition, and quantity, but all residuals must ultimately be dealt with. The toxicity of reaction products must be low enough that unwanted process residuals can be managed through aqueous discharge to a conventional wastewater treatment facility, disposed as solid waste in a landfill appropriate for the toxicity of the waste, released as allowable atmospheric emissions, or some combination of these three release routes. In legal terms, the concentrations and toxicities of the materials in aqueous, solid, slurry, or gaseous residual streams must fall below the limits set by the environmental permits needed to operate the agent-destruction facility and any downstream waste-management facilities.

One major challenge with some technologies is the management of large quantities of aqueous residuals. On-site management of aqueous residuals requires deciding either to change Army regulations to allow discharge directly to a wastewater treatment facility or to continue to evaporate the water and discharge it as an atmospheric emission, as is done in the baseline system. (The residual material remaining after evaporation is treated as a solid-waste stream.) Some of the hydrogen atoms originating from the chemical agents will ultimately bond with oxygen to form water so that, even with aggressive water recycling, some form of water release will be required. The extent of water recycling will affect cost.

A second major issue is the point at which process residuals can be transferred to off-site, private sector facilities for subsequent management. This question

---

[3]As stated in Chapter 1, the date mandated by Congress for the destruction of the stockpile is December 31, 2004. However, the latest date for the destruction of the stockpile according to the CWC will be 10 years from treaty ratification. Because the treaty has not yet been ratified, the latest date by which the stockpile must be destroyed may change. Congress may elect to amend the law so that the dates coincide. Until that occurs, however, the Army will continue to work toward the 2004 date.

requires consideration of appropriate waste management options (aqueous discharge, solidification or stabilization, landfill disposal, thermal destruction, etc.) for individual waste streams, the capability of private sector facilities to meet regulatory requirements and to process residual waste streams, the criteria for releasing process residuals to the private sector for treatment or disposal, and the technological capacity of available private sector facilities.

The process residual streams from alternative systems need to be compared in terms of both the composition of the stream and the intended management of it. An appropriate basis for this comparison begins with the mass balances for the overall process and for major chemical elements, such as nitrogen, sulfur, chlorine, phosphorus, and carbon. (Mass balance data that were available to the panel are summarized in Chapters 4 through 8.)

Process Stability, Reliability, and Robustness

Process stability, reliability, and robustness are key goals. Achieving them depends on many factors, a few of which are described here.

The batches of agent fed to a destruction process will vary in agent purity and in the composition of impurities as a result of variability in the conditions of their production and storage. For example, some containers of HD contain solids, which may make them difficult to feed through a system designed to handle liquid agent. The process must function effectively and reliably in spite of such variations in the process feed, i.e., the process must be sufficiently reliable that it can effectively destroy agent despite a range of variability in the chemical and physical composition of the feed material.

Operating conditions that can result in process instabilities, such as temperature or pressure excursions that can lead to catastrophic failure, must be avoided. Such conditions can include extreme operating conditions (e.g., high pressures, temperatures, or reaction rates) and corrosive reactants, residuals, or process environments.

Control strategies and process flexibility must permit the process to be controlled effectively even in the event of an upset such as a power failure or loss of agitation. The selected process must also provide for the decontamination and management of storage containers and other contaminated metal parts.

Process Monitoring

Implementing an alternative technology requires techniques to monitor the concentrations of agent and of reaction products in liquid, slurry, or solid process streams. Sampling procedures, response times, and required detection limits must be defined. The monitoring requirements for alternative processes may be quite different from the requirements for the baseline system. A critical issue is whether new monitoring techniques, not commercially available, are required, and if so, what the schedule for developing these techniques would be.

Energy and Natural Resource Requirements

The consumption of resources such as energy and water must be considered in selecting a technology, especially for locations where these resources may be limited. Resource constraints do not appear to be an issue at either Aberdeen or Newport, but a high demand for power or water, for example, may have secondary effects that need to be understood.

Scale-Up Requirements

Implementation of an alternative technology will require demonstrating the process with near-full-scale equipment prior to full implementation. The equipment required to demonstrate a process may differ for HD and VX. In addition, the scales at which the technologies under consideration have demonstrated the processing of agent are quite different, as is the scale at which these technologies have been used for other applications. Consequently, the engineering development required to scale-up the process will differ for each technology.

Applicability for Treating Other Wastes

Use of an alternative technology that is broadly applicable to treating common industrial wastes (including hazardous waste) is a concern to some in the communities near stockpile sites who fear the facility could be readily converted for treating additional wastes imported from off site, once stockpile destruction is completed. Thus, selection of a technology that would result in a versatile waste destruction facility may increase fears that the facility will not be decommissioned after the stockpile is destroyed. A contrary view, also held by some members of the communities, is that versatility could be a virtue at a site such as Aberdeen, which contains numerous hazardous wastes, other than the unitary agent stockpile, that also require disposal.

PROCESS SAFETY

Process safety encompasses concerns about worker safety, community health risks, and environmental protection. Evaluating process safety therefore includes assessing in-plant safety and health risks, risks to community safety and health, and risks to the environment. For each of these major risk categories, the evaluation should include the consequences of a release of chemical agent and of nonagent, toxic process residuals. Important contributing factors to the overall risk in each category include the risks from storing and handling agent in containers prior to processing, as well as the risk of releases from the destruction process itself.

The discussion below covers, in broad outline, the full range of risk factor evaluation and of risk assessment, preliminary and quantitative, that must be done in the course of developing an alternative technology through pilot-testing and on to construction of a full-scale operational facility.

For this particular study, time constraints and the immaturity and status of design of the candidate technologies precluded making quantitative risk assessments.

However, the panel was able to:

1. make a qualitative evaluation of whether each technology can be operated safely, given the current state of development (assuming adequate attention is paid to the intrinsic safety issues for each technology)
2. identify the intrinsic safety issues for each technology and evaluate the current treatment of these issues by the TPCs
3. provide focus for a future comprehensive, quantitative risk assessment prior to implementation

To avoid confusion, the following discussion refers to the activity of the panel as "evaluating risk factors" and reserves the terminology of "assessing risk" for the future detailed risk assessments. As explained in Chapter 1, the panel insisted that the Army obtain preliminary accidental-release risk assessments for the alternative technologies as input to the decision to be made by the Defense Acquisition Board on pilot-testing one or more alternative technologies. The panel's view of the scope appropriate to these very preliminary and qualitative assessments is discussed below, under Risk Assessments prior to the Pilot-Testing Decision.

In-Plant Safety and Health Risks

In-plant safety and health risks depend on the nature and magnitude of hazards within the processing facility. The panel's preliminary evaluation of an alternative technology included the following components of this risk category: the risk of catastrophic failure and agent release, the risk of exposing workers to agent, the risk of worker exposure to other hazardous chemicals used in or produced during the process, and the risks from hazardous process conditions. These risks are affected by (1) the hazard inventory (agent; stored thermal, mechanical, and chemical energy; and reactive chemicals), (2) process-intrinsic safety (safety features engineered into the process design), and (3) worker controls (e.g., in-plant monitoring for worker exposure, maintenance procedures, and campaign duration).

Risk to Community Safety, Health, and the Environment

Although the consequences associated with risks to community safety and health differ from the consequences of risks to the environment, the release factors that cause the risks are generally similar enough to treat both categories together, at least at this stage in evaluating alternative technologies. The release factors include not only those that can cause acute exposure to agent or toxic process residuals but also those that cause latent health effects or gradual environmental damage from long-term, low-level emissions and discharges.

Concerns about both kinds of exposure have led many citizens in the communities near stockpile sites to favor process designs with a "test-prior-to-release" requirement for all process residuals. This testing must be capable of detecting very low-level, continuing concentrations of a hazardous material, as well as one-of-a-kind, brief releases at high concentration.

Factors to consider in evaluating risks to the community and environment (and in detailed risk assessments) include all handling and processing throughout the projected period of facility operations, the limited scale and finite time of stockpile-destruction operations at each site, and hazards from off-site disposal of residuals. The specific components of risk, many of which require detailed risk assessment to identify and estimate realistically, include:

- risks from agent release and exposure during the destruction process or from storage and handling prior to destruction
- risk of latent health effects from exposure to nonagent releases from the destruction process (realistic information on this risk requires site-specific health effects assessments)
- risks from managing process residuals, whether off site or on site, after the destruction process (again, context-specific risk assessment is needed to provide realistic information useful to decision-making)
- risks to the community or environment associated with the total environmental burden (burden as quantified by total residual process streams that are released to the environment), including the potential impact on natural resources (agriculture, bodies of water, etc.) from aqueous discharges, atmospheric emissions, or solid-waste management

The first and third bullets in this list may require special consideration in future detailed risk assessments. One such consideration includes consequences for emergency preparedness or emergency response—for example, the extent of the area that would be affected by an accidental release of agent or of toxic nonagent materials.

Risks to the community and environment from agent storage have been cited as a reason for prompt destruction of the stockpile (NRC, 1994b). These storage risks have been the focus of ongoing debate in communities near several stockpile sites. The storage risks that vary with the agent-destruction system, whether that system uses an alternative technology or is the baseline system, depend primarily on the duration of storage and therefore on the overall schedule for each option. Actions that can reduce storage risk at individual sites, other than shortening the storage time, are for the most part independent of the technology for stockpile destruction. The Army is currently assessing the storage risks at all stockpile sites in the continental United States and may consider reconfiguring individual stockpiles based on the results of the evaluation.

Risk Assessments prior to the Pilot-Testing Decision

Before any technology is implemented at a stockpile site, two site-specific risk assessments will be required: a comprehensive quantitative risk assessment in which the likelihood of events leading to the unintended release of agent or toxic materials and the consequences of such a release are analyzed, followed by a health and environmental risk assessment in which the potential consequences of accidental or continuing low-level exposure of the community or the environment are assessed. These assessments cannot be properly performed until after pilot-testing of a technology and detailed engineering planning of the full-scale facility. However, the AltTech panel believes that a preliminary, comparative assessment of risks associated with the alternative technologies is necessary for a decision to recommend a technology for pilot demonstration. If the pilot demonstration is successful and the alternative technology is selected for full-scale implementation, the two more-rigorous, site-specific risk assessments must be completed before a full-scale facility is built and agent destruction operations begin.

As noted above, the panel encouraged the Army to support a preliminary accidental release risk assessment before the pilot-testing decision. A preliminary assessment for each alternative technology should be prepared as input to the decision on whether to pilot-test one or more of them. This assessment should include the kinds of accidental release scenarios that can reasonably be envisioned during the operation of the technology, a measure of the probability of various accidental release scenarios and their likely magnitude (the probability measure could be qualitative), a measure of the impact of potential accidental release scenarios on worker health and safety, and a preliminary assessment of the impact of a release on public health and the environment.

SCHEDULE

To compare the effect of alternative technologies on the implementation schedule for stockpile destruction, the panel needed estimated schedules for each alternative technology at each potential site. These technology-specific schedules had to include time ranges for technology development, pilot-scale evaluation, and full-scale implementation and operation. The panel requested schedules from the TPCs and the Army indicating major milestones—and the assumptions made in estimating them—for (1) laboratory and bench-scale development, if applicable; (2) pilot plant design, construction, and, operation, with subsequent analysis of pilot-plant data; and (3) design of the full-scale plant, acquisition of equipment, and the construction, startup, operation, and decommissioning of the full-scale facility.

Public opposition, regulatory review, and permitting requirements can cause significant delays in the implementation schedule, but informed public acceptance and support can help to overcome regulatory or statutory hurdles. The actual time required to implement a system and eliminate the stockpile will not only affect compliance with the CWC but will also significantly affect the overall risk at each site, because storage risk depends on the duration of storage.

The panel met with members of the communities near the Newport and Aberdeen sites, with representatives of the Indiana and Maryland CACs, and with state regulators to solicit information and learn how these groups see issues affecting the implementation of each alternative technology. In particular, regulators were asked to provide information on technology-specific permitting requirements. CACs and local communities were asked to discuss their specific concerns about the technologies selected for evaluation and their views on criteria that should be used in the evaluation.

ROLE OF EVALUATION FACTORS IN THE STUDY

The factors and subfactors described in this chapter provided the framework for the panel's assessments and evaluations. For example, the framework of factors was used as the outline for the information to be gathered and presented in the detailed individual assessments of the alternative technologies (Chapters 4 through 8). The framework was also used to generate the detailed questionnaires that were sent to the TPCs and regulators (Appendix J). The framework was also the basis for the public forums and for the panel's discussions with CACs (Chapter 9). Following the information-gathering stage, the panel refined the framework of factors and subfactors to derive specific evaluation criteria for comparing alternative technologies (Chapter 10).

3

Framework for Assessing Alternative Technologies

This chapter describes the framework and procedures within which the evaluation factors described in Chapter 2 were used to carry out the work of the AltTech Panel. The first section describes the framework as it was used to produce data-gathering questionnaires. The second section explains the basis for a supplementary consideration that arose during the study—the potential for the off-site treatment of process residuals.

FRAMEWORK FOR THE QUESTIONNAIRES

Because of the short duration of this study, the strategy for gathering data was critical. In particular, a framework for the information needed to address the evaluation factors (see Chapter 2) had to be ready prior to requesting information and making site visits to the Army and the TPCs. Because the panel had limited time for direct meetings with the TPCs and the Army, the panel provided advance notice of the type of information required. The evaluation factors were converted into a "Questionnaire for Technology Assessment" (see Appendix J), which was sent to each TPC and to the Army. Another reason for the questionnaire was to ensure that the proponents had fully considered all aspects of their technologies in the written responses, which the panel would later use to assess the technologies. The panel's evaluation was based on the completed questionnaires, additional data obtained in the course of the site visits, and information from follow-up questions and discussions.

The panel formed itself into technology assessment teams of approximately four members each, based on the expertise of the individual members. Each team was responsible for organizing the site visits to gather data on one technology, for analysis and evaluation of the data, and for the initial draft of the analytical chapter on that technology. The assessment teams reported on the status of their findings and evaluations at the full panel meetings. The analytical chapters were subsequently reviewed, revised, and approved by the full panel. The assessment teams also made follow-up trips and telephone calls as necessary to obtain needed information.

The panel found that the TPCs and the Army were very responsive to the checklist questions. Several data iterations ensued until the cutoff date of April 4, 1996. The absence of data in some responses to checklist items helped the panel and the respondents to focus further efforts where they were most needed. The discussion of each technology in the analytical chapters (Chapters 4 through 8) follows the questionnaire framework, which consisted of the following categories: process description; scientific principles; technology status; operational requirements and conditions; materials of construction; process stability, reliability, and robustness; operations and maintenance; utility requirements; scale-up requirements; facility decommissioning; process safety; and schedule. Each submission was required to provide a total solution to chemical demilitarization at the two sites, including handling and processing containers, treating dunnage and decontamination solutions, as well as destroying chemical agents.

Process Description. A detailed process description was needed so that the panel could understand the overall approach to agent destruction. The panel asked that the description include all available drawings and other materials needed for the panel to evaluate all components proposed as part of a pilot system.

Scientific Principles. To facilitate understanding of the basic physical and chemical principles underlying the technology, the panel asked for complete disclosure of all expected chemical reactions and end products.

Technology Status. The panel was interested in the degree of maturation and proof-of-concept demonstrations of the technology. Technology status proved challenging to evaluate because of the ongoing development of the technologies while the study was under way.

Operational Requirements and Considerations. This category addressed how the process would operate under actual conditions. Operational requirements included all process instrumentation and controls, material and energy balances, and the methodology and locations for disposing of process residuals. Operational considerations included how the bulk containers of agent would be moved from the storage location to the treatment facility; how the agent would be decanted, fed into the process, and treated; how remaining agent and agent heels in the ton containers, as well as the ton containers themselves, would be treated; and how process residuals would be managed, including the treatment and disposition of drained ton containers.

Materials of Construction. In addition to the materials to be used in constructing the facility, this category included questions about process streams, environmental chemistry, qualification of materials for use in the proposed facility, failure modes, material monitoring and inspection, and the previous experience of the TPC in operating the technology at processing rates and operating conditions (e.g., temperature, pressure, and materials) similar to those required for a pilot-scale demonstration of agent destruction.

Process Stability, Reliability, and Robustness. Process stability included consideration of potential deviations from "normal" operations that could lead to uncontrolled reactions or catastrophic failure of the facility. Reliability included information about the reliability of the equipment, such as whether it is in common use in the chemical industry and its performance under comparable operating conditions.

Operations and Maintenance. Issues of interest in operations included staffing and training requirements for operating a facility, the TPC's operational experience with the technology, operational safeguards and control systems, and startup/shutdown procedures. Under maintenance, the panel was interested in maintenance procedures and manuals, down-time expectations, documentation that maintenance was done, equipment replacement procedures, and maintenance staffing requirements.

Utility Requirements. The panel asked for the electrical, water, and fuel requirements for each process. Utility requirements only become a significant consideration if local sources would be unable to meet demand during an agent-destruction campaign.

Scale-Up Requirements. The panel asked at what scale each technology had already been demonstrated and with what feed materials. Other questions concerned the extent to which the process, or parts of it, had been demonstrated commercially, how process streams would increase in mass and volume, and whether scale-up might affect design of the chemical reaction vessel or other key components.

Facility Decommissioning. The agent-destruction facility will be decommissioned after the stockpile is destroyed. The panel asked about the process by which the facilities would be removed and the extent of site remediation needed.

Process Safety. Process safety issues include the potential risks of catastrophic failure and agent release, in-plant risks and hazards to workers, and the risks to the neighboring community and the environment from agent or other hazardous chemicals, whether from long-term, low-level exposures during normal operation or from brief but higher-level exposures after an accidental release.

Schedule. Because the storage risk to the community remains until the stockpile is destroyed, the panel sought to determine the time required to design, construct, and evaluate a pilot plant and the time for construction and systemization of a full-scale facility.

OFF-SITE TRANSPORT, STORAGE, AND PROCESSING OF PROCESS RESIDUALS

Internal Army procedures require special approval for off-site shipment, storage, or processing of wastes derived from agent processing. At the time of the *Commerce Business Daily* announcement of the Army's interest in alternative technologies (see Chapter 1), the program manager for the CSDP (Chemical Stockpile Disposal Program), who is often referred to as the program manager for chemical demilitarization (PMCD), had limited the requests for such approval to individual cases of shipping, storing, or processing contaminated (or possibly contaminated) materials. Examples included contaminated wastes from laboratory work on agents (analyses, investigation of destruction processes, etc.), potentially contaminated salts from the brine reduction systems at the Johnston Atoll and Tooele stockpile sites, and decontaminated personal protective suits from these sites. Special approvals have also been obtained by other parts of the Army for shipping ton containers decontaminated to a 3X status at the Rock Island Arsenal, Illinois, to be melted down, tested, and released for general reuse. Because of the limited conditions under which special approval had been sought or given in the past, the *Commerce Business Daily* announcement requested information only on technologies that would *not* require the off-site shipment of contaminated wastes, except for ton containers treated to 3X condition.

After the announcement and the start of the AltTech Panel's work, the Army recognized that there might be a programmatic advantage to off-site waste treatment by one or more licensed commercial treatment, storage, and disposal facilities (TSDFs) that have both extensive experience in handling hazardous wastes and the facilities to do so. However, uncertainty remained about the capabilities of commercial TSDFs, their willingness to accept the process residuals from an agent-destruction facility, and the costs for their services. Accordingly, the Army conducted a study to characterize the probable residuals from the neutralization processes (for which it had data to specify the residuals) and to determine the likelihood that they would be acceptable for subsequent treatment or disposal, or both. The Army then conducted a survey to acquire information on the general feasibility of and costs associated with various types of off-site shipment and disposal of process residuals.

Although the report on the results of the study and survey is only in draft form (U.S. Army, 1996c) and the Army is continuing to evaluate further details of off-site shipping, the initial results indicated that process residuals probably would be acceptable to several off-site facilities and several commercial facilities are interested in performing such services. The Army also obtained cost information from this survey, but the cost information was not considered by the AltTech Panel in the technical evaluation of alternative technologies.

The CSDP staff has since taken further action by requesting and receiving approval to ship the following items for off-site disposal (U.S. Army, 1996d):

- solid wastes generated from laboratory and monitoring operations: paper; plastic; glass; metal; wood; absorbents; and personal protective equipment (PPE) including gloves, boots, outer garments, and self-contained breathing apparatuses
- liquid wastes from laboratory and monitoring operations: decontamination solutions, acids, alkaline solutions, flammable liquids, rinse solutions, and analytical solutions
- plant wastes: filters (pre-filters, high-efficiency particulate-arresting filters, charcoal filters), PPE, dunnage, spill debris (rags, absorbents, plastic bags, and plastic sheets), brine salts from the pollution abatement systems, demister packing, ash from the furnace systems, and pieces of utility and process equipment

Although this list does not include all process residuals, it does include a number of components that might ease the burden on several of the alternative technologies being evaluated and sets the stage for possible future approval of off-site shipment, storage, or processing of other plant wastes. Although the Army study of this option has not yet been completed and the Army has not yet formally changed its policies, the panel found nothing in the available documentation that would preclude it.

The panel recognizes that procedures will have to be developed, such as setting standards and defining best practices for off-site shipping and treatment. Particulars include the maximum allowable residual concentrations of agent and other toxic components in various residuals, the methods for measuring and verifying the actual concentrations, and pathway constraints to ensure the safety of workers, the public, and the environment. Procedures will also have to be developed to allow verification that all precursors in the process residuals have been destroyed at the off-site location.

In light of this information, and at the direction of the Army as sponsor of the study, the AltTech panel agreed to expand the evaluation framework to include consideration of the off-site shipment and processing of wastes (see Appendix D). The reader should remember, however, that the technologies submitted by the other TPCs represented "total solutions" to chemical demilitarization and included methods for processing ton containers, decontamination solutions, and dunnage, as well as the destruction of chemical agents. Because the Army may not have discussed the implications of a change in Army policy with the TPCs, no modified concept design packages were received from them by the April 4, 1996, deadline. However, because submissions by the Army did include off-site shipment and treatment for hydrolysate from the neutralization processes, these options were considered and are addressed in Chapters 7 and 8.

4

Catalytic Extraction Process Technology

PROCESS DESCRIPTION

Catalytic Extraction Processing™ (CEP™) is a proprietary technology patented by its developer, Molten Metal Technology, Inc., and licensed to M4 Environmental L.P. for specified U.S. governmental applications.[1] M4 Environmental L.P. joined with several other firms to prepare the submission on CEP in response to the Army request for information on alternative technologies.[2] Hereafter in this chapter, M4 Environmental L.P. and its supporting firms will be referred to as the technology proponent company (TPC). In addition to processing of HD and VX, the submission included processing of the steel ton containers and all dunnage generated in the course of demilitarization operations at the two sites. Destruction of HD and VX by CEP is accomplished in a series of unit operations after the ton containers have been opened and the contents transferred to interim storage tanks.

CEP has been designated by the U.S. Environmental Protection Agency (EPA) as a nonincineration technology. The distinction between incineration (or combustion) and CEP is based upon reaction mechanisms as well as end products. Combustion, which occurs by means of a series of gaseous, reactive intermediates (free radicals), requires high temperature, intimate mixing, adequate residence time, and excess oxygen to achieve high destruction efficiency. CEP, by contrast, is conducted mainly within a molten metal bath at high temperature and low oxygen potential. The products of combustion are in high oxidation states (e.g., CO_2, H_2O), whereas products of CEP are in reduced states (e.g., CO, H_2).

[1]M4 Environmental L.P. is a 50/50 limited partnership of a subsidiary of Lockheed Martin and a subsidiary of Molten Metal Technology, Inc.

[2]The other firms participating in the submission are Bechtel National, Inc., Fluor Daniel, Inc., and Battelle Memorial Institute.

Technology Overview

A CEP reactor, which is called a catalytic processing unit (CPU), contains a bath of molten metal, typically iron or nickel. For treating chemical warfare agents, the TPC has decided that two CPUs are required. Each CPU is a steel pressure vessel containing a molten metal bath and an optional slag or flux cover. In CEP, these reactors are typically operated in the temperature range of 1425°C to 1650°C (2600°F to 3000°F). The vessel is lined with refractory materials selected to provide thermal insulation and resistance to corrosion, erosion, and penetration by components of the bath. An electric induction coil, embedded within the refractory lining surrounding the metal bath, provides the energy to melt the metal charge and maintain the temperature of the bath during processing. The CPU headspace, which is several times the height of the molten metal bath, provides physical space to allow disengagement of the offgas from the molten metal and slag. One or more tapping ports through the vessel sidewall allow recovery of metal and slag phases with minimal interruption of operation. One CPU is fitted with a side chamber that can be heated by its own induction coil to melt ton containers. The molten metal flows from the side chamber into the main bath of the CPU. The TPC plans to feed dunnage, placed in steel containers, directly into the metal bath.

The feed material and the cofeeds of oxygen and methane can be injected into the molten metal bath either through a lance entering the top of the bath or through one or more bottom-entering *tuyeres*. (The TPC has used top-entering lances in numerous bench-scale CPUs.) A tuyere consists of three concentric metal tubes cast into a removable refractory block that is bolted into the bottom of the CPU. The TPC proposes using the tuyere injection of liquid agent and cofeed gases for chemical demilitarization.

Feed material, which may be liquid, gas, finely divided entrained solids, or a pumpable slurry, is metered, mixed, and pumped through the central tube of the tuyere at moderately high pressure, less than 10 atmospheres. Oxygen, in stoichiometric proportion to convert all carbon in the feed and the methane cofeed to carbon monoxide, is metered into the next annulus at high velocity to induce turbulence, mixing with the feed stream, and formation of a jet that rapidly breaks up into small bubbles. A small amount of methane is fed through the outer annulus to cool the tuyere.

An inert gas is injected automatically into each of the feed lines as needed to make up the difference between the total flow required in each line and the setpoint flow of each feed component (agent, oxygen, and methane). During startup and shutdown, the inert gas alone is pumped through all feed lines to prevent molten metal from entering and plugging the tuyere.

According to the TPC's description of the process, when feed material is injected into the bath along with oxygen and methane, the molecular entities in the feed material are decomposed by catalysis into their component elements. These elements dissolve in the metal and form intermediates by bonding chemically with the metal. By appropriate selection of process conditions, the dissolved elements with high solubility in the metal (e.g., carbon, sulfur, and phosphorus) can either be retained in the metal bath up to their saturation limit or induced to react with less soluble elements (e.g., hydrogen, oxygen, and chlorine) to form gaseous products—principally H_2, CO, HCl, and H_2S with minor amounts of H_2O, and CO_2. These gaseous products then form bubbles, which ascend and exit the bath. According to the TPC, because CEP is carried out at low oxygen potential and decomposes feed molecules to elements regardless of their starting molecular structure, the process provides neither pathways nor precursors for the formation of oxides of nitrogen or sulfur or the formation of dioxins and furans.

The TPC has reported that it expects the process residuals from treating VX or HD, the ton containers, and dunnage to be ferrous alloys, aqueous hydrochloric acid, elemental sulfur, and a synthesis gas. The TPC also has reported that markets for the alloys, hydrochloric acid, and sulfur have been identified. The synthesis gas is combusted, along with natural gas, in an on-site gas turbine generator to provide electricity used in the process. A small amount of slag or ceramic (less than 5 percent of total solid product mass) is also produced and must be disposed of as waste. The panel agrees with the TPC that this slag is likely to pass the U.S. Environmental Protection Agency Toxic Characteristic Leaching Procedure (TCLP) test. (Unless it is delisted, however, it could still be classified as hazardous waste because it is derived from agent.)

Chemical Demilitarization Process

According to the submitted design, chemical demilitarization operations are to be conducted in a central processing building of approximately 13,000 square feet. The building is partitioned into distinct areas by function (Figure 4-1). Precautionary safety measures confine agent to small areas, reduce the possibility of cross contamination, and reduce requirements for heating, ventilation, and air conditioning (HVAC); high efficiency particulate-arresting filters; carbon filters; and agent monitoring equipment.

Ton containers are opened in area 100 and, if necessary for interim storage, cleaned to 3X condition. Dunnage from daily operations is compacted and packaged in small metal containers in the same area. The equipment and techniques used to handle ton containers, including the punch-and-drain process, vacuum transfer of agent and decontamination liquids to interim storage tanks, safe airlock passage, cascaded HVAC, double-containment envelopes, and low-pressure injection—are based on the equipment and techniques used in the baseline system facilities at Johnston Atoll in the Pacific Ocean and at Tooele, Utah. The only significant change is the addition of an aspirated, self-cleaning gland surrounding the punch to mitigate spillage of agent when the container is penetrated.

The two CPUs, designated CPU-1 and CPU-2, are located in area 200. The gas handling train (GHT) and facilities for product recovery are located in Area 300. Area 500 is devoted to product-gas utilization; products of CEP are stored in area 700; utilities are located in area 800; and area 1000 houses the emergency relief system. The CPUs and the equipment in the product recovery areas are of modular design, which will allow the TPC to use the same CPUs and product recovery equipment at the Aberdeen site to process HD and, afterward, at the Newport site to process VX.

For processing either agent, CPU-2 contains molten iron and processes all ton containers and dunnage. Emptied ton containers are fed by horizontal indexing conveyors and coordinated, double-door, cascade-ventilated airlocks to the premelting side chamber of CPU-2. The steel ton containers melt, and the organics, including all remaining gels, solids, and surface agent residuals, are pyrolyzed. Pyrolysis products and molten

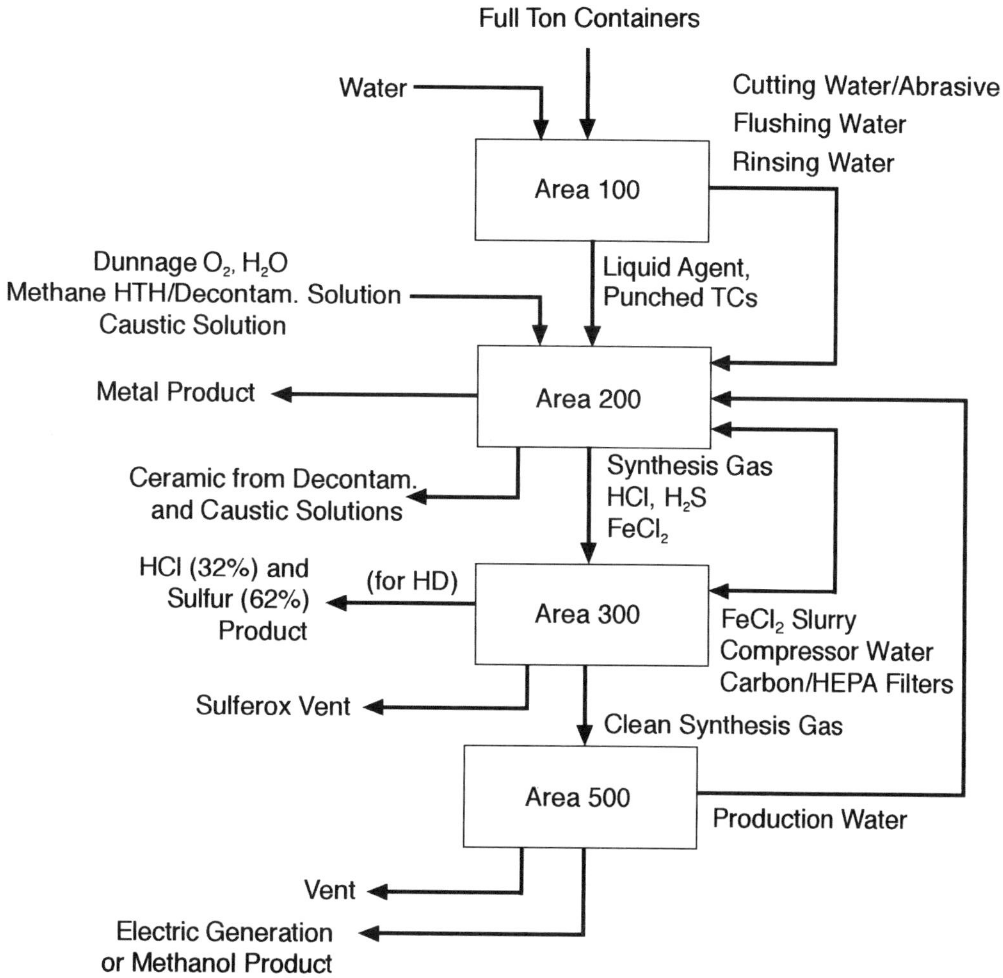

FIGURE 4-1 Primary agent and residue process flows for a chemical demilitarization CEP facility. Area 700 (product storage), Area 800 (utilities), and Area 1000 (emergency relief system) are not shown. Source: M4 Environmental L.P., 1996b.

metal then enter CPU-2 through a side chute above the level of the molten bath. The TPC states that dunnage canisters will be fed directly into CPU-2. If the ton containers are melted as they are emptied, at the proposed processing rate of VX (169 kg/hour) they will add about 725 kg of metal to the bath every 5 hours. This quantity of metal will increase the bath height about 8 cm, necessitating tapping the bath at approximately 10-hour intervals to maintain an optimum level. The metal tap, which will probably be located at the desired bath height, will be opened by heating it to melt the metallic or slag plug. The tap will be closed by cooling it to solidify a metal or slag plug.

Different strategies are required for processing HD (Figure 4-2) and VX (Figure 4-3). In the HD strategy, liquid agent is injected by tuyere into CPU-1, which uses a molten nickel bath to reduce the formation and carry-over of metal chlorides. Chlorine is released from the bath as HCl. Sulfur from the HD accumulates in the bath to a concentration of about 27 percent, a concentration at which sulfur is released from the bath as H_2S. The offgas from CPU-2, which originates from processing the ton containers, any residue in them, and dunnage is quenched with water, pressurized, and injected into CPU-1 to ensure complete reaction of any products of incomplete conversion. Product gases from CPU-1 are quenched with water, filtered, and scrubbed with water to recover aqueous HCl. At this point, the offgas consists primarily of H_2S, CO, and H_2. The H_2S is subsequently converted to elemental sulfur using the commercial SulFerox™ process. The remaining gases, principally H_2 and CO, form the synthesis gas, which is pressurized and stored in one of three

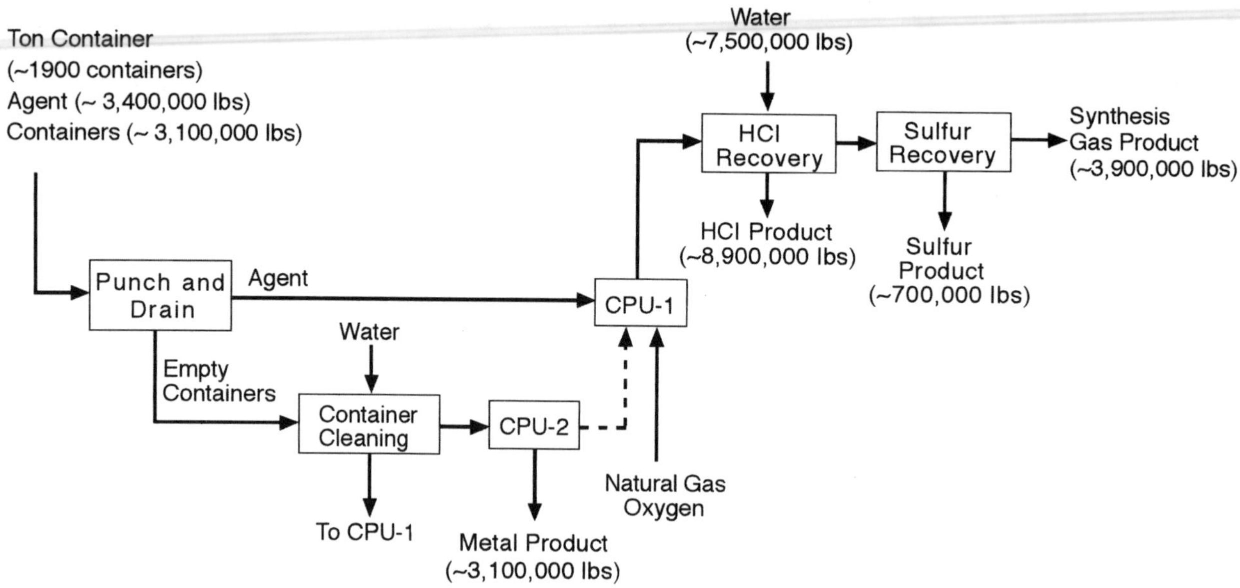

FIGURE 4-2 High level block diagram for the destruction of HD by CEP. Source: M4 Environmental L.P., 1996b.

tanks with a capacity of 4 m^3 each. After a filled tank has been analyzed for agent and other toxics, the gas is combusted in a gas turbine electric generator.

In the VX strategy, CPU-2 is the primary reactor for processing agent. Both sulfur and phosphorus from the VX are held in solution in the molten iron and recovered as an Fe-S-P alloy when CPU-2 is tapped to control the bath level. The offgas from CPU-2 is conditioned as described above for HD and injected into CPU-1, which in this case contains an iron bath and functions as a polishing reactor to ensure the destruction of remaining agent or other organics. The offgas from CPU-1 is quenched with water and filtered to yield the synthesis gas of CO and H_2. Trace amounts of HCN in the product gas are decomposed by catalysis to H_2, N_2, and carbon. The VX strategy uses the same approach as the HD strategy for storing and analyzing the synthesis gas prior to combustion.

In both treatment strategies, aqueous cleaning and decontamination solutions, including particulates and condensates recovered as water-base slurries from cooling and cleaning the CPU offgases, will probably be injected into CPU-2 for destruction, so that all slag-forming components are kept in the same CPU. Slag formed by the interaction of debris entering with the emptied ton containers, lime-based decontamination solutions, and dunnage can be removed in the same way molten metal is removed.

Should the need arise, the facility design includes the capability of opening a ton container with a high-pressure water jet containing abrasive particles. A water spray then removes the gels, residues, and remaining agent, and calcium-based decontamination solution is used to clean the container to 3X condition. The resulting finely divided aqueous slurry can be removed from the cleaning area by aspiration, transported by vacuum pumping to temporary storage, and injected into one of the CPUs for processing to the same residuals as other cleaning solutions and slurries. The use of a water jet, of course, would require suitable enclosure and capture/treatment of effluent from the spray operation.

If a situation arises in which liquids or gases from vessels, piping, or either CPU are vented by means of pressure relief devices, the facility design includes standby equipment to quench the vented material and absorb acid gases. Any residual agent or H_2S is combusted in a standby boiler prior to releasing the gaseous residual to the atmosphere.

SCIENTIFIC PRINCIPLES

The TPC and the developer of CEP describe the molten metal bath as a dissociation catalyst for molecular entities in feed materials, a solvent for elemental

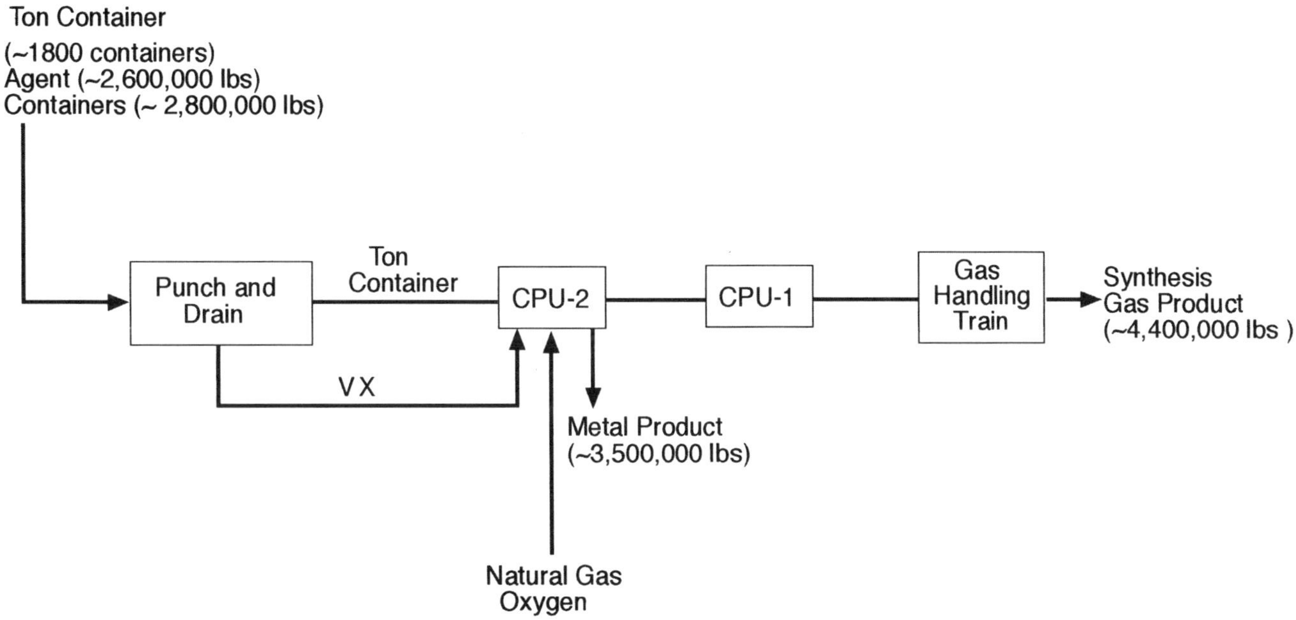

FIGURE 4-3 High level block diagram for the destruction of VX by CEP. Source: M4 Environmental L.P., 1996b.

fragments, and a medium for product synthesis. The TPC divides the process conceptually into stages comprising catalytic dissociation of the feed, formation of elemental intermediates with the solvent metal, product synthesis by interaction of elemental intermediates, and partitioning of products among metal, slag, and gas phases. A recent publication by technologists who work for the developer of CEP states, "the CEP unit is not acting as a thermal treatment device in that temperature is not the primary means to change the physical and chemical composition of the feed material . . ." (Nagel et al., 1996, p. 2158).

The above description does not address initial thermal and gas-phase reactions in the overall sequence of events between the introduction of feeds and the release of final products. Although bench-scale tests of the process have demonstrated that the process can destroy agent as required by the Army, analysis by the AltTech Panel indicates that the actual conditions are probably more complex than this description implies. The panel's review indicates that a complete description of the scientific principles underlying CEP requires discussion of several additional phenomena, including gas-phase reactions among agent, oxygen, and methane in the inlet jet immediately following tuyere injection; interactions of these gases and intermediate products with metal vapor inside bubbles; and boundary reactions between bubble components and the surrounding metal. Accordingly, the following discussion attempts to provide a more detailed description of the probable scientific principles and further develops details of the probable processes involved.

The TPC notes that the submitted design reflects many years of experience in the steel industry with injecting gases into molten steel baths by the use of similar tuyere inlets. However, experience in the steel industry relates primarily to the injection of gases for the purpose of changing the composition of the bath. The escape of a small surplus of these gases from the bath surface is of little concern other than as an economic loss. Thus, there is no long-established precedent from industrial experience for the complete reaction of injected gases with a molten metal bath to the very low level of residuals required for agent destruction. The panel is not aware of industrial experience with injecting liquids into a molten metal bath.

Dissociation and Reaction of Tuyere-Injected Materials

In the CEP, a liquid agent or other feed to be destroyed, inert carrier gas, oxygen in stoichiometric proportion to oxidize all carbon in feeds and cofeeds to CO, and methane are injected by tuyere at moderately high pressure (less than 10 atmospheres) and high velocity into the molten metal bath. The injected materials form a jet that extends several tuyere diameters into the bath. The high velocity of the oxygen gas stream causes turbulence and contributes to entrainment of metal vapor and droplets within the jet. These effects of the initial momentum quickly dissipate, and the jet breaks into bubbles that rise through the molten metal because of their buoyancy. Subdivision of larger bubbles increases the total surface-contact area and increases the collision frequency between gas molecules and the molten metal. As the bubbles rise to the surface, they continue to change in size for several reasons. They tend to increase in size as the ferrostatic head decreases; they tend to decrease as gaseous intermediates are absorbed into the molten metal; and they tend to increase as product gases released from the molten metal migrate back into them. Some very small bubbles may also form through the nucleation of gases produced in the molten metal and then grow as they agglomerate with other bubbles or accumulate more gas released from metal.

Radiant heat transfer from the hot metal to the aspirated liquid droplets and gas bubbles is extraordinarily rapid at the high temperature of the bath because the rate of radiant heat transfer is proportional to the fourth power of the absolute temperature. For example, a hypothetical sphere 100 μm in diameter will receive energy at 1600°C at the rate of 5×10^{-3} calories per second, which is sufficient to vaporize a like volume of liquid agent and heat the resultant vapor, as multiple 100-μm bubbles, to 1000°C in less than 50 milliseconds. The panel's judgment is that partial degradation of agent and gas-phase reaction between agent or agent fragments and oxygen is very likely under these circumstances. A significant fraction of the feed probably undergoes partial oxidation, and the products of partial oxidation then interact with the molten metal to form intermediates. The panel also concludes that oxidation is probably not complete and should not be termed combustion, even though reactions proceed stepwise by molecular collisions among gas-phase intermediates.

Increasing the effective pressure of the bubbles increases the gas density and therefore the collision frequency between bubble contents and the molten metal. Thus, increasing the operating pressure of the CPU or increasing the bath depth increases the rates of reactions in the bubbles. The TPC has ascertained that the processing rate for a given reactor increases significantly with an increase in operating pressure.

An important issue is whether there is opportunity for back reactions to form complex organic compounds from intermediates. The assumption that the opportunity is negligible is important to the TPC's statement that no detectable recombinant dioxins or furans are produced. However, it is possible and thermodynamically feasible to produce HCN in the conditions of the CEP bath when processing VX. In the original submission from the TPC, the inert gas was specified to be nitrogen. The TPC has subsequently considered using argon for this purge/make-up gas. For processing HD at least, using argon instead of nitrogen would resolve the issue of HCN formation by removing any source of nitrogen. Although the extent of HCN production can be controlled to very small concentrations, the fact that it does occur indicates that the claim that no detectable recombinant dioxins or furans (i.e., complex compounds) are produced does not apply to simple compounds like HCN.

Dissolution kinetics are also important to the formation of intermediates. For example, hydrogen is sparingly soluble in molten iron, and when organic compounds containing hydrogen are injected into molten iron, hydrogen gas evolves from the bath while the carbon dissolves in the metal. It is also reasonable to expect that the initial bubbles formed by the break-up of the jet contain H_2. (If nitrogen were used as the inert make-up gas, N_2 would also be a significant component of the initial bubbles.)

Catalysis by the Bath and the Formation of Intermediates

There is ample evidence in the peer-reviewed literature to support the TPC's position that the molten metal bath serves as a true catalyst by decreasing the activation energy for dissociation of organic molecules, participating in the formation of intermediates, and increasing the efficiency of product formation without itself undergoing change (Satterfield, 1991). Given the formation of intermediates, their relative solubilities in the metal are another factor to consider, particularly for the VX strategy, in which some elements are to be retained in the bath while others exit as offgas.

The panel estimated the solubility of VX components in the bath and the time required to saturate the bath under processing conditions of 1600°C and the proposed feed rate (Table 4-1). Columns 2 and 3 list the saturation solubility (in parts per million by weight) and the total weight of elements in the bath, based on a reasonable

TABLE 4-1 Calculated Solubility of VX and Cofeed Elements in Iron at 1600°C and Time to Saturate the Iron Bath at Processing Conditions

Element	Solubility in Bath ppm	Solubility in Bath kg[a]	Feed Rate kg/h[b]	Time to Saturate Bath h
C	54,000[c]	442	87.4	5.05
H_2	25[d]	0.20	17.8	0.011
P_2	110,000[c]	892	19.6	45.5
O_2	1,290[c]	10.6	130.7	0.081
S_2	110,000[c]	892	20.3	43.9
N_2	88[d]	0.14	8.9	0.016

Notes
[a]Bath assumed to contain 8,163 kg iron; contribution of dissolved elements was not considered.
[b]Feed rates: 169 kg/h VX agent; 110 kg/h oxygen; and 5 kg/h methane.
[c]From Massalski, 1986, pages 842 (C), 1746 (P), and 1762 (S).
[d]From Rao, 1985, pages 438 (H_2) and 463 (N_2).

assumption of the partial pressures of the gases derived from the feeds. Column 5 lists the time required to saturate the bath at the elemental feed rate given in column 4, which is derived from the molecular composition of the feed and cofeeds and their feed rates. These values are only computational estimates; numerous simplifying assumptions were needed, and interactions among bath components were ignored. However, the calculations do illustrate the following points.

Bath Saturation Point for Retained Elements. Because the solubilities in molten iron of carbon, phosphorus, and sulfur are significant, amounting to 5.4, 11, and 11 wt pct, respectively, considerable time is required to saturate the bath with these elements. The TPC's strategy for VX calls for controlling the release of phosphorus and sulfur gases (preventing breakthrough) by keeping the bath below saturation. The strategy is to remove alloyed bath metal at intervals by tapping, while adding molten iron by processing ton containers. Once the bath reaches saturation for phosphorus or sulfur, the ton containers must be processed at a rate sufficient to supply enough new iron to alloy all the phosphorus and sulfur in the agent feed. The calculated values in column 2 of the table indicate that the amount of iron in a ton container, 636 kg, will dissolve only about 69 kg of sulfur and a similar quantity of phosphorus. The 682 kg of VX within a ton container contains about 82 kg of sulfur and 79 kg of phosphorus. Although these calculations are based on numerous simplifying assumptions, they indicate that synchronizing the addition of iron to the bath with the agent feed rate will be critical in avoiding the breakthrough of sulfur and phosphorus into the offgas. In particular, these computations indicate that the TPC's suggestion of stockpiling ton containers for treatment at a later date while processing VX is not an option unless there is a significant alternative iron feed.

Hydrogen and Nitrogen. The solubilities of hydrogen and nitrogen in molten iron are extremely low, and Table 4-1 suggests that the bath will become saturated with these elements in less than 1 minute. Although the bath, when in continuous operation for processing VX, is likely to be saturated with hydrogen and nitrogen, the kinetics indicate that significant proportions of hydrogen and nitrogen in the feed may not pass through metallic intermediates but may form gas bubbles directly. Supersaturation of the bath as a whole with these and other sparingly soluble elements is likely because the feed materials are introduced into the bath at the bottom, where the ferrostatic head is greatest.

Oxygen. The solubility of oxygen in molten iron is much greater than hydrogen or nitrogen but far less than carbon, sulfur, or phosphorus. The calculated time of less than 5 minutes for the bath to become saturated reflects the high feed rate. The solubility of oxygen favors the formation of an iron-oxygen intermediate.

These calculations indicate all components in the feeds and cofeeds are soluble enough to support the TPC's description of the formation of elemental intermediates. Given the formation of elemental intermediates, product synthesis can occur by chemical reaction among those intermediates.

Partitioning of Products among Metal, Slag, and Gas Phases

To some extent, the process residuals from CEP can be customized by adding appropriate cofeeds or controlling operating conditions. As noted above, the design specifies that oxygen cofeed is provided in stoichiometric proportion to convert carbon in the feed material and the methane cofeed to CO at the desired carbon concentration and temperature of the bath. The oxygen stoichiometry determines the ratio of CO to CO_2 in the product gas, and this ratio is monitored as a process control on the oxygen feed rate. Hydrogen appears as H_2 in the product gas because the oxygen potential in the bath is less than the potential required to form significant amounts of H_2O. Similarly, SO_2 and NO_x formation are thermodynamically unfavorable.

For processing HD, sulfur can be recovered in the gas phase by allowing sulfur in the bath to increase to a saturation concentration above which the formation of H_2S from H_2 and the Fe-S intermediate is thermodynamically favored. Or, sulfur can be recovered as an alloy element by tapping bath metal from the CPU before the saturation concentration is reached, as the TPC proposes to do for processing VX. The chemistry of phosphorus, although more complicated, is similar in that phosphorus can be obtained as an iron alloy by tapping the metal before the saturation concentration is reached. The panel notes, however, that although CEP has been performed extensively with iron baths containing carbon, sulfur, and chlorine, to the panel knowledge it has not been performed with iron baths containing phosphorus in addition to carbon and sulfur.

Metals such as aluminum, calcium, and silicon that form oxides that are more stable than CO at the operating temperature will be oxidized and will accumulate in the slag phase (as Al_2O_3, CaO, and SiO_2, respectively). Cofeeds may be required to ensure the slag is sufficiently fluid. For example, silica and lime are appropriate cofeeds if the feed material contains appreciable aluminum or alumina. Metals whose oxides are less stable than CO will either accumulate in the molten metal (Co, Cr, Cu, Ni, Mn) or exit the bath as vapor (Cd, Pb, Zn).

Iron is the preferred bath metal for processing VX. However, if iron were used to process HD, there would be substantial formation and carryover of $FeCl_2$ vapor, which would form a dust in the downstream systems, requiring a more extensive dust removal strategy than the particle filters included in the current design. The use of a nickel bath for processing HD reduces this problem because $NiCl_2$ is less stable than HCl and does not form to a significant extent. Nearly all of the chlorine from the HD forms HCl and is recovered in the aqueous scrubber. Under the same processing conditions, a nickel bath will become saturated with sulfur in about the same time as an iron bath of equal mass and will become saturated with carbon in less than half the time of an iron bath.

Process Modeling

The most important consideration to the panel, in light of the short residence time of bubbles in the bath, is whether agent or significant fragments of agent can avoid decomposition by remaining in or migrating to a bubble and passing unreacted through the bath. An analysis of the probability and consequences of the requisite reactions at the molecular level would involve complicated computations dependent on numerous assumptions. Instead, it is customary in such circumstances to use engineering models that work from both basic principles and experimental data to provide an approximation adequate for design purposes. The TPC has done extensive experimentation and modeling to understand bubble formation, break-up dynamics, and the operating limits of CEP performance. The models used by the TPC indicate that the process depends heavily on three factors: (1) bubble size, with the critical largest-bubble diameter being on the order of a fraction of an inch (the actual size is proprietary); (2) residence time, with the typical single-path residence time being a fraction of a second (actual time is proprietary); and (3) an energy dissipation term that reflects the degree to which metal vapor and droplets inside the bubbles increase the gas-metal contact.

Although these models were developed and used by the TPC, the panel did not review or evaluate them in detail for this report. Rather, the panel has relied upon the TPC's representations that the model results correlate well with the very high DRE (destruction removal efficiency) values that were achieved in the experimental and commercial-scale demonstration reactors to which the models were applied. The TPC has stated that it intends to use a residence time that provides a design safety factor of at least 10 to assure the destruction of VX or HD agent to at least the required six 9's DRE (99.9999 percent).

Conclusions on the Underlying Science

The TPC's explanation of CEP performance is based upon accepted free energy principles.[3] The panel believes the engineering design models used to design the system have been based upon solid scientific data. The panel did not, however, review these models in detail.

The TPC's original submission did not include equipment for holding the synthesis gas until analysis had ensured the complete destruction of agent or other toxic components prior to combusting the gas in a gas turbine or using it in some other way. However, in response to the concerns of communities near the storage sites, the TPC has subsequently changed the design to include three 4-m^3 storage tanks, in parallel, in the synthesis gas line prior to the gas turbine. Each tank has the capacity to store 15 minutes of anticipated output of synthesis gas pressurized to 20 atmospheres, gauge (300 psig). This storage capacity allows the synthesis gas to be analyzed before it is used as a fuel and the emissions are released to the atmosphere.

The proposed design for a chemical demilitarization facility is undergoing continuous development as the TPC accumulates operating experience in other applications. The opinion of the AltTech Panel is that the process is adequately understood and satisfactorily engineered at this time to process either HD or VX successfully and safely, when operated properly, to meet the required six 9's DRE.

TECHNOLOGY STATUS

The information available to the panel on CEP operational units is summarized in Table 4-2. As of early 1996, the TPC reported more than 15,000 hours of molten-metal test experience with its reactors. Much of this experience was in tests on the 10 to 15 bench-scale units at the TPC's Fall River site. The nominal bath size of these units is 4 to 9 kg.

Fall River Demonstration Unit

The Fall River Demonstration Unit (Demo Unit) is the largest operational CPU. As of April 1996, the longest period of continuous, commercial-scale operation in this unit while processing liquid or gaseous organics was 120 hours, during which 1,680 kg of feed was processed. The associated on-stream factor was between 50 and 80 percent, depending on experimental requirements.[4] The TPC plans to use an on-stream factor of about 82 percent for the CPUs for destroying HD and VX at Aberdeen and Newport.

The TPC also reports that the Demo Unit was used to demonstrate the long-term operability, reliability, and product performance of CEP as a contractual milestone prior to an agreement with a major chemical manufacturer to build a commercial facility. The 93-hour test included a switch-over from injecting solid feed material (biosludge) to injecting heavily chlorinated liquid organic material (RCRA waste F024). The TPC reports that the results of this test surpassed more than 40 performance criteria (for environmental protection, product quality, reliability, operability, feed injection, etc.) established by the customer, Hoechst Celanese. The reported test results included an on-stream factor up to 90 percent, mass balance closures at 100 percent, and feed injection rates that met commercial-operation requirements. The TPC reported that steady-state operational requirements were met and surpassed (validated by on-site customer evaluations), as demonstrated by the steady-state production of high-quality synthesis gas that met the customer's on-site recycling requirements.

---

[3]The panel wishes to thank Dr. Nev A. Gokcen, former supervisor (retired), Thermodynamics Laboratory, Albany Research Center, Bureau of Mines, for his help in discussing the applicability of the free-energy equations used by the TPC as taken from Table C-3 (p. 892) of *Stoichiometry and Thermodynamics of Metallurgical Processes* (Rao, 1958). The text identifies the equations as the "standard free energy change between the Raoultian to the 1-wt.% standard state."

[4]The on-stream factor, or availability, is defined for this chapter as the number of days per 360-day year a facility is fully operational.

TABLE 4-2 Status of CEP Units from Bench Scale to Commercial Scale[a]

Location	Reactor Units	Nominal Metal Bath Size (kg molten metal)	Development Scale	Comments
Fall River, Massachusetts	10–15 CPUs	4–9	bench	Much of TPC's bath operating experience is with these experimental units.
	APU-10	450	pilot	Repeated continuous runs of >100 hours each. Tuyere injection of liquid chlorinated organic feed.
	Variable Pressure Reactor	68	pilot	Demonstrated hot metal operation for >700 hours. Automated heating to maintain bath temperature.
	Demo Unit	2,700	commercial size	Used for demonstrating CEP at commercial scale.
Quantum-CEP Oak Ridge, Tennessee	RPU-1	45	bench	Used for depleted uranium hexafluoride. Panel observed unit in operation.
	RPU-2 (2 units)	~9 (per unit)	bench	Used for treatability studies. Panel observed unit in operation.
	RPU-3	450	pilot	Has performed more than 15 small-scale tests and a 27-hour pilot test.
	RPU-4 "Combo"	1,360	commercial size	Bath size expandable to 3,200 kg. Under construction for summer 1996 startup. To be used to demonstrate CEP at commercial scale.
SEG-Q-CEP Oak Ridge, Tennessee	2 units	up to ~900	commercial size	For batch-mode volume reduction of radioactive ion-exchange resins. Processed >27,000 kg of resins as of May 1996.

[a]Table data based on information from Valenti, 1996, and M4 Environmental L.P., 1996b.

Oak Ridge Facilities

The Quantum-CEP reactor units at the TPC's Oak Ridge site are referred to as RPUs (radioactive processing units). Members of the AltTech Panel observed the bench-scale units at Oak Ridge in operation during site visits.

The SEG/Quantum-CEP units are located at a separate site in Oak Ridge and are designed for batch-mode commercial operations. Each campaign will consist of a 36-hour startup, 3 to 5 days of injection of radioactive ion-exchange resins, and a 36-hour shutdown, for a total campaign duration of 6 to 8 days. During the panel's site visit in March 1996, the SEG facility at Oak Ridge was still in scale-up activities using nonradioactive resins, prior to commercial operation. As of May 1996, the facility was reported to have processed more than 27,000 kg of ion-exchange resins. The TPC reported that a peak throughput rate of 150 percent of design had been achieved and that equipment upgrades were being made.

Agent Testing

Battelle/Columbus Laboratory (a member of the team that prepared the TPC submissions) has tested agent destruction in a bench-scale CEP unit. The TPC has issued a news release reporting a "destruction percentage" of eight 9's (99.999999 percent) for processing HD and VX (M4 Environmental L.P., 1996a). From the AltTech Panel's preliminary review of the full report on these tests, the panel concludes that the tests demonstrated that the CEP technology can destroy agent to at least the six 9's DRE required by the Army. Further implications of the test results for a full-scale operation are discussed below in the section on Scale-Up.

Summary of Technology Status

The development of the various subsystems required for a chemical demilitarization facility has been demonstrated by successfully injecting feed materials, generating process products, and achieving high on-stream factors at developmental facilities.

A wide range of materials has been processed, including polystyrene with graphite, ion-exchange resins, acetone, industrial biosolid waste, chlorotoluene with heavy organics, chlorobenzene, fuel oil with chlorotoluene, dimethyl acetamide with heavy organics, benzonitrile, diazinon, diazinon with sulfur, and surplus metal components. These materials have been in various physical forms, including liquids, slurries, fine solids, and bulk solids. Various feed-addition systems, including configurations with a top-entering lance or a bottom-entering tuyere, have been studied. Successful tuyere injections of liquids, slurries, and fine solids have been demonstrated in which the injection rates and the reactor design were optimized for steady-state operations. Injection rates comparable with commercial levels have been demonstrated at both the demonstration-scale and advanced processing units.

Bulk additions of metal components, scrap metals, and wood have been demonstrated at feed rates comparable to commercial scale and with successful conversion of materials. The TPC's design for processing bulk solids uses two reactors. The receiving unit includes a premelting chamber for melting and volatilization. The second unit is used to polish the offgas from the first unit.

Panel Summary of Technology Status

As of May 1996, the TPC has accumulated considerable test experience with CEP technology, as described above, and is gaining commercial experience. However, the TPE does not yet have extended, continuous commercial experience with CPUs of commercial size.

PROCESS OPERATION

Process Description

The TPC provided the following process diagrams, which will be referred to in this and ubsequent sections as needed:

- Block flow diagram for CEP facility (Figure 4-4)
- CEP process flow diagram for VX feed injection system into CPU-2 with premelting chamber for ton containers (Figure 4-5)
- CEP process flow diagram for VX CPU-2 offgas treatment (Figure 4-6)
- CEP process flow diagram for VX CPU-1 gas handling train (Figure 4-7)
- CEP process flow diagram for VX relief system (Figure 4-8)
- CPU block diagram and material balances for HD treatment (Figure 4-9)
- CPU block diagram and material balances for VX treatment (Figure 4-10)
- CEP heat and material balances for VX gas handling (Table 4-3)

Agent Detoxification

Residual Agent

Based on tests using HD, VX, and agent surrogates as CEP feed materials, the TPC anticipates a DRE for each agent in excess of six 9's (99.9999 percent). If, as the result of equipment failure, operator error, or some other circumstance, residual agent remains in the synthesis gas emerging from the gas handling train (see Figure 4-7), it can be detected in the hold-up tanks before the gas is released to the energy recovery system for combustion. If analysis of a tank detects the presence of agent above the six 9's DRE limit, the contents can

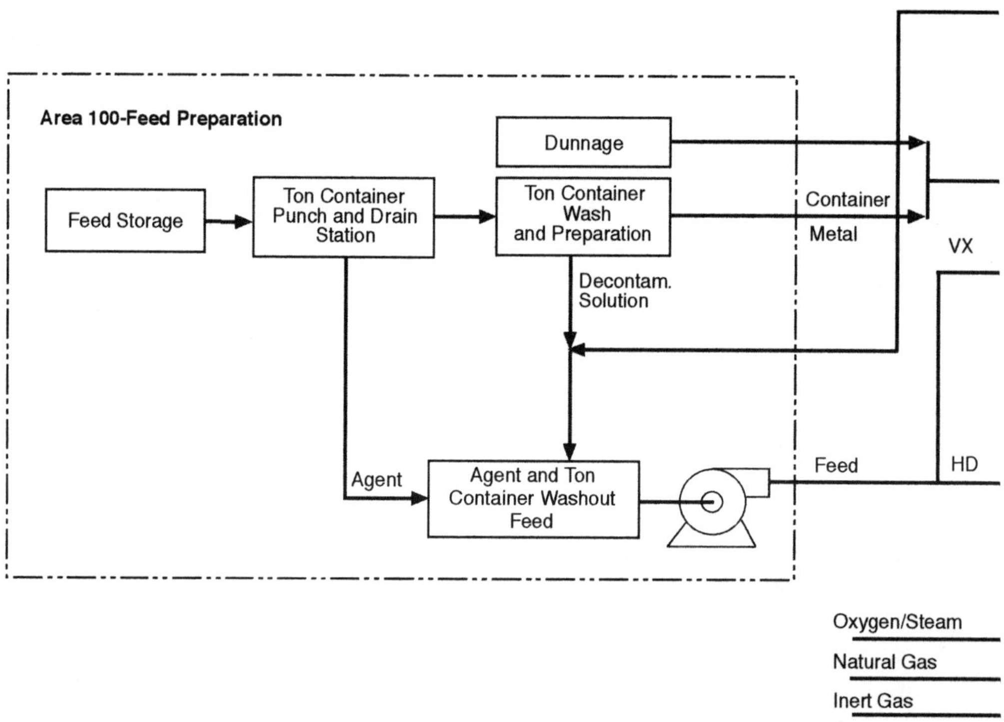

FIGURE 4-4 Block flow diagram for CEP facility. Source: M4 Environmental L.P., 1996b.

be recycled to the appropriate CPU for retreatment. Neither the TPC nor the panel expects that agent or other off-specification gases will be emitted from the process.

In a case requiring venting gases from the CPUs, piping, or other vessels by way of the pressure relief system (Figure 4-8), the on-line caustic scrubbers would further destroy any agent that might potentially

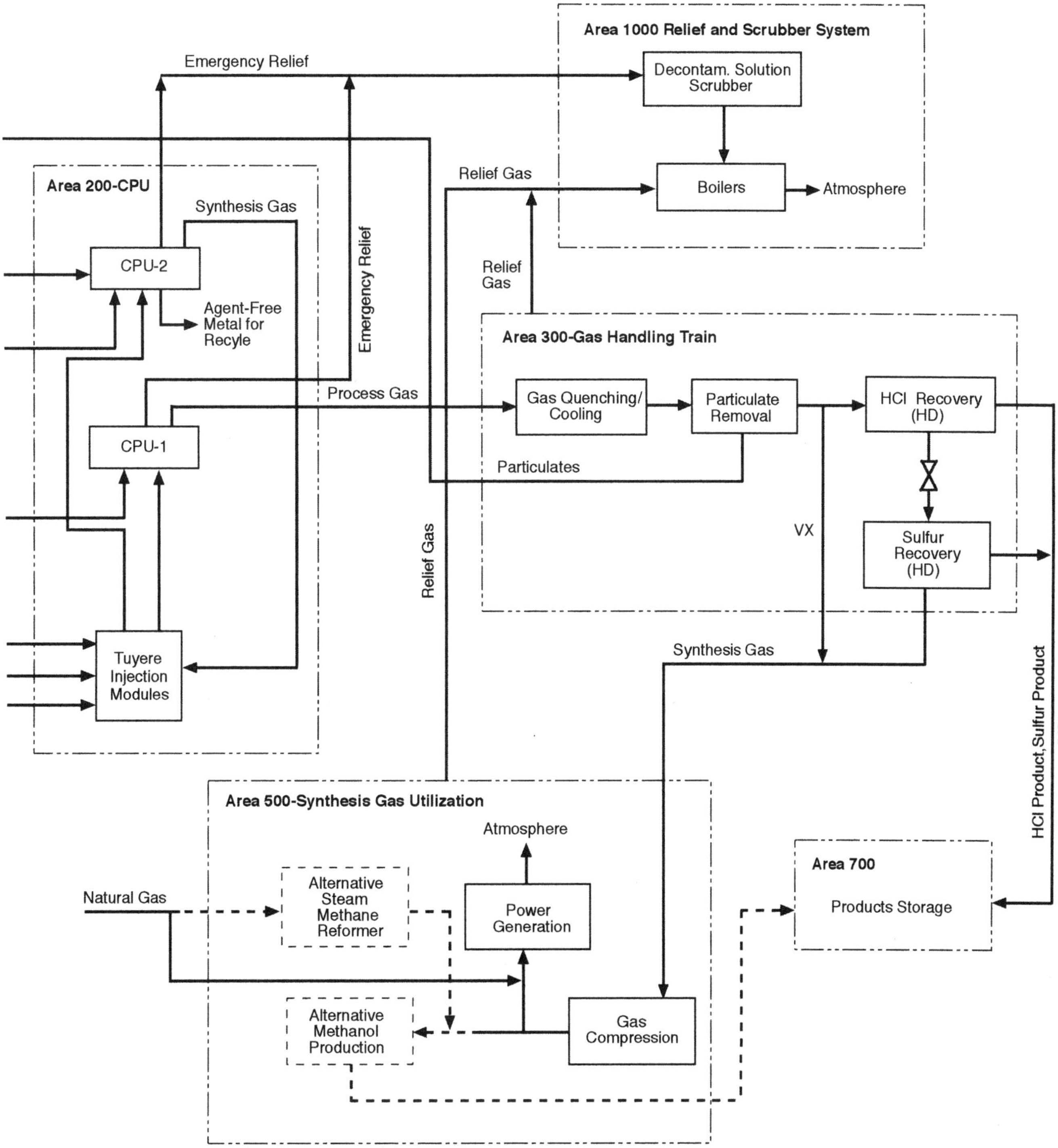

enter the relief header downstream of the reactors. (The exact level of destruction is not known, but it would be more like a 3X condition than a 5X condition, if agent did in fact exit the CPU.) Only under unusual circumstances would the relief system be exercised. If it is, the only residuals would be the scrubber liquor wastes, which would not contain agent above the 3X level.

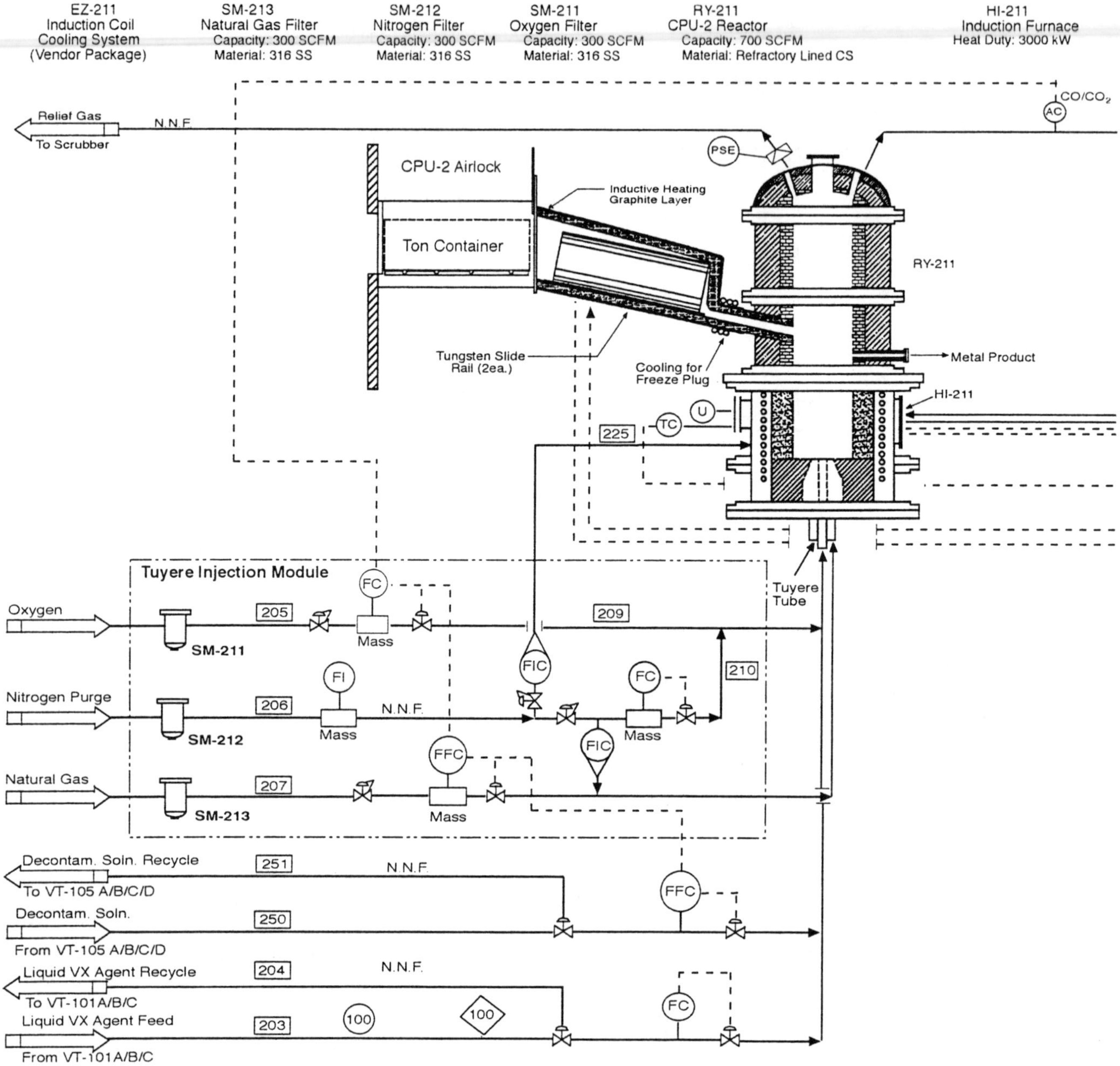

FIGURE 4-5 CEP process flow diagram for VX feed injection system into CPU-2, with premelting chamber for ton containers. Source: M4 Environmental L.P., 1996b.

Reversibility of Reactions to Reform Agent

None of the process reactions is reversible to the extent that agent could be reformed. The formation of chemical warfare agents as unintended by-products in the product stream from CEP treatment of HD or VX is not possible under the proposed operating conditions. The reaction paths and conditions required for the production of HD or VX from species in the product gas stream will not be present in an operating CEP plant.

Toxicity of Process Residuals

The solid, liquid, and gaseous residuals from the process are discussed below in the section on Residual Streams. The process as designed does not produce

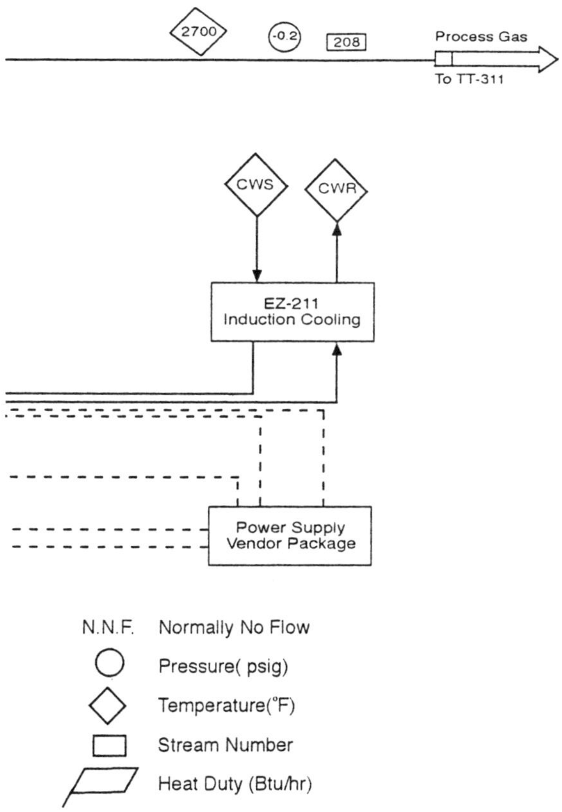

residuals with toxicities that are known to be hazardous to human health or the environment.

Cleaning Out Ton Containers

It is not necessary to remove all residual agent from the ton containers prior to their destruction by CEP. The procedure presented by the TPC ensures detoxification to the Army 5X standard because the containers are melted, a treatment at more severe conditions than the conditions required by the 5X standard. Analysis of ton containers prior to processing is not necessary, provided they are not stored prior to CEP treatment. (Interim storage of emptied containers would require cleaning to the 3X standard.) The molten metal and slag phases from CPU-2 will be cast into ingot or slag molds, as

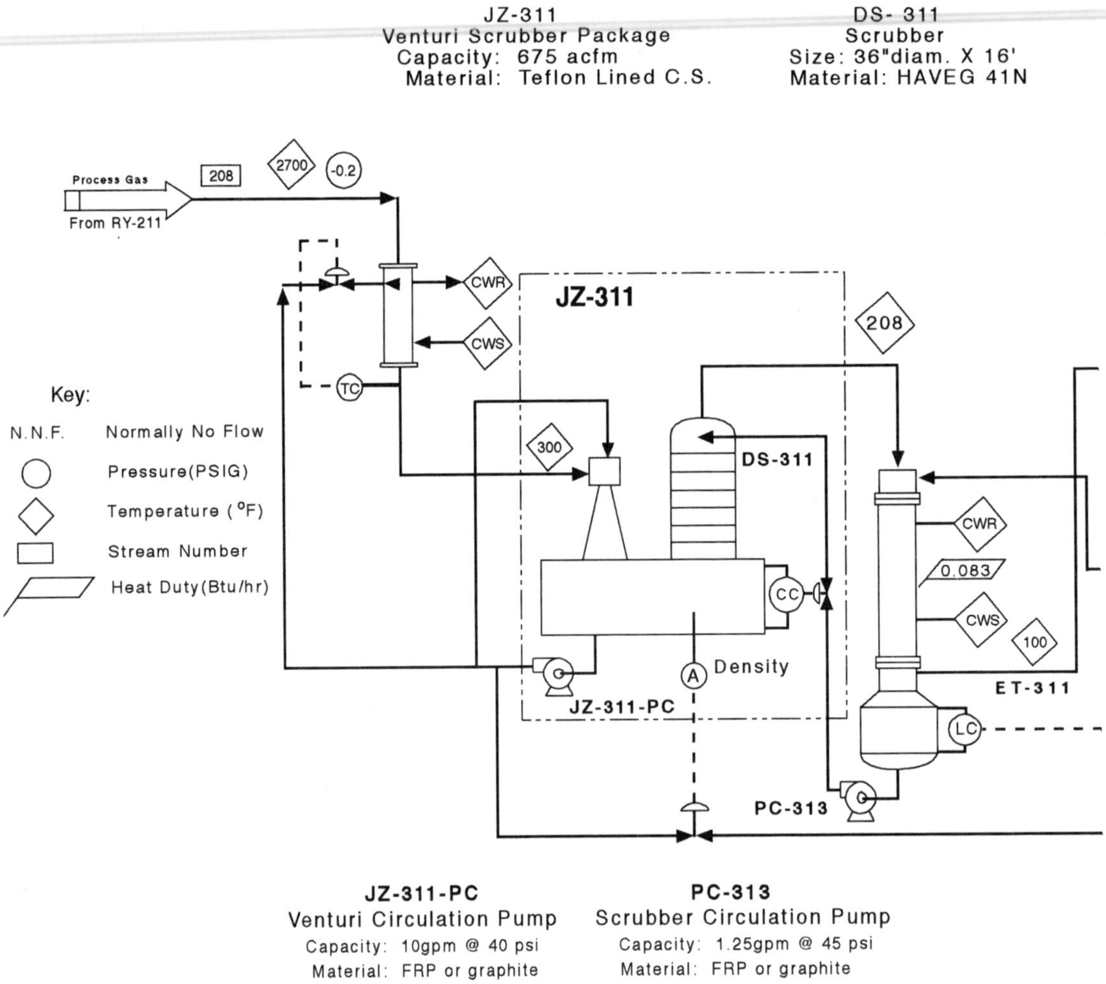

FIGURE 4-6 CEP process flow diagram for VX CPU-2 offgas treatment. Source: M4 Environmental L.P., 1996b.

appropriate. The metals will be offered for sale, and the slags will be committed to an appropriate landfill, as determined by TCLP testing.

Operational Modes

Substantial time is required to heat the CEP system, including the CPUs and the gas handling trains, to operating temperature or to cool the system from operating to ambient temperature. Therefore, it is preferable to operate a CEP facility continuously, 24 hours per day, for extended periods. The units can be kept in a shut-down-but-ready mode if electrical power to the induction coils keeps the bath near operating temperature and if the tuyeres are kept open by maintaining flows of inert gas through the feed lines in place of the agent, oxygen, and methane feeds.

Startup and Shutdown

As explained above, it is preferable to operate a CEP facility continuously, 24 hours a day. Startup and shutdown typically cause the greatest wear on the process equipment. Although operating the system for only 8 hours a day is technically possible, it is not a reasonable approach. Startup of the CPUs requires:

- opening the vessel and filling it with a weighed quantity of iron or nickel spheres (or other metal shapes)
- installing the gas-fired headspace heater
- starting the systems for handling offgas from each CPU
- starting inert gas flow through the tuyeres to keep them open and cool as the bath metal heats and melts

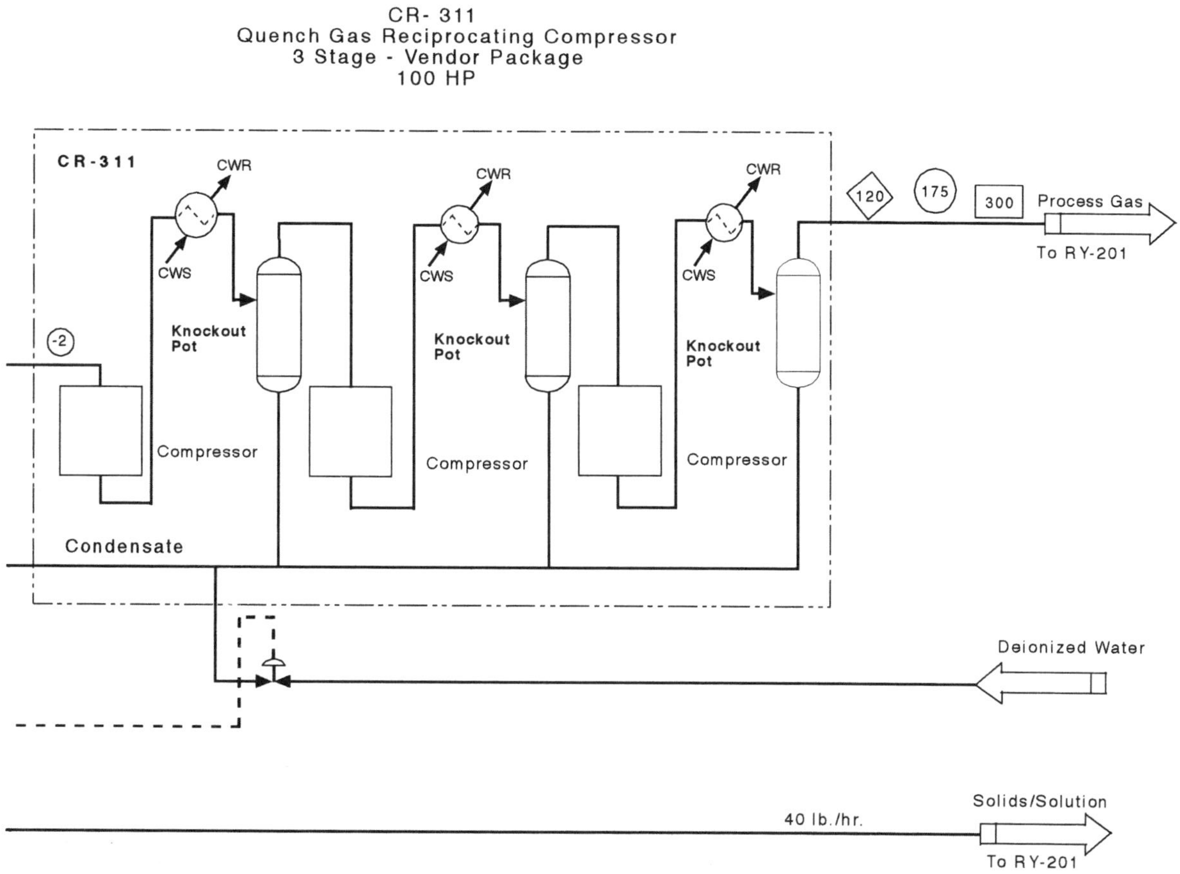

- preheating the CPUs with a gas-fired heater through the critical metal melting stage
- inserting additional metal, if required to adjust metal level
- stopping and removing the preheater and closing the reactor vessel
- turning on electrical heaters to gradually heat the downstream equipment for the gas handling systems (to avoid too rapid heating of the Haveg™ or other special materials in the HCl recovery area)
- switching from inert gas feed to feed streams of methane, oxygen, and finally agent

Shutdown to a hot standby mode requires gradual substitution of an inert gas for agent, oxygen, and methane to keep the tuyeres open; readjustment of the electrical power to keep the baths molten; and maintaining the gas handling trains for both CPUs at operating temperature. Restart from hot standby is the reverse of this shutdown procedure.

Moving to a cold shutdown from a hot standby mode requires that the metal and slag be drained and that the CPUs be allowed to cool. Failure to drain the units would require breaking out the solidified metal and replacing the refractory.

CPU-2 Operation

The configuration and operation of the CPUs are similar except that CPU-2 has a side chamber to melt the ton containers. Emptied ton containers, which may contain agent residues, enter this premelting chamber on CPU-2 by means of horizontal indexing conveyors and coordinated double-door, cascade-ventilated airlocks. The chamber is purged with inert gas, and the chamber induction coil is activated to heat the

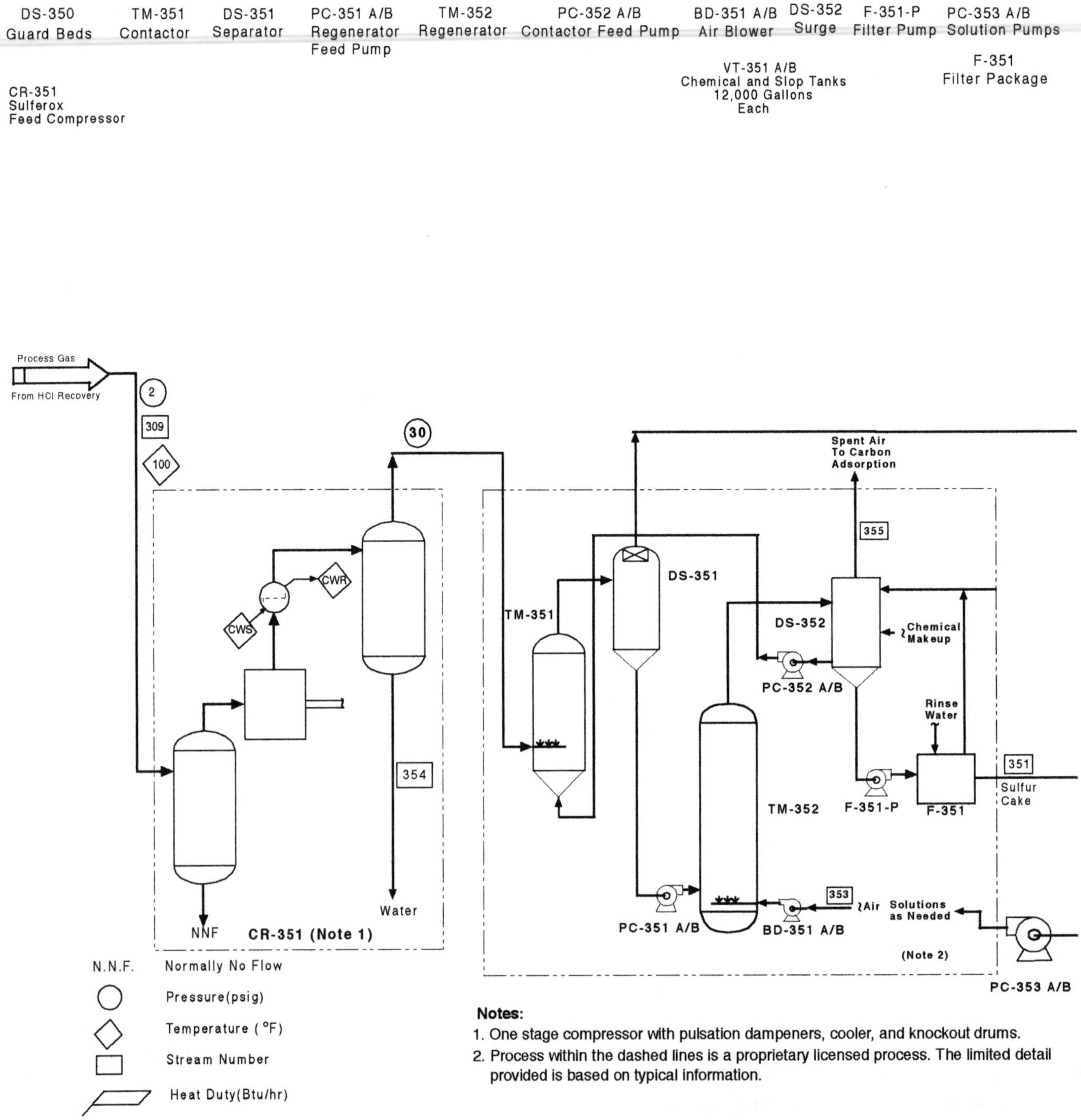

FIGURE 4-7 CEP process flow diagram for VX CPU-1 gas handling train. Source: M4 Environmental L.P., 1996c.

chamber and melt the metal. Visual observation through a viewport determines when melting is complete.

Molten metal is tapped from CPU-2 at intervals, as needed to maintain appropriate bath depth and remove Fe-S-P-C alloy (in the case of VX processing). The bath is tapped by opening a proprietary-design tapping nozzle on the side of the bath. The tap is opened by heating to melt the solidified metal plug. The molten alloy flows out into a mold. When the desired amount of alloy has been removed, the heating is replaced by cooling to solidify the molten alloy in the tap to form a

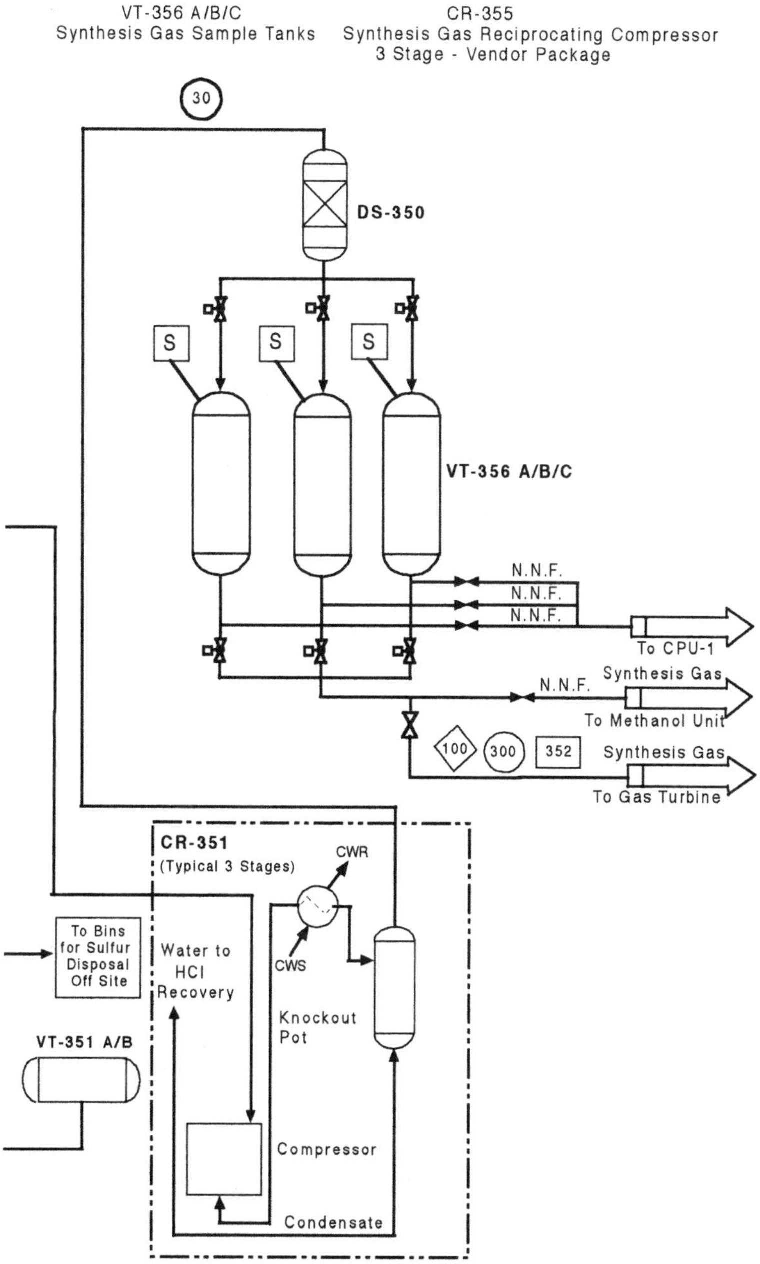

metal plug. Ceramic slag is similarly tapped at intervals, as required.

Feed Streams

This section discusses only the feed streams into the facility and not the internal process streams.

Agent

The design flow rate for chemical agent is set to achieve destruction of the stockpile at each site in a nominal one-year period. The HD design flow rate is 204 kg/h to CPU-1. The VX rate is 169 kg/h to CPU-2.

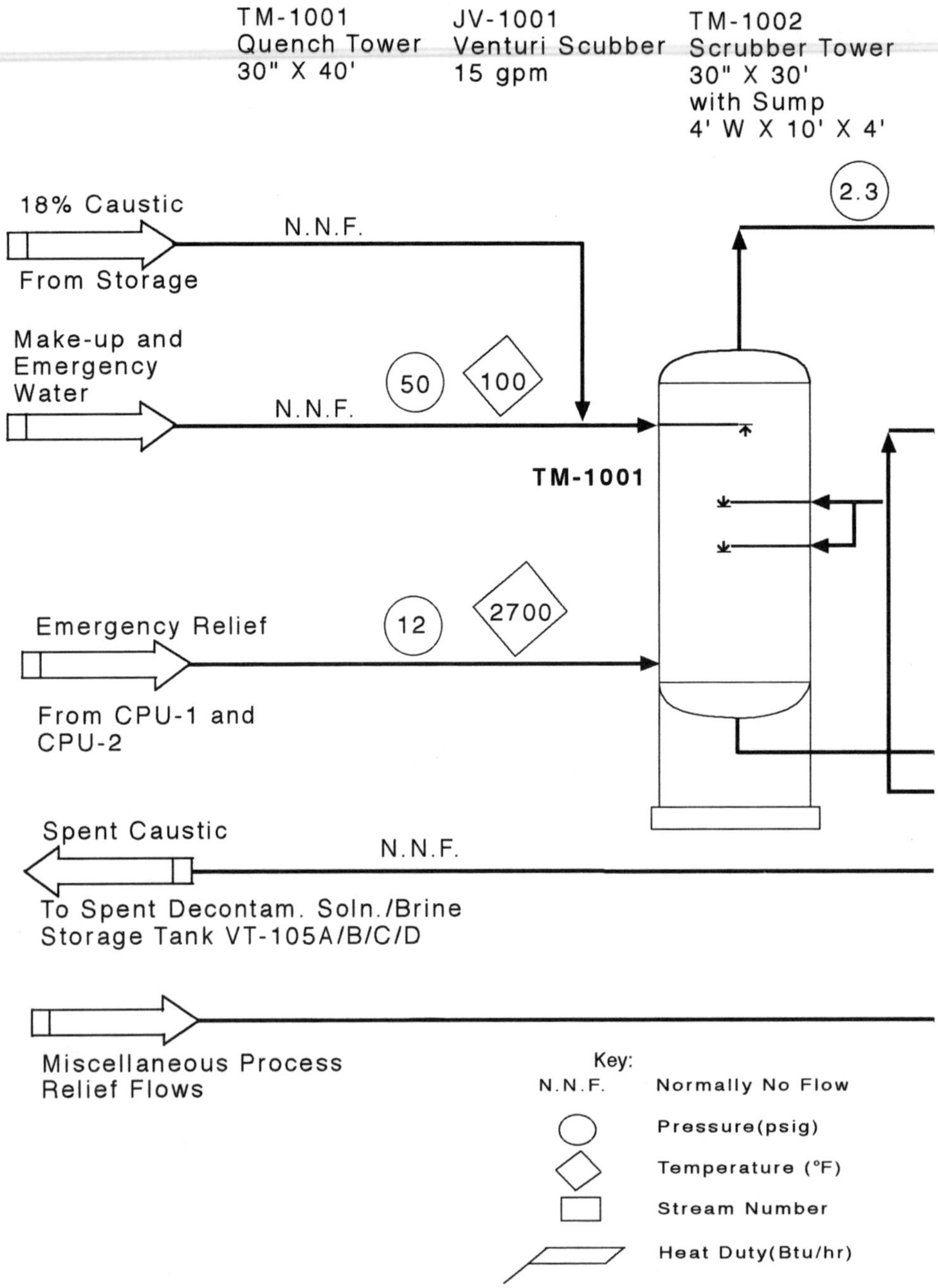

FIGURE 4-8 CEP process flow diagram for VX relief system. Source: M4 Environmental L.P., 1996b.

Metal

At cold startup, each of the CPUs is loaded with iron (or nickel for the HD CPU-1). For HD processing, there is no additional metal feed stream (other than metal from ton containers and dunnage canisters) unless the units are drained for maintenance or repair and then restarted. The same is true for VX processing, provided the addition of ton containers can be synchronized with the agent feed rate, as explained in the section on Catalysis by the Bath and the Formation of Intermediates.

Gases

Oxygen is used to oxidize the carbon in the agent and the methane to CO. An inert gas is injected automatically

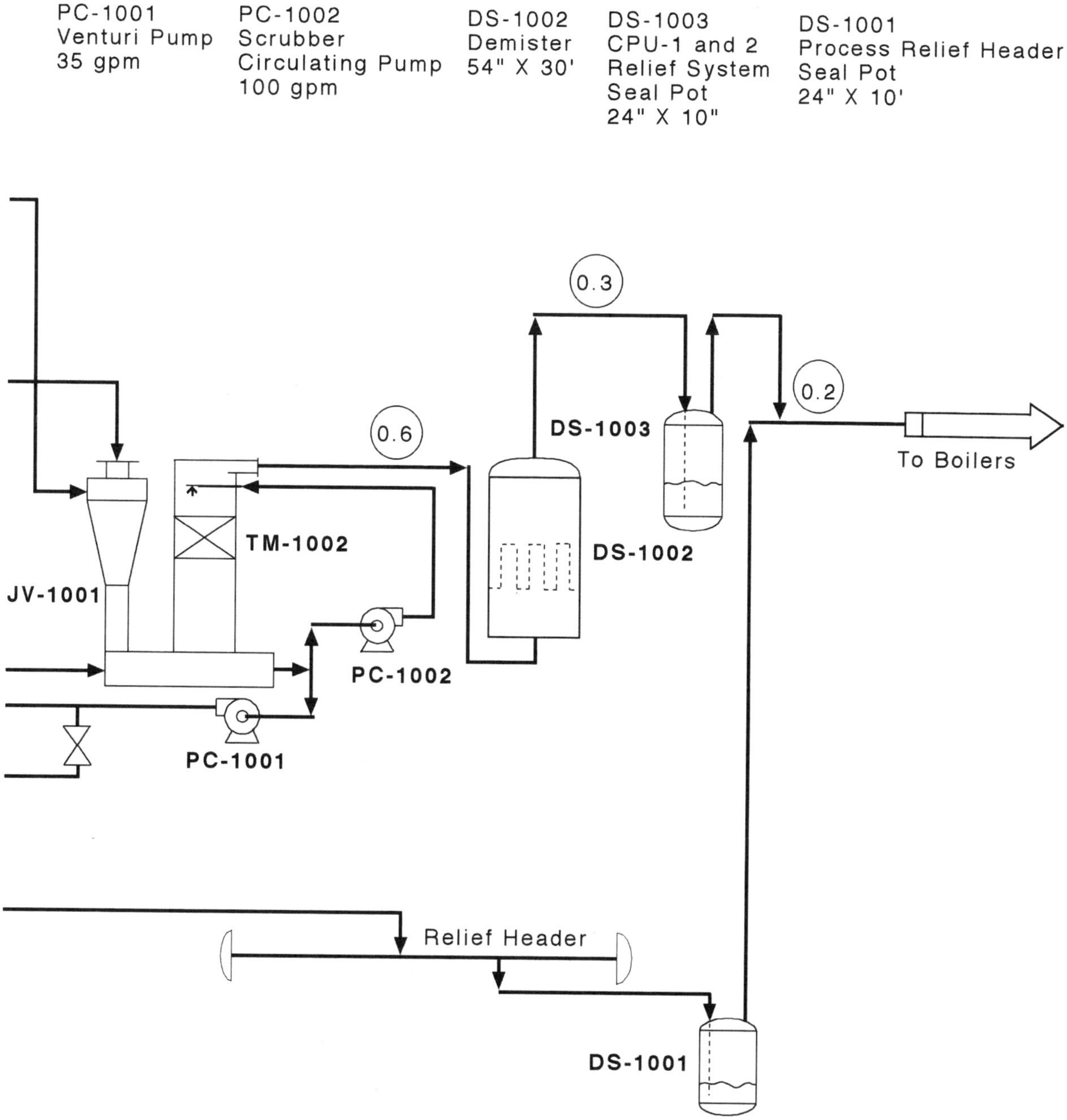

as needed into each feed line to make up the difference between the flow rate of the feed material and the desired total pressure in that line. The flow rates for these feeds are shown in Figures 4-9 and 4-10.

Gas Storage Units

Oxygen will be supplied from an off-site vendor. The on-site storage area will have standard oxygen safety systems. The TPC plans to use pipeline natural gas as the methane source, with no on-site storage.

Decontamination Solution

The TPC submissions do not specify the required quantity of decontamination solution, but it should be less than the amount required in the baseline system because CEP does not require decontamination of ton

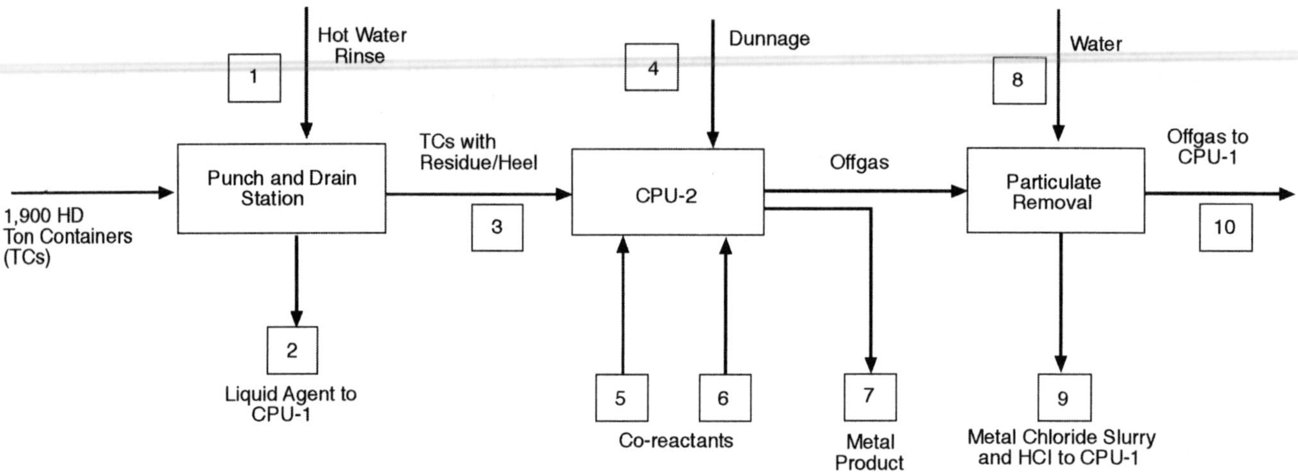

Component	Stream 1 Hot Water	Stream 2 Liquid HD	Stream 3 TCs/Resid.	Stream 4 Dunnage	Stream 5 Oxygen	Stream 6 Methane	Stream 7 Metal Prod.	Stream 8 Water	Stream 9 Slurry	Stream 10 Offgas
C		990,945	78,109	18,515		5,412				
H		165,699	7,886	1,904		1,804				
Cl		1,426.311	45,826							
S		666,045	50,426				50,426			
O_2				6,031	129,893					
P			1,216				1,216			
Fe			3,012,355				3,003,331			
Mn			15,200				15,200			
$FeCl_2$									20,481	
H_2O	158,500	158,500						96,945	96,965	
CO										237,960
H_2										10,597
HCl									35,348	
Total (lb)	158,500	3,407,500	3,211,000	26,450	129,893	7,216	3,070,173	96,945	152,298	248,557

FIGURE 4-9a CPU block diagram and material balances for HD treatment. Adapted from M4 Environmental L.P., 1996b.

containers. Decontamination solution would be used primarily to decontaminate the punch-and-drain equipment and, work area. Standard storage and mixing facilities for the decontamination solution will be used.

To avoid introducing sodium into the CPU-2 bath, the TPC prefers, according to its submissions, calcium-based decontamination solutions instead of the Army standard sodium-based solutions. Although there is experience in the use of calcium-based decontamination solutions, their effectiveness and acceptability to the Army have not been established.

Pretreatment Requirements

Cleaning the ton containers is not necessary in this process. If the Army requires precleaning of the ton containers for temporary storage, the high-pressure water-jet cleaning system will require a small amount of water (on the order of a few gallons per ton container) and iron abrasives. The drainage from the cleaning system will be pumped to temporary storage and ultimately processed in CPU-2.

Residual Streams

This section covers the residual streams coming out of the chemical demilitarization facility. It does not describe internal process product streams.

Mass Balance

The mass balances provided by the TPC for residuals from each agent are shown in Figures 4-5, 4-9, and 4-10

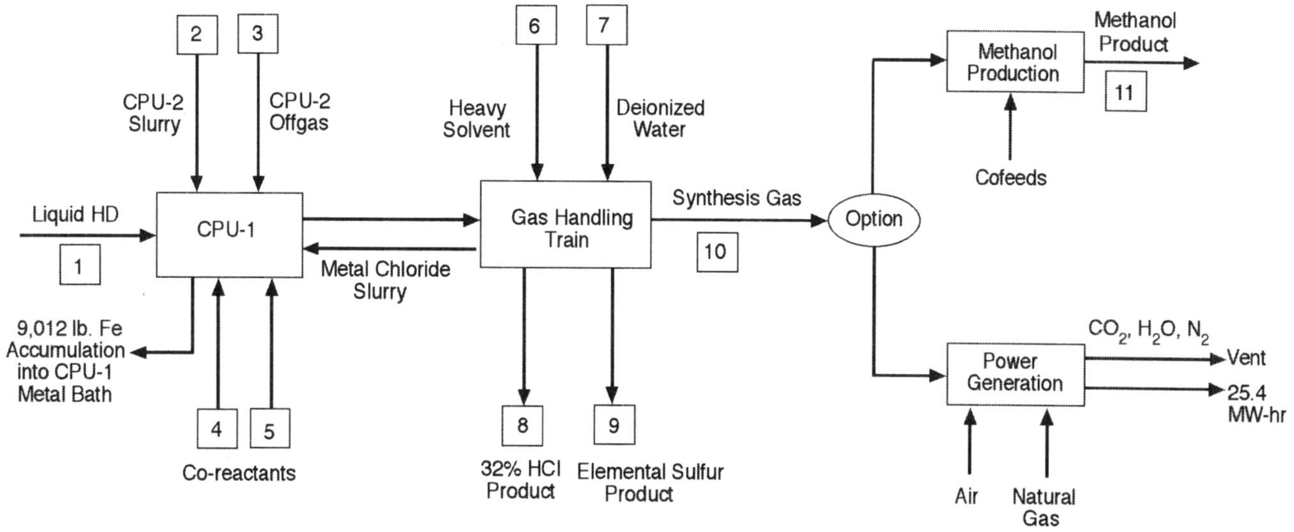

Component	Stream 1 Liquid HD	Stream 2 Slurry	Stream 3 CPU-2 Gas	Stream 4 O_2	Stream 5 Methane	Stream 6 Solvent	Stream 7 D.I. Water	Stream 8 Aq. HCl	Stream 9 Sulfur	Stream 10 Synthesis Gas	Stream 11 MeOH
C	990,945				76,151	393,349					
H	165,699				25,562	71,054					
Cl	1,426,311										
S	666,045								666,357		
O_2				1,718,615							
Fe											
$FeCl_2$		20,481									
H_2O	158,500	96,945					3,217,224	3,217,224			
CO			237,960							3,643,882	
H_2			10,597							218,739	
HCl		35,348						1,513,992			
H_2S											
MeOH											3,793,514
Total	3,407,500	152,744	248,557	1,718,615	101,713	464,403	3,217,224	4,731,216	666,357	3,862,621	3,793,514

FIGURE 4-9b CPU block diagram and material balances for HD treatment. Adapted from M4 Environmental L.P., 1996b.

and Table 4-3. There are no residuals from Area 100, the feed handling and punch-and-drain systems. All feed materials are eventually sent to Area 200, the CPU area, for processing. The residuals from Area 200 are the metal and slag phases that are tapped from the CPUs. The offgas from CPU-2 is fed to CPU-1. The offgas from CPU-1 goes to Area 300 for processing in the gas handling train.

Solids

HD and VX processing will produce about 1,360 and 1,590 metric tons per year, respectively, of metallis products. The TPC proposes to sell this material.

The only solid-waste residual will be approximately 62 metric tons per year of ceramic slag from processing decontamination solutions and dunnage. The ceramic slag will be placed in drums and shipped to a permitted hazardous-waste landfill. The TPC reports having had initial discussions with several commercial disposal firms regarding disposal of this material, as well as pursuing possibilities for marketing it. If sodium-based decontamination solution is used at the facility, the sodium will appear in the ceramic slag and alter its properties, including its solubility and strength.

H_2S in the offgas from processing HD will be converted to elemental sulfur and offered to the market.

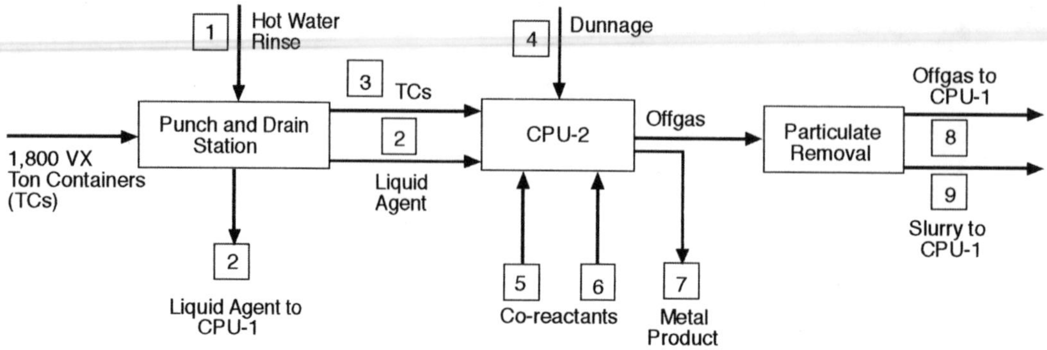

Component	Stream 1 Hot Water	Stream 2 Liquid VX	Stream 3 TCs	Stream 4 Dunnage	Stream 5 O_2	Stream 6 Methane	Stream 7 Metal Prod.	Stream 8 Offgas	Stream 9 Slurry
C		1,309,889	28,800	18,515		60,166			
H		260,154		1,904		20,055			
O		332,568		6,031	1,443,991				
P		316,476	1,152				317,628		
S		319,158	1,440				320,598		
N		142,146						142,146	
Fe		1,341	2,834,208				2,835,549		
Mn			14,400				14,400		
Ni,Cu		268					268		
H_2O	118,800	118,800							144,000
CO								3,305,466	
H_2								295,407	
Particulates									144,000
Total (lb)	118,800	2,800,800	2,880,000	26,450	1,443,991	80,221	3,488,443	3,743,019	288,000

FIGURE 4-10a CPU block diagram and material balances for VX treatment. Adapted from M4 Environmental L.P., 1996b.

Liquids

There are no continuous aqueous residual streams that will require disposal. Internal aqueous process streams, including spent decontamination solution, scrubbing liquors from the relief-system vent-gas, and spent liquors from the HCl and sulfur recovery processes, can be fed to the CPUs. The HCl from HD processing will be recovered as an aqueous solution that can be offered to the market.

Gases

The offgas from processing HD will include H_2, CO, HCl, H_2S, and trace components. The TPC anticipates that the offgas from processing VX will contain the same gases, except that HCl and H_2S will be present in trace quantities, at most. The panel expects that there will probably also be trace amounts of HCN. The HCl and H_2S from HD will be recovered as aqueous HCl solution and elemental sulfur, respectively.

The gases remaining after scrubbing, referred to by the TPC as synthesis gas or syngas (see Figure 4-9), will be burned along with natural gas in a gas turbine generator to supply in-plant electricity needs, subject to permit approval. The TPC projects that the effluent gas released to the atmosphere from the gas turbine will have the composition shown in Table 4-4. If combustion of the synthesis gas is not allowed, the TPC has stated that it will provide a methanol recovery module, which will recover hydrogen, carbon, and oxygen as liquid

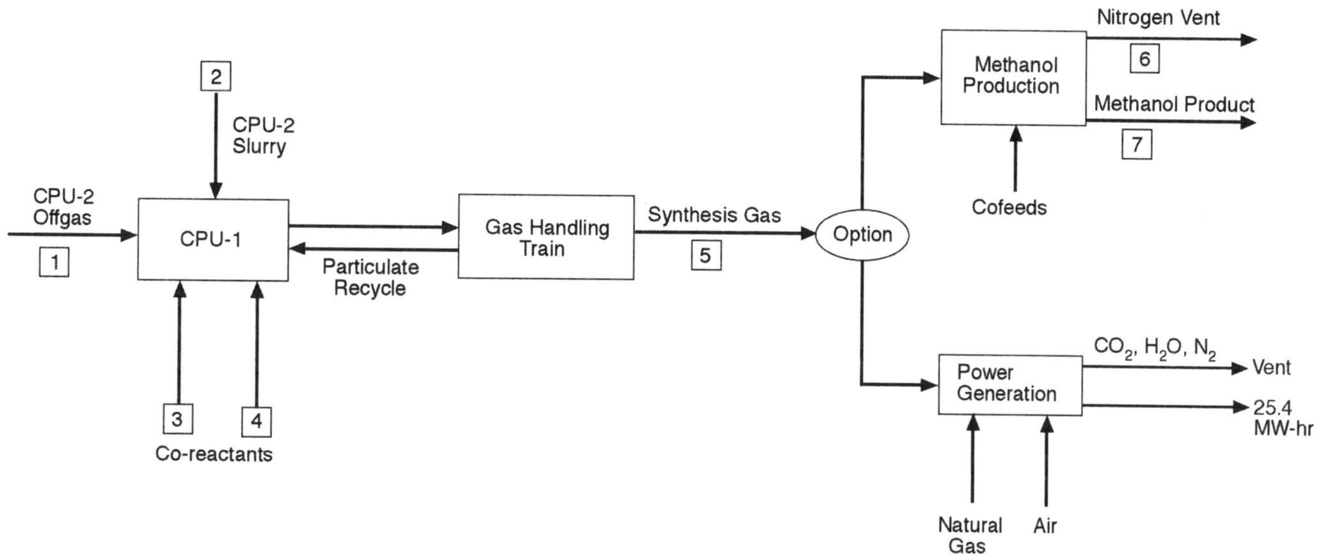

Component	Stream 1 CPU-1 OG	Stream 2 CPU-2 Slurry	Stream 3 O_2	Stream 4 Methane	Stream 5 Synthesis Gas	Stream 6 N_2 Vent	Stream 7 MeOH
C				96,003			
H				32,226			
O			norm. 0				
CO	3,873,868				4,097,759		
H_2	339,434				387,772		
N_2	142,146				142,146	142,146	
H_2O		144,000					
Particulates		144,000					
MeOH							4,161,641
Total	4,355,448	288,000	norm. 0	128,229	4,627,677	142,146	4,161,641

FIGURE 4-10b CPU block diagram and material balances for VX treatment. Adapted from M4 Environmental L.P., 1996b.

methanol. The panel has not analyzed the fate of trace gaseous components if methanol recovery is substituted for synthesis gas combustion.

There are also minor air emissions from the chelate regeneration equipment in the sulfur recovery system. This vent stream passes through an activated-carbon filter before being released to the atmosphere. During startup of the CPUs, intermittent combustion gases are produced by the headspace heater, which burns natural gas.

Nonprocess Wastes

Dunnage from daily operations will consist of PPE (personal protective equipment) including demilitarization protective ensembles, undergarments, suits, gloves, and boots that are no longer usable; rags used in maintenance and decontamination operations; and laboratory waste. The dunnage will be compacted, packaged into small metal containers, and fed to CPU-2 for destruction. The materials in the dunnage contribute to the ceramic slag and the offgas components, described above.

Off-Site Shipping and Processing Options

The CEP technology as submitted by the TPC to the Army is a "total solution" approach to chemical demilitarization. It includes methods for processing ton containers, decontamination solutions, and dunnage, as well as for destructive processing of chemical agents. Most

Table 4-3 CEP Heat and Material Balances for VX Gas Handling

Stream Number	229	301	302	310	312	313	352
Description	Reactor offgas from CPU-1	Quench offgas	Process gas feed to JV-301	Quench water to reactor offgas	Particulate slurry	Make-up water to DS-305	Synethesis gas to gas turbine
Phase	Vapor	Vapor	Vapor	Liquid	Slurry	Liquid	Vapor
Mass flow (lb./h)							
H_2	53.9	53.9	53.9	0.0	0.0	0.0	53.9
CO	569.1	569.1	569.1	0.0	0.0	0.0	569.1
H_2S	0.0	0.0	0.0	0.0	0.0	0.0	0.0
N_2	19.7	19.7	19.7	0.0	0.0	0.0	19.7
Water	0.0	353.3	768.1	768.1	0.0	19.0	19.0
CH_3OH	0.0	0.0	0.0	0.0	0.0	0.0	0.0
Particulates	20	20	0.0	0.0	20	0.0	0.0
Solvent	0.0	0.0	0.0	0.0	20	0.0	0.0
Total mass flow (lb./h)	662.7	1,016.1	1,410.8	768.1	40	19.0	661.7
Mole flow (lb. mole/h)	47.7	67.4	90.4	42.6		1.1	48.8
Volume flow (ft.^{3}/h)	54,520	40,378	24,723	12.0	12	0.30	6,557.0
Temperature (°F)	2,700	1,199	300	100	100	1,000	100
Pressure (psia)	29.7	29.7	29.7	46.7	20.7	35	44.7
Vapor fraction	1	1	1	0	0	0.0	1
Enthalpy (Btu x 10^6/h)	-0.028	-2.431	-5.253	-5.225		-0.129	-1.068

Source: Adapted from M4 Environmental, L.P., 1996b.

TABLE 4-4 Expected Composition of CEP Gas Streams prior to and after Combustion in a Gas Turbine Generator

Constituent	Gas Stream to Generator HD Offgas	VX Offgas	Generator Exhaust Gas
CO	12.4%	12.4%	19.5 ppmv
H_2	9.7%	15.3%	none
HCl	<0.5 ppmv	none	<0.5 ppmv
H_2S	<0.03 ppmv	none	<0.03 ppmv
SO_X	none detectable[b]	none detectable[b]	0.039 ppmv[c]
N_2[a]	1.17%	1.53%	none
NO_X	none detectable[b]	none detectable[b]	<130 ppmv[d]
HCN	none detectable[b]	none detectable[b]	not stated
Trace Organics	none detectable[b]	none detectable[b]	9.7 ppmv[e]

[a]TPC states that most of the nitrogen shown is typical of the natural gas combusted with the synthesis gas. TPC states that no nitrogen is introduced in the HD process and nitrogen from VX processing is approximately 0.36% prior to natural gas injection.
[b]TPC used the following lower detection limits: SO_X = 1 ppm; NO and NO_2 = 3 ppm; HCN = 0.01 ppm; trace organics = 0.1 ng 2, 3, 7, 8 TEQ/Nm^3.
[c]TPC based this value on typical sulfur concentration in natural gas.
[d]Expressed as NO_X corrected to 15 percent oxygen. TPC stated that, if required, this amount could be reduced to 42 ppmv by water injection.
[e]TPC based value on unburned hydrocarbons from the natural gas cofuel to the turbine generator.

Source: M4 Environmental L.P., 1996b.

of these feed materials are converted to useful products, including iron-based alloy, synthesis gas for power generation, aqueous HCl, and elemental sulfur.

INSTRUMENTATION AND CONTROL

The CEP design includes a distributed control system (DCS) for overall monitoring and control of material processing and related support systems. The control architecture for the CEP chemical demilitarization facility is an integrated DCS that provides executive control of the monitoring and process intervention required for safe and efficient operation in processing chemical agents. Two fully operational control systems will be installed. One actively controls and monitors the process; the second remains on active standby, monitoring the process and serving as a redundant system that can take over control operations if the primary system malfunctions or some other internal problem arises. The facility includes a local area network with an independent bus for control and communications.

Process instrumentation and controls are located throughout the central building and support areas for monitoring and controlling parameters such as tank and bath levels, flow rates, pressure, pH, temperature, motor current, weight, volume, and valve position. The sensor instrumentation for monitoring process parameters includes detectors, signal conditioning, transmitters, and other devices as required. Continuous, real-time control is provided for critical processes. The DCS interfaces with the process monitoring and control instrumentation through input/output devices, which are located throughout the facility to reduce the amount of cabling, the number of connections, and the number of cell penetrations. Ground-bus connections isolate the grounds for the instrumentation and control circuits from power grounds. Additional analytical instrumentation is used to monitor for agent releases in the central building.

Most of the systems and equipment to be controlled are located in various work cells of the central building. These in-cell systems have hermetic feedthroughs for wall-penetration assemblies that provide interfaces for equipment, components, or input/output devices.

Monitoring and control systems that perform safety functions are hard-wired and sufficiently redundant to meet the criteria for avoiding single-point failures. They are powered by an uninterruptible power supply consisting of batteries, with chargers and inverters to allow use of power from backup generators. The design basis for these systems includes protection against natural events (e.g., earthquakes or severe storms) and worst-case environmental conditions. Systems are designed with fail-safe circuits to meet these requirements. Each redundant system required to perform safety functions is physically and electrically separated from its counterpart and from non-safety-related circuits and components.

Part of the TPC's stated control strategy is to perform an analysis of the entire system during the detailed design phase to define the critical control systems that will be hard-wired. The hard-wired systems will include all safety systems and all systems necessary to ensure the safety of workers and the public and to protect the environment.

Operations will be directed and monitored from a master control room adjacent to the central building. The control room is isolated from areas that could become contaminated with agent. Video surveillance provides visual monitoring of the entire process, end to end.

The process monitoring and controlling requirements for the feeds to the CPU reactors include gas mass-flow controllers for the oxygen, inert gas, and natural gas streams and liquid flow controllers for agent and for solutions used to clean ton containers. Agent assays of the ton container contents will be performed by taking a grab sample from each container and analyzing it via GC/MS (gas chromatography followed by mass spectrometry) with a lower detection limit less than 0.1 µg/ml (100 ppb).

Key parameters for controlling the CPUs are bath temperature, bath composition, bath level, and containment monitoring. Monitoring and control for each of these are described below.

Bath Temperature Control

By varying the power to the induction coil, the bath temperature control system maintains the molten metal bath at a stable operating temperature (±28°C) at least 110°C above the liquidus temperature of the bath (temperature at which the bath metal is entirely molten). Based on the preliminary design submitted, the operating temperature of the CPU-1 bath for processing HD (nickel bath) is likely to be about 1425°C. The iron baths will operate at about 1500°C to 1650°C.

Two temperature-sensing systems are used for monitoring: an infrared lightpipe and thermocouples embedded in the CPU refractory material. The primary temperature sensor is the infrared lightpipe, which provides a continuous, non-invasive method for sensing bath temperature. The lightpipe, which transmits infrared radiation directly from the bath to a dual-wavelength pyrometer, provides fast response and, precise measurements, and requires minimum calibration.

The redundant system for controlling the bath temperature uses thermocouples embedded in the refractory wall combined with a proprietary, on-line control model that predicts the metal bath temperature during operation. The method is non-intrusive and robust for CEP processing conditions.

In addition to control of bath temperature, headspace temperature is kept high enough to avoid solidification of molten metal on surfaces.

Bath Composition Control

Control of the bath composition is necessary to obtain the required agent DRE, to produce offgas with the desired composition, and to maintain the structural integrity of the containment system. The carbon concentration in the bath is controlled by varying the oxygen flow rate and monitoring the composition of the offgas, specifically the ratio of CO to CO_2. The model used to infer carbon concentration from the composition of the offgas has been validated with actual measurements of bath carbon.

A contingency method of modeling the bath composition is based on the material balance for feed and product streams to and from the bath. The TPC has routinely estimated bath carbon concentration in its large Demo Unit CPU by using a feed-forward model and an offgas composition model. The basis of each model is a general steady-state carbon balance on the reactor. In the feed-forward model, composition is estimated using partitioning and thermodynamic models. Analysis data on offgas composition provide estimates for the second model. The results from these models are combined with feedback control based on the CO/CO_2 ratio to ensure an appropriate bath carbon concentration.

For VX processing, sulfur and phosphorus are controlled by adding iron from ton containers and tapping

Fe-S-P-C alloy from the bath, but monitoring procedures were not discussed.

Monitoring Bath Level

There is a bath-level monitoring system for each CPU. Each CPU is fitted with a side-mounted lightpipe that senses the bath temperature directly and provides an indirect indication of the bath height when compared with the bath temperature provided by the thermocouples in the refractory lining. In addition, a microwave level switch is used as a sensing system for maximum bath level.

Monitoring Containment

The CPU design provides for two linings of refractory to serve as the primary and secondary containment for the molten metal. The inner lining, called the working lining, is the primary containment. The outer lining is designed primarily as an insulating layer to lower the temperature at the outer steel vessel, but it also serves as a backup containment, capable of holding the bath long enough for the molten metal to be drained if the working lining is breached. In addition, portions of the outer steel vessel are water-cooled, which cools the adjacent refractory enough to freeze a layer of slag on the surface of the working lining, thereby prolonging its life.

The two systems for monitoring the integrity of the primary containment are embedded thermocouples and grid assemblies. These redundant monitoring systems give the operators an indication of normal refractory wear and warn of molten metal encroachment to the secondary containment. During normal operations, the primary monitoring system is the thermocouples embedded in the refractory. The temperature differences among thermocouples indirectly measure refractory wear from the temperature gradient across the working lining, which is directly proportional to thickness of the refractory.

The secondary level of monitoring the refractory containment consists of detection grids incorporated in the primary lining. Contact with molten metal opens a grid and provides a reliable indication of either localized or uniform deterioration of the working lining. Complete coverage of the refractory lining with grids, together with the embedded thermocouples, provides continuous monitoring of the refractory, thereby allowing sufficient time for a normal system shutdown in the event of excessive deterioration.

These containment monitoring systems have performed reliably in the units at the TPC's Fall River facility (see Table 4-2).

Monitoring Residual Streams

Solids

Metal ingots and ceramic slag can be analyzed by the EPA's TCLP test to verify compliance. Verifying that the metal ingots and ceramic slag do not contain agent within their internal matrices is difficult because any technique used to extract samples for analysis is also likely to destroy agent. However, this internal verification is probably not necessary because the conditions under which the ceramics and metal ingots are produced exceed the Army's definition of a 5X material (which is considered agent-free).

Gases

The TPC plans to install a continuous emission monitoring system to monitor gas effluent streams for O_2, CO_2, CO, NO_X, H_2, HCl, and H_2S. Similar monitoring systems have been proven and used extensively at the operating demilitarization facilities. The TPC states that it will review and incorporate lessons learned from these sites prior to specifying the final type of detector to be used for the emission monitoring system.

Provision for retaining synthesis gas for analysis prior to release for combustion has been added to the original design, as described above in Conclusions on the Underlying Science.

The depot area air monitoring system (DAAMS) and "mini" continuous air monitoring system (MINICAMS) used by the Army are sufficient for monitoring for agent inside the CEP facility and at the site perimeter. Gas chromatographs, mass spectrometers, and the continuous emission monitoring system are capable of analyzing the feed, internal process, and residual streams to meet regulatory and operational requirements.

Monitoring Synthesis Gas prior to Combustion

The TPC plans to choose among one of three analysis systems during the next stage of design. One is the automatic continuous air monitoring system (ACAMS), which is the standard Army monitoring system to existing agent-destruction facilities. The second is the MINICAMS, which is also used to monitor for agent at existing Army facilities. The third system is the TAGA 6000E (trace atmospheric gas analyzer), which has been tested at the Army Chemical Agent Munitions Disposal System. The ACAMS and MINICAMS use gas chromatography with flame photometric detectors and have response times of 3 to 5 minutes for the agent detection levels required. The TAGA 6000E has a response time of 15 seconds. The TPC plans to install several sensors for each of the three retention tanks, with a "voting logic" system to reduce the number of false positives. If the system logic determines that agent is present in a tank, the tank contents would be recycled to CPU-1 for reprocessing. The TPC's description makes no reference to testing the retained gas for constituents other than agent.

A preliminary analysis by the panel suggests that this three-tank design may not be adequate; at least one more tank may be required. At 20 atmospheres, gauge, each 4-m^3 tank holds 60 kg of synthesis gas. If a tank is found to be contaminated (call it Tank A), the contents must be fed back through CPU-1, along with cofeeds of oxygen and methane. The minimum mass to be "reprocessed" is thereby increased to about 64 kg, all of which reappears as offgas from CPU-1, assuming the bath is saturated with C, O, and H. The gas in the next tank to be filled (call it Tank B) must be presumed to be contaminated until that tank is filled and testing shows it is clean. If Tank B is contaminated, it cannot be used to hold the surge from Tank A. This leaves only the third tank (Tank C) to hold the 64 kg of gas from reprocessing Tank A. Tank C will be full before Tanks A and B are emptied. The fourth tank must be empty and ready to handle the overflow from reprocessing Tank A. When the fourth tank is full, Tank A can be refilled to handle overflow from Tank B, if it is contaminated.

Air in the Containment Building

In the submitted design, air inside the secondary containment building will be monitored using a variety of instruments to provide both real-time and time-weighted-average agent monitoring. A detailed agent-monitoring plan for a CEP demilitarization facility would be developed initially as part of the detailed design process before pilot-testing. The plan would be refined as the facility is constructed and commissioned. The general strategy for safety and environmental agent monitoring is much the same as the strategy used at the Johnston Atoll Chemical Agent Disposal System (JACADS) and Tooele Chemical Agent Disposal Facility (TOCDF), although the TPC states that less of the plant would require monitoring by virtue of the inherent safety features of CEP. In the central building, each enclosed room would be monitored by a near-real-time instrument and a DAAMS. The detection range and alarm level will be based on the hazard category (protective clothing level) for each room.

Near real-time monitoring could be provided by either the ACAMS or the MINICAMS, These instruments would be used to monitor for agent throughout the demilitarization facility at the following statutory levels: MPL (maximum permissible limit, a very high level), HLE (high-level exposure), TWA (time-weighted-average, a low level), and IDLH (immediately dangerous to life and health). According to the TPC, the MINICAMS provides additional flexibility in software functionality and the future availability of the ACAMS is uncertain, so the TPC currently considers the MINICAMS as the monitor of choice for near real-time monitoring. DAAMS, which is used at operating Army facilities, will be used to monitor the perimeter for very low levels of agent and, in the event of a MINICAMS or ACAMS alarm, to obtain longer-term samples to confirm whether agent was present.

STABILITY, RELIABILITY, AND ROBUSTNESS

Stability

Stability of CEP is discussed under the topics of out-of-control operations, stored energy, and catastrophic failures.

Out-of-Control Operation. The large mass of the metal bath provides commensurately large thermal inertia, which prevents a significant temperature excursion in the event of perturbations in the feed rate of agent or cofeeds. The bath mass provides a margin of safety for bath composition and feed rate and allows the CPU to operate over a relatively wide range of conditions.

Stored Energy. According to the TPC, the total stored energy of each iron bath is approximately 4 x 10^6 kJ. The nickel bath used for HD processing has two-thirds the mass of an iron bath and about 2.6 x 10^6 kJ of stored energy.

Catastrophic Failures. There are no identified process mechanisms, such as uncontrolled reactions, under *normal operating conditions* that could lead to a catastrophic failure of the facility. However, catastrophic accidents can always occur if the equipment fails—a break in a tuyere or tapping nozzle, for example—or if there is operator error, such as inserting an undrained ton container into the CPU-2 melting unit. In response to questions from the panel, the TPC has added several levels of operational controls to the design to prevent an accidental insertion of an undrained container.

An extended failure of electrical power would require a cold shutdown of the CPUs, with related problems whose severity would depend on the reliability of emergency standby power to open taps and drain the molten baths before they solidified (see Startup and Shutdown, above).

Reliability

Performance Record

The CPUs closely resemble the induction furnaces used in melting metal, as well as the TPC's several demonstration CPUs. Materials of construction were selected in light of process conditions and process-fluid characteristics. Allowances for stress and wear are incorporated to ensure adequate life and performance throughout the operational period.

The basic CPU design has been tested under severe conditions. Most of the front-end equipment is either the same as equipment in the Army baseline incineration system or closely resembles that equipment and is likely to be as reliable.

The offgas recovery units are based on proven commercial design but require some special features for processing the offgas from chemical agent destruction.

Backup Systems

In the event of an equipment failure in the oxygen supply, methane cooling gas supply, or offgas treatment, the system can stop the agent feed almost instantaneously. The CPUs can be held at hot standby condition indefinitely.

If the site has a single line of access to the electric power grid, an uninterruptible battery power system with a response time of a few milliseconds can maintain critical safety and control services until backup power can be brought on line. Essential services for a no-feed, hot standby condition can be provided by the gas-powered turbine generator used to recover energy from the synthesis gas. If a turbine generator is not installed, a diesel generator capable of a 10-minute response from cold start can be used to provide power for standby services.

Robustness

The CPUs can operate over a range of operating conditions. The thermal inertia of the bath is large enough that, with a loss of power, the bath takes approximately 2 hours to freeze. Responses to upsets and control mechanisms have already been described.

MATERIALS OF CONSTRUCTION

Systems and Materials

The block flow diagram for the facility in Figure 4-4 shows the layout and interconnects for process operations. The conceptual design for the facility was performed by competent engineering firms that are participants in the team that prepared the submissions. These firms have experience in designing chemical processing units and nuclear power plants, many of which have been in operation for years and have documented safety records. System design and material selection appear to be based upon sound engineering practice.

An inquiry from the panel led to one change in material selection from the original submission. The initial design specified tungsten for the slide rails inside the premelting chamber of CPU-2 to support the ton container during melting. The TPC changed the material to a refractory oxide after a question from the panel about the substantial solubility of tungsten in iron at the melting point of iron.

Materials Specifications

According to the TPC, the design follows the published specifications of the American Society of Mechanical Engineers (ASME) for piping materials, valve bodies and trims, shell-side and tube-side materials for heat exchangers, and impeller materials for pumps. The corrosion allowances and specifications for piping and components, including special materials requirements such as stress relief, also use the ASME recommendations for specific components.

Welding Specifications

Most of the piping, vessels, and other equipment in a CEP facility contain welds. Where equipment is welded to piping, the equipment is generally flanged and bolted to the welded piping spools. Structural steel used to support the piping and equipment is also typically welded. According to the design for an agent destruction facility, agent transfer lines from the storage tanks to the CPUs are double-walled piping; the annular space between the walls is monitored for low-level agent vapor, as an early indicator of a leak in the inner wall. Special stress-relief requirements, welding processes, filler metals, and gas shielding conform to standard welding specifications. These extensive specifications are normally tailored to the requirements of a project during the detailed engineering phase. The design states that welding procedures will follow the current ASME codes and applicable Military Standard, MIL-STD-1261C(MR).

Stress Relief

In the design generally, stress relief, where required, is based on details of the material, thickness, or service. Materials that often require stress relief regardless of thickness are martensitic steels containing 1 to 12 percent chromium. Carbon steel often requires stress relief above a certain thickness, per the applicable codes. For instance, ASME Section VIII for vessels requires stress relief when carbon steel is thicker than 1.5 inches (3.8 cm), and ASME Standard B31.3 for piping requires stress relief when carbon steel is thicker than 0.75 inches (1.9 cm).

Stress relief for service generally applies when the material would be susceptible to stress corrosion cracking, such as when carbon steel is in contact with caustic or amine solutions or when stainless steel is in contact with chloride or sulfide solutions. Operating temperature is often an important variable in determining if stress corrosion cracking may occur. For the solutions listed above, the temperature range of concern is from 38°C to 66°C. In the TPC's design, these solutions listed above are either at room temperature or an appropriate lining is specified.

Weld Inspection

According to the submitted design, the minimum amount of weld inspection will be to an appropriate industry code, typically ASME Section VIII for vessels, ASME/ANSI B31.3 for piping, and AWS D1.1 for structural steel. (ANSI refers to codes approved by the American National Standards Institute; AWS refers to codes approved by the American Welding Society.) This degree of inspection requires spot radiography and hydrotesting for the majority of welds of equipment and piping. For the double-walled agent transfer line, large vapor lines, and refractory-lined piping and equipment, hydrotesting will not be practical, so 100 percent radiographic testing will be performed. The TPC states that a reputable third party will conduct the weld inspections and evaluate results. The TPC will furnish welding specifications with the detailed design to provide information on inspection methods and criteria. Weld inspections will be conducted in accordance with paragraphs 5.1.4 through 5.1.4.4 (magnetic particle inspection, radiographic inspection, dye penetrant inspection, and ultrasonic inspection) of MIL-STD-1261C (MR). A report will be issued in accordance with Data Item Description DI-THJM-81194.

Environmental Chemistry and Conditions

Nominal Internal Environmental Conditions

The CEP processing conditions described here are based on the submitted design, which is preliminary and subject to revision during further design and development. For processing HD, the nominal chemical environment in CPU-2, where ton containers and dunnage are processed, is a molten iron phase containing a controlled concentration of carbon and a gas phase consisting of H_2, CO, H_2S, and HCl. Table 4-5 gives the nominal composition for elements other than carbon. The nominal composition of the metal phase in CPU-1 for HD processing is nickel containing about 2 percent carbon. Temperatures in both CPUs are in the range of 1425°C to 1650°C, at an absolute pressure of about 2 atmospheres in CPU-1 and 1 atmosphere in CPU-2.

TABLE 4-5 Nominal Composition of CPU-2 Metal Phase (weight percent)

Element	HD	VX
Sulfur	1.64	9.11
Phosphorus	0.04	9.19
Iron	97.82	81.28
Manganese	0.50	0.41
Nickel, copper	0.008	0.008

For processing VX, the bath in CPU-1 is iron with carbon controlled in the range of 1 to 2 percent. The nominal composition of the metal phase in CPU-2 has higher concentrations of sulfur and phosphorus than in the CPU-2 bath for HD (Table 4-5).

For processing either agent, the chemical and physical environment of the quench, absorber, and compressor between the two CPUs is the gas phase from CPU-2. This gas consists mainly of CO, H_2, and H_2S. In HD processing, some HCl will be present from residuals in the ton containers and from spent process solutions. Temperatures in this area range from about 1500°C exiting CPU-2 to 38°C at the suction of the compressor; absolute pressures range from 1 atmosphere as the gas leaves CPU-2 to about 10 atmospheres at the discharge of the compressor. The temperatures for quenching and cleaning CPU-2 offgas range from 260°C for the offgas at the inlet to the absorber to 38°C after the cooler and about 66°C in the bottom of the absorber.

The gas handling train operates at low pressure, about 1 atmosphere, gauge. For HD processing, the offgas from CPU-1 will be scrubbed in the HCl recovery section to absorb HCl gas in water and recover it as HCl solution. H_2S in the offgas is converted to elemental sulfur. For VX processing, the HCl and sulfur recovery systems are not required because VX does not contain chlorine, and the sulfur is retained in the iron bath of CPU-2. The offgas is scrubbed with water, compressed, stored for analysis, and sent to the gas utilization unit (e.g., gas turbine or methanol recovery). Typical flow rates in the gas handling train during HD destruction are shown in Table 4-6.

Nominal External Environments

Design for exterior environments generally depends on whether the equipment is inside or outside a building, whether heat is being transferred, or whether protection of personnel or equipment is required. In the CEP design as submitted, the environment inside the central building will be protected from weather and maintained at a comfortable temperature. Atmospheric contaminants

TABLE 4-6 Flow Rates in the Gas Handling Train for HD Processing

Gas Handling Service or Equipment	Flow Rate
Reactor offgas	750 acfm (354 l/s)
HCl product	2.5 gpm (9.5 l/min.)
Quench water to reactor offgas	1.4 gpm (5.3 l/min.)
Primary HCl recovery column overhead	250 acfm (118 l/s)
Recycle liquid to primary column	1.6 gpm (6 l/min.)
Primary column pump-around	7.3 gpm (28 l/min.)
Makeup water to secondary column	1.2 gpm (4.5 l/min.)
Offgas to sulfur recovery	220 acfm (104 l/s)

are not expected to be a controlling condition for the design at either site because the piping, equipment, and structures are protected from the weather. Equipment and piping will be insulated either for heat conservation or for protection of personnel (maximum surface temperature 66°C) and equipment. Heat conservation requirements, which will be determined during detailed design, will be based on the cost of heat loss or on the need to provide a stable internal temperature to prevent undesirable swings in process controls. Insulation and heat tracing will be used to prevent freezing in areas where the ambient temperature could fall below freezing and the contents of the piping and equipment could freeze.

The TPC has stated that, for the design of the pilot-test facility, the exterior environments for the piping and components—temperature extremes, relative humidity, atmospheric contamination, and leached chemicals—will be approximated by ambient conditions for the nearest city for which data are available. For the final design, the TPC plans to use conditions at the sites. These conditions enter into the specifications and design basis of various items of equipment as well as the structural facilities. For example, the ambient wet- and dry-bulb temperatures are used to set the design cooling water temperature and to specify the capacity of the cooling tower. The rainfall, snowfall, and wind velocity are important to the design of all buildings, other outdoor structures, and surface drainage. The seismic zone will be determined during detailed engineering and taken into account in the design of structures.

Ambient air composition is important if the small amounts of certain substances, such as carbon dioxide and ammonia, that may be present in air are significant to the process. For CEP these components have no significant impact on the design as long as they are not present in concentrations harmful to humans. Air is used in CEP for combustion air to the gas turbine generator, startup burners, and the relief-system boiler; for blowing (oxidative regeneration of) the SulFerox solution; and for evaporative cooling of water in the cooling tower. None of these uses is sensitive to minor impurities.

Crevices, Surface, and Bottom Deposits

The TPC states that its construction practice is to minimize all crevices, deposits, sources of galvanic corrosion and other design features that can increase corrosive conditions. The detailed design will be reviewed for this purpose by materials specialists on the TPC team. Corrosion in crevices can occur in aqueous electrolytic services. In this design, most of these services are being handled with Haveg, impregnated graphite, or plastic-lined carbon steel, which prevents of crevice corrosion. The industry codes and the TPC's standard practice is to use butt welding for all piping instead of socket welding. Galvanic couples will be avoided in electrolytic services except where the area ratios are such that corrosion is expected to be minimal. (For example, alloy valve trim is specified in carbon steel piping but galvanic corrosion is minimized because the surface area of the trim is much smaller than the area of the carbon steel valve-body and piping.) If underdeposit corrosion is a risk, either larger corrosion allowances will be specified on the bottom head or boot or upgraded alloys, coating, or lining will be specified. The TPC plans to assess the risk and take adequate design precautions based on past experiences with similar services.

Heat Transfer Surfaces, Heat Fluxes, and Crevice Geometries at Tube Supports

Reactor Vessel Shell. Heat flux in the CEP design is limited to that which will produce an external metal temperature of approximately 150°C. This heat flux is in the range of 200 to 500 Btu/h/ft^2 (2,300 to 5,700 kJ/h/m^2) of external surface.

Reactor Containment. The entire reactor is lined with several overlapping courses of refractory brick. Where the bricks meet, some molten metal, slag, or gas can penetrate between them, but this penetration is stopped by the next layer. Molten material freezes as the temperature drops through the refractory, sealing the interstices from further penetration.

Reactor Internals. The bath refractory is surrounded by an induction coil that heats the bath metal. The coil is internally water cooled. This technology is in widespread use in the steel industry.

Reactor Offgas Piping. Hot offgas in the gas handling train is transferred in a jacketed pipe, which is designed to be cooled with water to maintain the pipe temperature within the maximum temperature limit for carbon steel. Insulation is provided to protect personnel.

Heat Flux in Crevice Geometries. Crevices are particularly prone to corrosion when the heat flux in the vicinity of the crevices creates an enduring temperature differential at the crevice surfaces. For example, tube-to-tube-sheet joints in heat exchangers are prone to corrosion, particularly the crevice in the back of the tube-sheet. In most designs, the tubes are not rolled to the full width of the tube-sheet, which results in this crevice. Corrosion at this crevice is a concern especially with stainless steel tubes. Because there are no heaters or fired furnaces in the present design, no problems of this type are anticipated and no special requirements have been specified for tube rolling. In HCl environments where corrosion would be expected to be severe, the design specifies graphite block exchangers that do not use tube sheets or other constructions with crevices.

An important crevice that does exist in this system is the joint between the headspace refractory and the refractory containment of the metal bath. The panel learned that, during the early stage of testing the Demo Unit at Fall River, molten metal leaked out through this joint into the annular space that contains the induction coil and burned out the rubber hoses that supply cooling water to the coil. The TPC subsequently developed a proprietary means of sealing this joint that prevents such leakage. The leak did not create a safety hazard but did require a complete shutdown and replacement of the induction furnace.

Startup and Shutdown Procedures

Startup and shutdown procedures have already been described (in Startup and Shutdown in the Process Operations section). Detailed startup procedures, including hot and cold restart specifically for an agent destruction facility, will be developed in the detailed design phase, based on the existing general CEP operating manuals.

Deoxygenating and Heating Rate on Startup

The CPU is deoxygenated as part of normal startup. The procedure for deoxygenating on startup is to pass an inert gas through the CPUs and the downstream piping and equipment until oxygen levels, as determined by analysis, are well below the lower flammable limits of the expected offgas composition. One way to ensure that dead spaces are purged is to open all vents, drains, and bypasses with the inert gas flowing. A variation is to pressurize the system with inert gas and then vent down to atmospheric pressure, with pressurizing and venting repeated several times. Still another variation is to evacuate the system and then break the vacuum with the inert gas, with several repetitions. The TPC plans to decide which procedure to use in this facility during the detailed design phase and will incorporate it into the operating instructions.

The only critical equipment items sensitive to temperature change rate are the refractory lining of the CPUs and the special materials in the HCl recovery system, such as Haveg and graphite. A reasonable rate of temperature change for these items is 110°C/h. Heating rates will be specified in the operating instructions for the CPUs and for other equipment containing ceramic, graphite, or plastics such as Haveg.

Design Life of the Process Equipment

The process equipment is sized to process the entire inventory of HD at Aberdeen in 300 operating days and then to be relocated to Newport to process the entire inventory of VX in 300 operating days. A pre-operational period will be required to check out the equipment and controls and to train the operators. Therefore, the panel expects that the required operating life of the process equipment is less than 3 years, which is well within the normal design life of chemical processing equipment (generally 10 to 20 years). The TPC has stated that no attempt will be made to reduce quality and corrosion allowances because of the shorter life expectancy of this facility.

Certain parts of a plant of this kind may require replacement during a normal operational period. Examples are the refractory lining of the CPUs and parts of the HCl recovery section. Refractory life depends on many variables, such as temperatures, changes in temperature, compressive stresses, the corrosive action of slags, actions of different molten metal solutions, and actions of gases. In CEP reactors, changes in temperature are both gradual and controlled, thereby reducing the stress on the refractory linings. Injection forces are mediated by directing the jets from the tuyeres toward the center of the bath. Refractory life is therefore expected to be long enough for the relatively short duration of each agent campaign.

Qualification and Testing of Materials of Construction

The design states that selection of materials of construction will be based on equipment operating conditions and on corrosion and mechanical testing. Materials selection for the punch-and-drain system will be based on the baseline system and lessons learned from existing facilities that process agent. Refractory for the CPU linings will be selected on the basis of testing experience at the TPC's research facilities. The panel believes the refractory can be maintained to accommodate the projected one-year agent processing campaigns at each site; replacing the refractory will probably not be necessary.

Materials selection for the gas handling section will be based on the experience of the TPC partners with similar applications, in consultation with experts in the manufacture of chlorinated chemicals, and on corrosion testing of material coupons at the TPCs research facilities. This experience indicates that, with proper maintenance and operating procedures, these materials rarely fail within the first 10 years in service. The expected operating life of this facility of less than 3 years is therefore well within the anticipated usable life of the materials.

Potential Failure Modes for Materials and Components

This section describes only the experience and analytical work related to understanding the failure modes of materials and components in a CEP system. The TPC's general approach to identifying failure modes and hazards in CEP technology and in the design for an agent destruction facility is described below in the Failure and Hazards Analysis section under Operation and Maintenance.

Several systems in the design of the facility use materials and components designed for intrinsically safe modes of operation. First, the molten metal bath quickly dissociates the chemical agent, and this dissociation greatly reduces the chances of contamination downstream. Second, the tuyere line diameter and pressure are designed to limit the agent flow rate to a safe maximum. As a consequence, a valve failure, even in full-open mode, cannot cause a hazardous condition. Third, the reactor has three internal containments (two refractory linings and the steel vessel) and two external containments (the CPU module and the enclosed central-building) to reduce the potential for an off-site release.

The TPC states that, in addition to the hazard studies discussed below, the failure modes of the CPUs are understood from the TPC's nearly four years of experience at the Fall River facility. The principal failure modes affect reliability and economical performance but not safety. Careful design and operation are needed to avoid plugging the tuyere (which would prevent agent feed and cause downtime), excessive wear on the refractory (which would reduce on-stream time), loss of coolant to the induction furnace (which would cause downtime), and inadequate control of the process (which could lead to solidifying or skulling of metal or ceramic phase on the walls of the CPU and thus reduce on-stream time).

The failure modes in the gas handling train that are of some concern are loss of coolant in the offgas precooler (which could damage downstream equipment), solidifying of molten carryover from the CPU in the piping to the first quench, and corrosion in the offgas handling equipment.

Monitoring and Inspection

Monitoring methods for the bath temperature, composition, and containment, as incorporated in the CEP design, are described above in Process Instrumentation and Control. Offgas from CPU-1 will be cooled by water quenching. The temperature of the gas quench outlet will be measured and the flow rate of quench water adjusted to maintain the set-point temperature.

Inspection Frequency, Locations, and Observations

The TPC plans to base the frequency of inspection for the monitoring system on its general industry experience with corrosion and the Army's experience with corrosion at other agent destruction facilities. For example, probes for the continuous emission monitoring system last only a few days in high temperature, acidic environments, so they will be monitored on a daily schedule of preventive maintenance. The schedule for other monitoring locations with lower corrosion rates will be weekly or monthly.

The agent monitoring system itself will be used to warn of leaks in agent piping, fittings, valves, and pumps. All equipment used to deliver agent to the CPU will have double containment walls. The space between the primary and secondary containment walls will be monitored with DAAMS tubes, which will enable maintenance personnel to identify and repair leaking valves, fittings, etc., in the primary containment before the leak allows agent to escape the second containment.

The TPC plans to develop a maintenance control document as part of the detailed design phase. This document will include equipment maintenance schedules; parts lists for routine maintenance; lubrication requirements for each item of equipment; and maintenance procedure summaries specifying the frequency, purpose, references, prerequisites, and listings of all tasks and reviews. The documents will also include an instrument index and spares list, as well as preventive maintenance procedures for instruments, and will serve as a source book for miscellaneous maintenance items required for startup. Software will be used to record maintenance schedules and provide daily reminders and reports.

OPERATIONS AND MAINTENANCE

Operational Safeguards

All important variables such as temperatures, pressures, flow rates, and levels are measured, recorded, and alarmed throughout the system. Critical controls are provided with automatic alternatives if there is a safety risk or the possibility of damage to equipment. In areas of the plant that handle agent, the interstitial space in double-walled piping and equipment will be continuously monitored for agent, as a means of detecting leaks in the primary containment.

In the gas handling train, the quench water source has assured backup water sources, such as the firewater system. The backup water source ensures that hot offgas from the CPUs is cooled to prevent damage to the gas handling train.

The entire system is designed for operation via remote instrumentation, controls, and video cameras from a control center separate from the central building. The architecture of the DCS uses a centrally integrated executive protocol, which includes an emergency process-shutdown that is hard-wired and completely independent of the control computers and requires no human intervention.

The plant design adheres to approved safety principles for operations involving hazardous chemicals, including the following:

- All operations are designed to keep agent and agent-contaminated fluids inside the ton container, storage tank, or process piping at all times. Agent and agent-contaminated fluids are transferred from the collection point to nearby storage tanks by vacuum pumping techniques.
- The capacity and number of storage tanks for agent and agent-contaminated fluids are set to the minimum needed for the design throughput. Each tank is contained within a separate cell, and all cells are located together in the same area.
- Pumps for pressurizing the agent feed are located as close to the reactor as possible to minimize the length of piping that conveys pressurized agent to the CPU. The pump pressure is as low as possible consistent with maintaining reliable feed conditions under all operating conditions.
- Liquid agent and agent-contaminated fluids are transferred only through double-wall piping. The annulus is purged continuously with inert gas and monitored to detect the presence of agent.
- Pipes and ducts are welded and fully inspected. Bolted and sealed connections are used only where they are essential.
- In the event of a transfer-pump failure, agent or agent-contaminated fluid in the piping drains back into the source tank.
- All agent-involved pipes are sized and routed to allow unimpeded flow and minimize the chance of contamination traps.
- All components involved in pumping, storage, or piping of agent are mounted to be readily accessible for corrective maintenance and area housekeeping by personnel wearing appropriate safety gear.
- The areas around the CPUs are designed for convenient and secure access and are maintained at ambient temperature, to permit immediate emergency response via multiple routes for personnel in full protective clothing.
- The central building is partitioned in such a way that air monitors placed throughout the process areas can detect and verify agent leaks quickly and effectively.

Failure and Hazards Analysis

The TPC has performed several hazard and operability studies of CEP technology for the demonstration and commercial facilities described above. In addition, the TPC contracted with a third party to perform a hazard analysis specifically to support its submission for the chemical demilitarization program (M4 Environmental L.P., 1996e). This analysis, which used a failure modes and effects analysis (FMEA) approach, identified 1,129 failure events. Of these, 17 unique events for both facility sites were assigned a risk assessment code of 2, indicating that the risk was not acceptable. None of these code 2 risks involved exposure to chemical agent, and only one involved personal injury. The remaining 16 involved only a possible loss of processing capability because of damage to critical components in the gas handling train.

The TPC plans to conduct additional safety and hazards reviews during the design, engineering, and facility commissioning phases of development. The TPC states that, for these reviews, it will use methodologies and techniques developed by E.I. DuPont de Nemours and Company, Imperial Chemical Industries, and the Chemical Process Safety Institute that meet or exceed the requirements specified in the Occupational Safety and Health Administration regulations, Process Safety Management of Highly Hazardous Chemicals (29 CFR 1910.119).

The TPC also plans to implement a comprehensive health and safety program to establish best practices for ensuring safety. These practices include emergency response plans, plans for communicating information on chemical and radiological hazards, ALARA (as low as reasonably achievable) review procedures, safety training requirements, procedures for change management, and standard industrial safeguards. The TPC intends to document all operational procedures and practices, incident investigation reports, and compliance audits.

Maintenance

Routine Maintenance Requirements

For the feed preparation systems, feed systems, and balance of the plant (Areas 100 and 900), most of the routine maintenance after startup involves checking and adjusting for wear and tear of mechanisms and stops and replacing pressure seals and glands to prevent leakage of fluids and gases. Critical elements of the feed preparation equipment such as the punch tools, the probes for extracting liquid agent, and the water-jet cutting nozzles and cleaning heads need frequent replacement because they have high rates of wear.

Because operations at the two sites will be of short duration (about one year each) and the number of process cycles to be completed is fairly low (1,700 ton containers at each site, plus miscellaneous discrete items), the wear on the process equipment should be within acceptable limits.

An important aspect of routine maintenance will be calibration of instruments such as the ACAMS (automatic continuous air monitoring system) or MINICAMS. Because both of these instruments are gas chromatographs, they require a significant level of routine calibration and maintenance. The experience of one of the TPC partners in working with the instrumentation at the Tooele Chemical Disposal Facility gives the TPC team experience in setting up and operating a calibration and maintenance program for these and other agent-monitoring instruments.

Maintenance Manuals and Procedures

The TPC provides maintenance manuals and operating procedures for all its operating CEP units. Because the CEP facility for chemical demilitarization is still in the conceptual design phase, no facility-specific manuals or procedures have been developed yet. The TPC plans to develop a project maintenance manual covering preventive maintenance, lubrication, scheduled checks and inspections, cold test plans, and integrated test plans for startup. The manual will be prepared as the detailed design nears completion and will contain detailed procedures, checklists, and valve line-ups.

Documented Record of Performance

The feed preparation systems, feed systems, and most of the balance-of-plant systems (Areas 100 and 900) use equipment that is the same as or similar to equipment used in the Army baseline incineration system. Records of performance probably exist for this equipment, and one can reasonably assume that similar levels of operation and maintenance will apply when the equipment is used in the proposed CEP system.

Downtime Experience

Based on the TPC's experience to date, the TPC has allowed for approximately 60 days of maintenance and

300 days of continuous operation per operating year for each site (Aberdeen and Newport).

UTILITY REQUIREMENTS

Table 4-7 summarizes the TPC's stated utility requirements for a CEP agent-destruction facility. The numbers in the table represent steady-state processing of agent at the design rate (upper bound) of one ton container (liquid agent to CPU-1, empty container to CPU-2) approximately every 4 hours.

The principal utility requirements are natural gas and electric power. Note that the total electric power load of 1,510 kW shown in the table is a *net* load and includes a load-reducing contribution of 3,525 kW from cogeneration. Of the 33.35 x 10^6 Btu/hr (9,767 kW equivalent) of natural gas required at steady-state operation, 30.6 x 10^6 Btu/hr (8,962 kW equivalent), or 92 percent, is used for cogenerating electric power. The energy contribution to cogeneration from the synthesis gas is estimated at about 2 x 10^6 Btu/hr (586 kW equivalent).

For electric power, the maximum operating load of about 7,500 kW (not shown in Table 4-7) occurs when starting up the two CPUs together and lasts a maximum of 2 days. During CPU startup, there is also additional demand for natural gas to fuel the headspace heaters.

The water requirement is minor, consisting of makeup for a small offgas scrubber, makeup for a small cooling tower, and use by personnel. The total average requirement is estimated at 10 gallons (38 liters) per minute.

SCALE-UP REQUIREMENTS

The discussion of scale-up requirements for CEP is divided into issues related to scaling up the equipment and issues related to how processes are likely to perform when carried out at a larger scale.

Equipment Scale-Up

Front End and Back End Equipment

The development of all process operations and equipment at the front end of the process, as well as the back end of the plant, is well advanced. The same or similar equipment is used either in the Army's baseline program or in industry at the scale required for an agent-destruction facility. For example, the punch-and-drain equipment for ton containers has operated successfully at the JACADS chemical demilitarization facility.

CPU Equipment

The state of development of the CPU and related equipment is described above in the Technology Status section. The Demo Unit is a commercial-scale reactor with a metal bath size of 2,700 kg. The three iron CPUs in the CEP conceptual design submitted to the Army are about 8,200 kg each; the nickel bath is about 5,350 kg. Based on these preliminary estimates of nominal bath size, a scale-up of approximately 3:1 from the largest CPU in operation is required. In the judgment of the panel, the TPC has sufficient experience and understanding of CEP technology to perform the scale-up of bath size successfully.

The TPC has told the panel that it plans to use multiple tuyeres in each of the CPUs. Basic oxygen furnaces in the steel industry use many more tuyeres than are under consideration for this process. (At a meeting with the panel in January 1996, a TPC representative said that 16 to 20 tuyeres per furnace is common in the steel industry.) The TPC is continuing to validate the use of multiple tuyeres in an agent-destruction CPU, and confirmation on an appropriate number of tuyeres will be part of a final engineering design.

The design concepts for the premelting chamber to melt ton containers and for the system for feeding dunnage (in steel canisters) into the CPU-2 bath do not, to the panel's knowledge, have similarly close industrial counterparts. The TPC has conducted a demonstration program to test the processing of scrap metal, as a surrogate for some solid-waste feed streams of interest to the U.S. Department of Energy. However, the premelting chamber as suggested for the chemical demilitarization facility will require extensive development and demonstration. The TPC's reported experience to date includes a demonstration test in which six marine-location markers supplied by the Department of Defense were enclosed in cylindrical steel containers 0.8 m long and 9 cm in diameter. The containers were fed one by one into a molten metal bath through a gland in the top of the CPU. This test lends some credence to the submitted method for processing dunnage by loading it into cylindrical steel canisters 1 m long by 30 cm in diameter and feeding the canisters into CPU-2.

TABLE 4-7 Summary of Utility Requirements for a CEP Facility

Unit	Description	Plant Air (scfm)	Instrument Air (scfm)	Breathing Air (scfm)	Nitrogen (scfm)	Oxygen (scfm)	Fuels: natural Gas (Btu x 10^6h)	Fuels: Fuel Oil (Btu x 10^6h)	Chemical: Type	Chemical: Gallon	Water Make-U (gal./min.)
100	Feed preparation	120 @ 200 psig + 30 @ 90 psig	10 @ 90 psig	100 @ 90 psig	40 @ 50 psig				10% HTH decontam. soln.	10 to 40 per ton container = 5 to 20 gal./h	0.5
200	Catalytic processing	20 @ 90 psig	20 @ 90 psig	100 @ 90 psig	150 @ 200 psig[a]	50 @ 200 psig	0.6				
300	Gas handling train	20 @ 90 psig	10 @ 90 psig	50 @ 90 psig					10% HTH decontam. soln.	600[a]	
500	Power generation	10 @ 90 psig	10 @ 90 psig				30.6				
700	Product storage	20 @ 90 psig	20 @ 90 psig								
900	Infrastructure	100 @ 90 psig	5 @ 90 psig	50 @ 90 psig	10 @ 50 psig		2	1.2[b]	10% HTH decontam. soln.	600[a]	9.5
1000	Relief system	10 @ 90 psig	75 @ 90 psig	150 @ 90 psig			0.15		10% HTH decontam. soln.	600[a]	
Totals		120 @ 200 psig +210 @ 90 psig	75 @ 90 psig	300 @ 90 psig	150 @ 200 psig + 50 @ 50 psig	50 @ 200 psig	33.35	1.2[b]			10.0

Unit	Description	Steam[c]: High Pressure (lb./h)	Steam[c]: Low Pressure (lb./h)	Boiler Feed Water: High Pressure (lb./h)	Boiler Feed Water: Low Pressure (lb./h)	Condensate: High Pressure (lb./h)	Condensate: Low Pressure (lb./h)	Electric[e] Power (KW)	Cooling[d] Water (gal./min.)	Demineralized Water (gal./min.)	Domestic Water (gal./min.)	Sanitary Sewage (gal./min.)
100	Feed preparation		20 @ 15 psig		0.05							
200	Catalytic processing							4,000	750			
300	Gas handling train							225	210	5		
500	Power generation	(1,720) @ 435 psig		4.2				(3,525)	151			
700	Product storage							5				
900	Infrastructure		780 @ 15 psig		1.95		0.5	800	379		5	5
1000	Relief system							5				
Totals		(1,720) @ 435 psig	800 @ 15 psig	4.2	2	0	0.5	1,510	1,500	5	5	5

[a] For intermittent use.
[b] Additional fuel oil will be required for a short period, to power diesel generators in case of an electrical power outage.
[c] For steam ( ) indicates quantity produced.
[d] Cooling water supplied at 80°F and returns at 100°F.
[e] Connected electric load is 6000 KW, essential load is 3500 KW, UPS load is 150 KW.

Performance Scale-Up

Front End and Back End Performance

All the processes in areas 100 and 900 have been demonstrated in the Army baseline system with live agent at scales similar to the scale for an operational CEP facility, except for the optional high-pressure water-jet systems for cutting open and cleaning ton containers. The panel expects the water-jet systems will work as proposed because they are commercial systems that have worked well on similar materials under extremely harsh conditions over long periods of time.

CPU Performance

The TPC has done extensive experimentation and modeling of CPU performance to understand bubble formation, breakup dynamics, and the operating limits of molten metal baths. As described in the Process Modeling section, this modeling work has identified three key factors in CPU performance to be bubble size, residence time, and energy dissipation by gas-metal mixing and gas-metal contact within gas bubbles. The TPC states that the modeling results correlate well with DRE values achieved in actual tests. The design for the full-scale baths is stated to provide a residence time with at least a tenfold safety factor over the residence time required to meet the requirement of at least six 9's (99.9999 percent) DRE.

Testing Agent Surrogates in CEP

The TPC tested destruction of an HD surrogate, half-mustard gas (HMG, 2-chloroethyl ethyl sulfide). The result was a DRE of at least nine 9's for conversion of HMG to synthesis gas, HCl, Fe-S alloy, and H_2S. The DRE calculation was limited by the amount of agent processed and the lower detection limit of the analytical method.

In another test, diazinon, which is structurally similar to VX, was reported to have been converted to synthesis gas, with the phosphorus and sulfur from the diazinon retained in the metal phase as an Fe-S-P alloy. Analysis of the offgas was conducted in accordance with EPA method TO-14. By this method, no C_2 or higher hydrocarbons were detected at the lower detection limits, which are in the part-per-billion range. Third-party analyses confirmed that no hazardous organic constituents were present in the ceramic or metal alloy products, which also passed the TCLP test for RCRA metals. The TPC states that the results verify that these solid products are nontoxic and potentially marketable.

The AltTech Panel agrees with the TPC's interpretation of these tests as showing that the technology can destroy agent. The AltTech Panel sees no reason to expect the *qualitative* aspects of these test results to be different when the process is scaled up. The major conversion products and the partitioning between gaseous and condensed-phase are expected to be the same. The panel also believes the tests provide a strong *preliminary* indication that the residuals from a carefully designed CEP process to destroy chemical agents are likely to be nontoxic and safe for release to the environment or to commercial use, as the TPC anticipates.

However, the panel cautions that the particular *quantitative* results obtained in these tests on surrogates, such as a particular DRE value or the nondetection of trace products at parts-per-billion concentrations in residuals, *should not be directly extrapolated to full-scale operation* unless information on certain key scaling parameters is provided. In the case of CEP test results, an important scaling parameter is one that the panel has named the *specific processing rate*, which for convenience can be defined as the amount of agent (in kilograms) processed per hour, per unit size of the bath (measured, for example, in 1,000 kg of molten metal). The closer the specific processing rate of a test is to the specific processing rate projected for a full-scale operation, the more confidence one can place in extrapolating quantitative test results. In the case of the tests on agent surrogates, the panel did not receive data from which specific processing rates could be calculated. Therefore, the quantitative results obtained under full-scale operation could be better or worse than these bench-scale test results with agent surrogates.

Testing Actual Agent in CEP

As noted in the Agent Testing section of Technology Status, the TPC has tested actual HD and VX agent in a bench-scale CPU at Battelle/Columbus Laboratories. The panel received the full report on these tests in early June 1996. The report states the agent destruction efficiency of the bench unit as eight 9's (99.999999 percent) for HD and VX. Based on the panel's preliminary review of the report, it appears to be more accurate to call this result a DRE because the offgas passed through at least one filter before it was tested.

The panel obtained sufficient data on the tests on actual agents to calculate specific processing rates for comparison with the rates for the full-scale system (Table 4-8). (The latter were computed from the design feed rates of agent and the nominal bath size.) Of several bath compositions tested for each agent, the panel used the results from the bath composition closest to that of the full-scale bath under steady-state operation. The bench-scale tests used a single top-entering lance to feed agent into the bath, whereas the design for a full-scale facility has bottom-entering tuyeres.

As the table shows, these bench-scale tests of agent destruction were run at significantly lower specific processing rates than the rates the TPC has designed for a full-scale facility. In the panel's judgment, with the admonition stated above about extrapolating quantitative results from small-scale tests to performance of a full-scale operating facility, the implicit scaling factor in the specific processing rate for VX of 2.6:1 is within acceptable engineering practice. In making this judgment, the panel has taken into account the TPC's stated design safety margin of 10:1 in bath residence time and the reported test result of eight 9's DRE, which implies a performance margin beyond the required six 9's DRE. The panel cautions that the implicit scaling factor in the specific processing rate for HD of 5.4:1 leads to even greater uncertainty in extrapolating the bench-scale DRE to full-scale performance.

The panel believes that the TPC understands the complexity of scaling quantitative performance measures such as DRE from bench-scale tests to full-scale operations. However, the panel would prefer DRE data for VX and especially for HD from bench-scale tests conducted at specific processing rates closer to the rates for the full-scale design.

UNIT OPERATIONS

This section summarizes the unit operations in CEP treatment of chemical agents for the Aberdeen and Newport sites, including unit operations required to treat secondary process streams and residuals prior to disposal. A unit operation is a combination of equipment that accomplishes one specific step in a process. Table 4-9 lists the unit operations for CEP by process area.

PROCESS SAFETY

Process-safety risk factors for a CEP agent-destruction facility can be divided into two categories: factors related to handling agent prior to its introduction into the CPUs and factors related to the molten bath technology.

The risk factors inherent in the handling of agent prior to entry into the CPUs include storage risk, transportation risk, and the risk from the punch-and-drain operation. These risk factors are common to all the agent-destruction technologies reviewed in this report, but they can be exacerbated or ameliorated by aspects of a specific technology. For example, how quickly a facility using the technology can reach operational status or the rate at which the agent can be processed with that technology can alter the storage risk by changing the length of time that the agent must be stored. The CEP technology is well advanced, and the design calls for processing the agent at each site in one year. Both of these technology-specific features help in reducing storage risk. As another example, the capability in the CEP design for treating emptied ton containers to the equivalent of 5X condition by melting and processing them immediately reduces the risk from handling the containers. The process-safety risk factors inherent in CEP include issues associated with high-temperature molten

TABLE 4-8 Specific Processing Rates of Bench Tests Relative to Full-Scale Design Rates

Agent Tested	Bath Composition	Specific Processing Rate (kg agent/hour/1,000 kg bath metal): Bench Test	Full Scale (design)	Scaling Factor (full scale/bench)
HD	Ni + 2% C	7	38	5.4
VX	Fe + 7% P + 7% S + C	8	21	2.6

TABLE 4-9 CEP Unit Operations by Process Area

Area 100, Container and Dunnage Transportation and Handling	Feed storage (ton containers) Punch and drain station Ton container wash and preparation Dunnage handling and preparation Liquid (agent and container-washout) storage and feed
Area 200, CPUs	CPU-1 Premelting chamber to CPU-2 CPU-2 CPU-2 offgas quench, scrub, particulate removal, and compressor
Area 300, Gas Handling Train	Gas quench and particulate removal HCl recovery Sulfur recovery
Area 500, Synthesis Gas Utilization	Gas compression and retention/analysis Power generation Steam-methane reformer (option for methanol recovery)[a] Methanol production (option for methanol recovery)[a]
Area 700, Products Storage	Sulfur product storage HCl product storage Methanol product storage (option for methanol recovery)[a]
Area 800, Utilities	Inert gas storage and feed Oxygen storage and feed Natural gas feed Air—plant air and instrument air Water—plant, potable, cooling, boiler feed, and chilled Steam—generation and condensate handling Electricity Diesel power backup
Area 1000, Relief and Scrubber System	Scrubber (decontamination solution) Boilers

[a]These unit operations are only present if synthesis gas is converted to methanol instead of being burned to generate power. Under the methanol option, the power generation unit process would not be installed.

baths such as the integrity of the refractory confinement, the proximity of the molten bath to water cooling coils (raising the possibility of steam explosions), the behavior of the tuyeres, and the instrumentation for monitoring the refractory confinement. In the panel's judgment, none of these factors presents an insurmountable impediment to the safety of the process. Many of the risk factors have already been addressed by the TPC in the hazard analysis it conducted for design of a chemical demilitarization facility (discussed above under Failure and Hazards Analysis) or on the basis of the TPC's research and operational experience with CEP.

The panel was satisfied that the TPC had adequately addressed several issues the panel had raised during site visits regarding integrity of the refractory. The panel found no evidence of scenarios involving a loss of electrical power, loss of cooling, failures of pumps or valves, breaks in agent lines from inadvertent overpressurization, or inadvertent temperature transients that would lead to off-site releases of agent or toxic process products. Pessimistic scenarios for a coincident loss of normal power, loss of backup power, and loss of cooling result in the solidification of the molten metal bath in place without significant release to the atmosphere.

Based on the panel's preliminary and qualitative evaluation, the most significant off-site risk appears to be associated with risk factors inherent in handling agent prior to the CEP process. In particular, the principal risk factors appear to involve mishaps during the punch-and-drain operation or damage from airplane crashes or other external events to holding tanks where agent is stored before being fed to the main reactor. The subsections on process safety below address the risk factors specific to CEP technology. However, the panel believes that none of these factors seriously challenges the safety of the facility.

Safety Issues Related to Off Site Releases

The following issues should be addressed fully and clearly in a final CEP process design.

Integrity of the Refractory. The work by the TPC on the integrity of the refractory must be included in the safety documentation for a final CEP design. The TPC has done much work to avoid gas-jet impingement on the refractory lining of the CPU and to select refractory materials for the lining that resist gas permeation, thermal degradation, corrosion, erosion, and penetration by components of the molten metal and slag.

Integrity of the Agent-Bearing Components. This issue was explored briefly by the panel, and no significant issues were uncovered. However, because certain parts of the design are still preliminary, the panel encourages the TPC to pursue its stated plans for continuing, comprehensive safety and hazard analyses as part of the development process. Particularly important is further exploration of scenarios involving failures of piping or components. (Failure could be caused by thermal attack by molten material, system overpressure, subtle system interactions, or other causes.)

Cooling Offgas Piping. Scenarios involving a failure to cool the offgas piping should be explored. This is probably not an issue, but at the time of the panel's review, the consequences of such scenarios were not clear.

Buildup of Combustible Gases. The TPC's design as submitted prevents a buildup of combustible gases in the vicinity of the system by maintaining a high ventilation rate. Assurances should be made that combustible gas buildup cannot occur and that the high ventilation rate does not compromise the design capability to contain leakage of agent.

Worker Safety Issues

There are a number of worker safety issues associated with high-temperature molten baths, high-temperature corrosives in the scrubbers, and secondary containment (concerning both inadvertent leaks and maintenance activities). These risk factors need to be addressed in the final operational design, and realistic emergency responses need to be spelled out.

Specific Characteristics that Reduce Risk Inherent in the Design

Because of the natural temperature gradient in the CPU refractory material, the molten material will solidify before it gets very far into the refractory. This self-sealing feature helps keep the molten metal away from the water-filled induction coils and thus reduces the possibility of a steam explosion.

A loss of electrical power, of cooling water to the heat exchanger, or of the cooling for pumps could result in the molten metal solidifying in place. Although solidification would be an operational problem if it were to occur, it is not a safety issue.

SCHEDULE

Figure 4-11 is the latest schedule submitted to the AltTech Panel from the TPC for the major activities and milestones in a chemical demilitarization program to use CEP technology at the Aberdeen and Newport sites. Table 4-10 is the panel's analysis, based on the TPC schedule, of activities on the critical path to completion of the program, their duration, and the cumulative time from start of the program to the end of that activity. An important aspect of the TPC's concept as submitted to the Army is that the same CEP equipment would be installed first at Aberdeen for HD destruction, then moved to Newport and installed there for VX destruction. Advantages and disadvantages of this approach are discussed below.

Another key aspect of the design is that the TPC's preferred approach, after a go-ahead from the Army to begin work, is to move directly to design of a facility

Activity Description	Early Start	Early Finish
Milestones and Design		
Design Requirements Defined	11/1/96	
Design (Aberdeen/Newport)	11/1/96	6/29/01
Start Construction	1/30/98	
Destruction of Agent @ Aberdeen Complete		10/9/00
Aberdeen Complete	4/16/01	
Destruction of Agent @ Newport Complete		8/25/03
Newport Complete	2/27/04	
Aberdeen		
Equipment Specs. and Procurement	8/18/97	10/30/98
Site Prep. and Installation	1/30/98	2/3/99
Final EIS (Aberdeen)	1/30/98	
Aberdeen Startup and Test-Limited Agent Operation	2/4/99	9/6/99
System Modifications	4/7/99	10/6/99
Full-Scale HD Treatment	10/7/99	10/9/00
Decontam./Decommision APG Facility (Phase I)	10/10/00	4/13/01
Newport		
Equipment Specs. and Procurement	8/18/97	6/29/01
Final EIS (Newport)	8/30/00	
Newport Site Prep. and Facilitization	9/8/00	4/11/01
CEP VX Installation and Checkout	4/16/01	11/15/01
Newport Startup and Test	11/16/01	6/19/02
System Modifications	1/21/02	7/22/02
Full-Scale VX Treatment	7/23/02	8/25/03
Decontam./Decommision Newport Facility (Phase I)	8/26/03	2/26/04
Regulatory Milestones (Aberdeen)		
Prepare/Obtain Air Construction Permit	11/1/96	1/30/98
Prepare/Obtain Recycle Determination	11/1/96	1/30/98
Submit Operating Permit	6/30/98	
Operating Permit Approval	9/30/98	

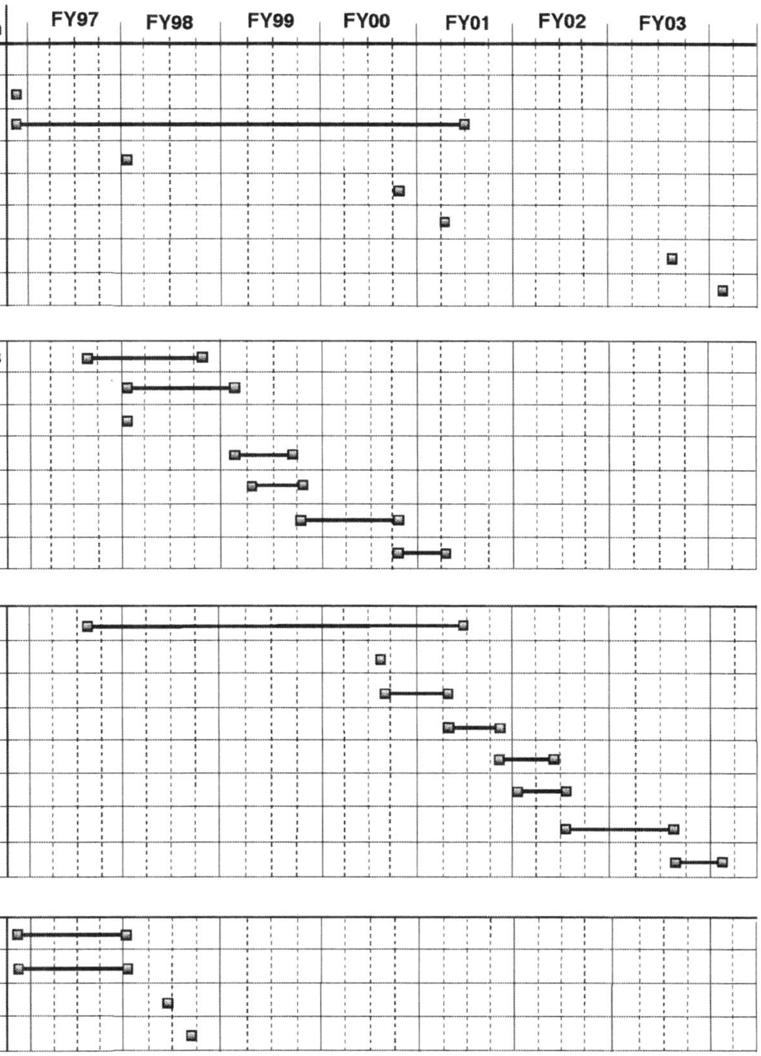

FIGURE 4-11 CEP program schedule and phasing concept. Source: M4 Environmental L.P., 1996d.

with full-scale CPUs for the next stage of development. A facility at that scale is more conventionally referred to as a demonstration plant than a pilot plant. To indicate how the schedule relates to the Defense Acquisition Board's decision to proceed with pilot-scale development, the panel will refer to this next stage as pilot/demonstration. The facility for this pilot/demonstration phase at each site will be equipped with enough gas-handling capability to ensure protection of human health and the environment, but the full gashandling train will not be installed until full-scale operation.

The TPC foresees no scale-up effort required to move from pilot testing to full-scale processing. The panel cautions, however, that although use of full-scale equipment at the pilot/demonstration stage means that no *equipment scale-up* will be required, whether *performance scale-up* is needed depends on how closely the final stages of pilot testing resemble the process conditions for full-scale, continuous operation. The pilot/demonstration activities will entail a good deal of work, including systemization with agent surrogates, preoperational surveys, an operational readiness evaluation, and similar requirements prior to full-scale operation. Provided that the TPC continues testing and develops an adequate design basis prior to construction of the pilot/demonstration facility (that is, resolves remaining issues such as demonstrating the premelting chamber, scaling the bath to the larger size required, resolving the number and placement of tuyeres, and demonstrating process performance at the design

TABLE 4-10 Critical Activities in the Program Schedule

Activity	Duration (months)	Cumulative (months)
1. Prepare and obtain regulatory permits, etc., for Aberdeen	15	15
2. Aberdeen construction (site prep. and installation)	12.2	27.2
3. Aberdeen pilot/demonstration (startup, test, and system modifications)	8	35.2
4. Aberdeen full-scale HD operations	12	47.2
5. Newport construction (site prep. and installation)[a]	14.25	60.45
6. Newport pilot/demonstration (startup, test, and system modifications)	8.25	68.7
7. Newport full-scale VX operation	13	81.7

[a]Newport construction overlaps one month with Aberdeen full-scale operation.

specific-processing rates), the panel believes that 8 months can suffice for performance scale-up and required startup activities.

The full-scale operation at each site is designed to be continuous, 24 hours per day, at the agent feed rates specified above in the Feed Streams section. The scrubbed offgas is either combusted with natural gas in a gas turbine generator to produce electricity for the plant or converted to methanol. At this stage, process residuals would be placed on the commercial market. The design as submitted is not clear about how process residuals would be handled during the earlier pilot/demonstration stage.

The TPC has stated that the submitted design provides sufficient throughput to allow all agent, ton containers, and dunnage to be destroyed in 12 months from the start of full-scale operation at Aberdeen and in 13 months from the start of full-scale operation at Newport (M4 Environmental L.P., 1996d). Assuming that construction at Aberdeen can be approved by January 30, 1998, the TPC anticipates that the program for both sites will be completed before the end of 2003, more than a year before the Army deadline of December 31, 2004. The AltTech Panel believes that the TPC's goal of completing the destruction of each stockpile in 12 to 13 months after commencing full-scale operation is achievable, if the throughput rates assumed in the submission can be sustained for the duration of the operation.

In the panel's judgment, the time allotted for pilot/demonstration activities at Newport is essential. The VX configuration uses the same equipment but a different set of processing parameters and constraints, as well as handling a different agent and a different partitioning of chemical elements to product phases.

After processing HD at Aberdeen has been completed, the CEP systems will be decontaminated, decommissioned, and relocated to Newport for processing VX. The TPC believes this plan for reusing equipment is a cost-effective and time-saving solution for destroying agent stockpiles at multiple sites. The panel agrees that there are advantages to sequential operations but cautions that there are also risks to the schedule. A significant delay in the Aberdeen schedule could delay the agent destruction schedule at Newport. In fact, any delay in one of the activities along the critical path can delay subsequent activities.

For example, the submitted schedule reflects early and vigorous efforts to complete the required reviews and secure necessary approvals. The TPC estimates that a permit for construction of a plant producing atmospheric emissions can be obtained in Maryland within 15 months of project start. The panel notes that this relatively short time for permitting may depend on the TPC acquiring a recycle waiver from RCRA permitting requirements. If the permitting process takes longer and construction is delayed, the schedule does have about

15 months of slippage time at the end to still meet the Army deadline.

The panel notes in passing that the time shown in Figure 4-11 for decontamination and decommissioning is probably only the time required to decontaminate and decommission the CEP systems. (The schedule refers to the activity as phase 1 of decontamination and decommissioning.) Additional time will probably be required for decontaminating and decommissioning the central building and the associated infrastructure.

5

Mediated Electrochemical Oxidation Silver II

PROCESS DESCRIPTION

Silver II is a patented electrochemical process. It was originally developed in 1987 by AEA Technology at Dounreay, Scotland, as a means for destroying solid and liquid radioactive organic waste streams from the U.K. Fast Reactor fuel development program. AEA Technology submitted the Silver II technology to the Army for consideration as an alternative technology for agent destruction at the Aberdeen and Newport sites and will therefore be referred to as the TPC (technology proponent company) for the Silver II process in the remainder of this report.

Most of the TPC's effort to date has been dedicated to operation of a 4-kW pilot plant for destroying inactive fuel solvent composed of 10 percent tributyl phosphate in kerosene. In addition, laboratory tests conducted at Dounreay since 1987 have demonstrated destruction of 68 organic compounds encountered in industrial wastes, including HD (distilled S-mustard), VX, and GB (another unitary chemical nerve agent).

Figure 5-1 is a schematic diagram of the heart of the Silver II process as described by the TPC for destruction of VX and mustard. The core reactions take place in two separate 180-kW, electrochemical cells (model ICI FM21), which are connected in parallel through a 360-kW power supply. Each FM21 cell comprises 45 anode-cathode compartments, each 10 mm wide by 240 mm high; each electrode is separated by a Nafion[1] membrane, which is permeable to cations and water but impermeable to anions (Figure 5-2). The anode-cathode chambers are connected in parallel, each pair requiring a normal operating current of 2,000 A at a nominal 2 volts DC. Thus, the 360-kW power supply unit for a standard module must provide a total of 90 kA and 180 kW to each of the two cells that make up the module. The aggregate volume of all the anode-cathode chambers within a cell is 2.5 m^3.

At the start of operation, the composition of the anolyte is approximately 8 molar in nitric acid, 0.5 molar in silver nitrate, and 0.02 to 0.03 molar in agent. The catholyte is 4 molar nitric acid.

When power is applied to the cell, Ag(I) ions are oxidized at the anode to the highly reactive Ag(II). The Ag(II) species has been shown to exist in the form of $AgNO_3$+ ions (Po et al., 1968), which impart a brown color to the solution in the absence of organics. In the presence of organics, $AgNO_3$+ ions oxidize water into intermediates such as hydroxyl radicals that rapidly oxidize the organic species. Simultaneously, Ag(II) is

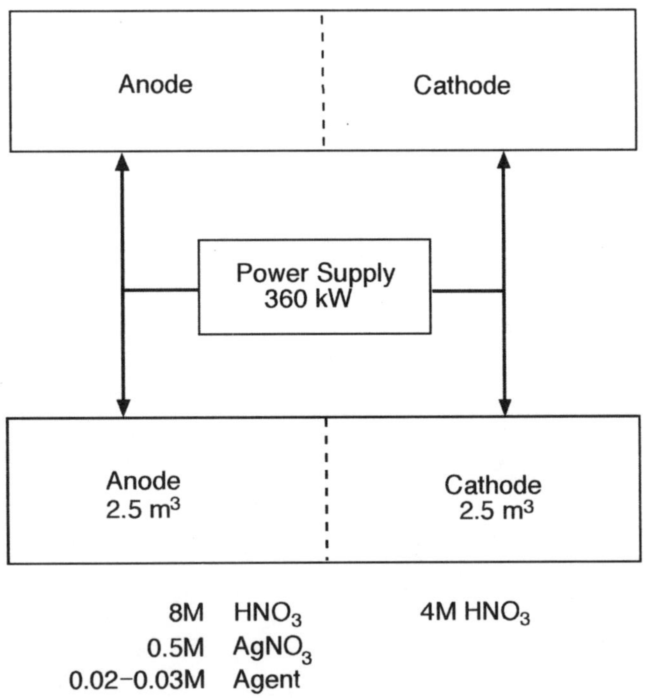

FIGURE 5-1 Schematic diagram of the basic cell module for mediated electrochemical oxidation.

[1]Nafion is a perfluorosulfonic acid polymer developed by E.I. Du Pont de Nemours and Company.

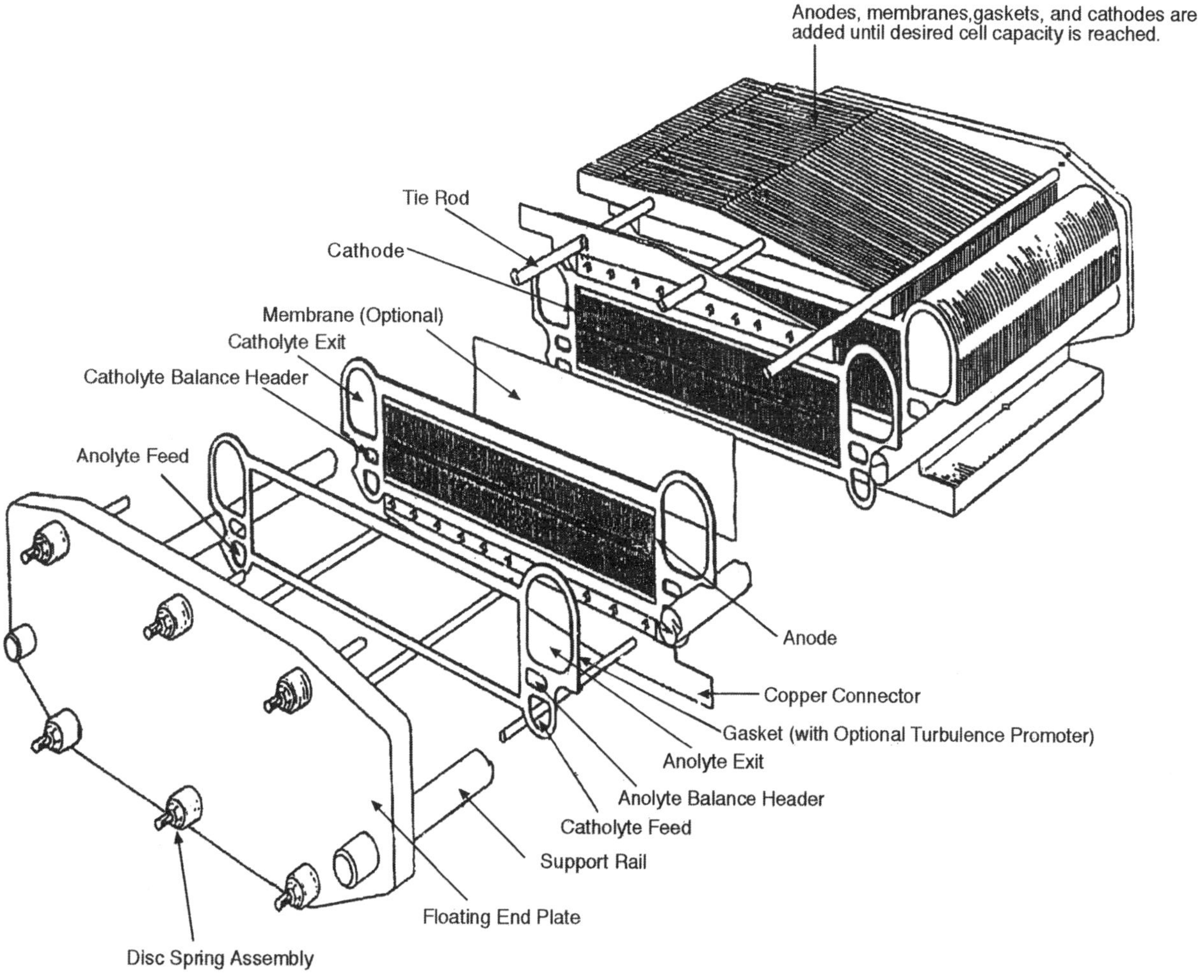

FIGURE 5-2 Exploded view of the FM21 electrochemical cell. Source: AEA Technology.

reduced back to Ag(I), which migrates back to the anode surface where it is reoxidized to Ag(II). Silver therefore serves as an electron transfer intermediate that is not consumed in the process. However, when chloride ions or organic chlorides are present, as in HD, Ag(I) precipitates as AgCl.

The anticipated overall anode reactions for VX and HD are as follows:

$$\text{VX: } C_{11}H_{26}SNPO_2 + 31H_2O = 11CO_2 + H_3PO_4 + H_2SO_4 + HNO_3 + 82H^+ + 82e^-$$

$$\text{HD: } C_4H_8SCl_2 + 12H_2O = 4CO_2 + H_2SO_4 + 2HCl + 28H^+ + 28e^-$$

Some CO will form as well, by analogous reactions, but laboratory tests have shown that carbon is converted primarily to CO_2. Hydrated protons (hydronium ions, H_3O^+) move across the membrane toward the cathode, where the primary reaction is reduction of nitric acid to nitrous acid:

$$HNO_3 + 2H^+ + 2e = HNO_2 + H_2O$$

Nitrous acid will partially decompose to NO gas, nitric acid, and water. In the laboratory tests observed by the AltTech Panel, the gas leaving the cathode compartment had the characteristic red-brown color of NO_2, which can form by oxidation of NO in the gas phase when O_2 is present.

The overall cell reactions are:

$$\text{VX: } C_{11}H_{26}SNPO_2 + 40HNO_3 = 11CO_2 + H_3PO_4 + H_2SO_4 + 41HNO_2 + 10H_2O$$

$$\text{HD: } C_4H_8SCl_2 + 14HNO_3 = 4CO_2 + H_2SO_4 + 2HCl + 14HNO_2 + 2H_2O$$

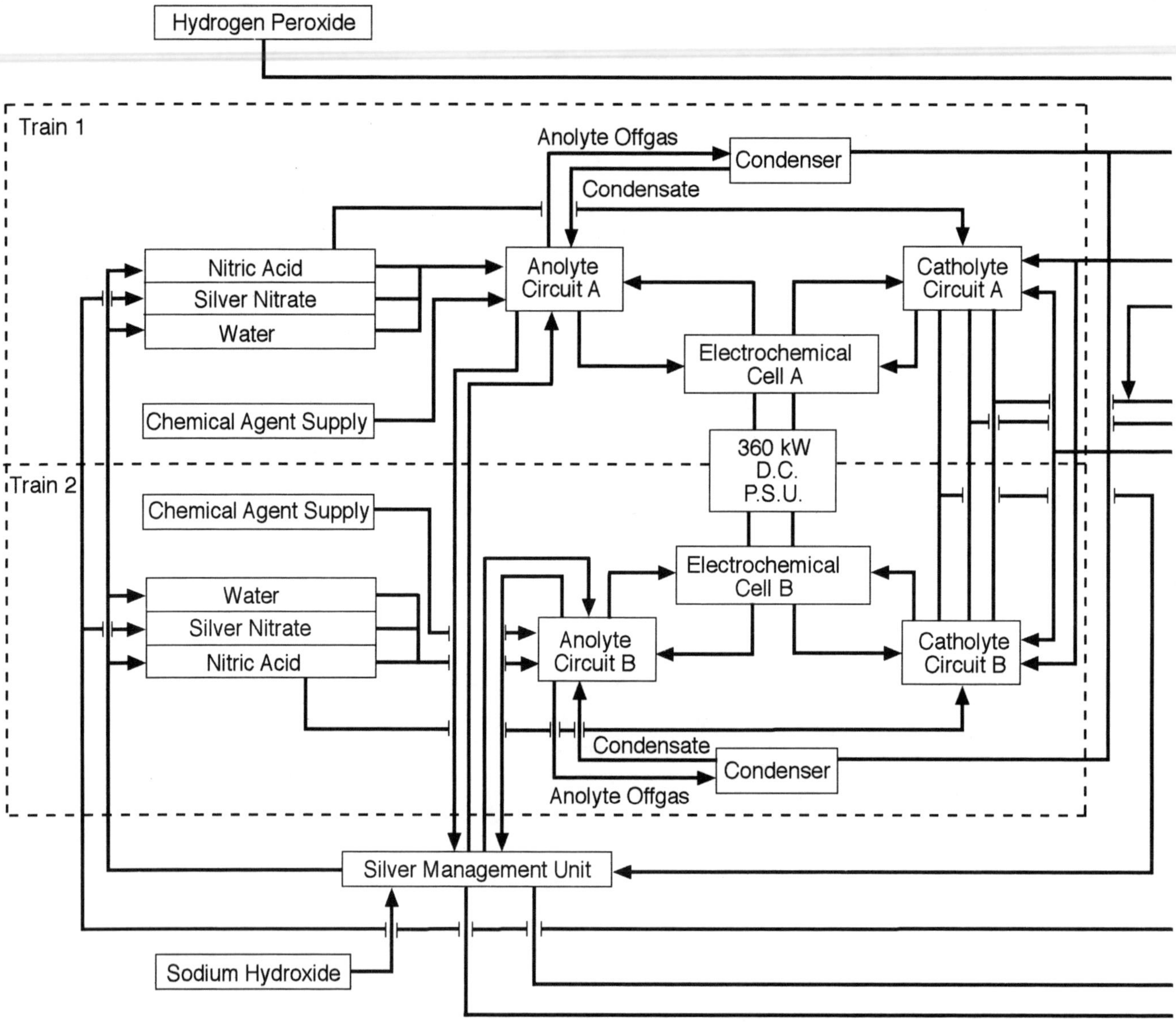

FIGURE 5-3 Block flow diagram of the Silver II process total system. Source: AEA Technology.

The reaction products are treated in subsequent steps outside the cell to reoxidize HNO_2 to HNO_3 and to neutralize the acids to their corresponding sodium salts. Therefore, the net reactions are as follows:

VX: $C_{11}H_{26}SNPO_2 + 20.5\ O_2 + 6NaOH = 11CO_2 + Na_3PO_4 + Na_2SO_4 + NaNO_3\ 16H_2O$

HD: $C_4H_8SCl_2 + 7O_2 + 4NaOH = 4CO_2 + Na_2SO_4 + 2NaCl + 6H_2O$

The overall reactions are similar to the overall reactions for incineration of VX and HD, but they occur at low temperature (less than 90°C) and close to atmospheric pressure. In both processes, carbon is released to the gas phase primarily as CO_2. In the electrochemical process, the sulfur, phosphorus, and chlorine components of the agent appear in the final effluent as hydrated anions in aqueous solution (sodium is the principal cation). This solution can be analyzed and treated further, if necessary, prior to release. In combustion processes like the baseline incineration system, these elements yield gases (assuming oxidation is complete), which must be removed in a treatment train, but the treated process gas stream is difficult to analyze prior to release to the atmosphere.

Three additional reactions that can occur will affect the energy efficiency of the process. First, Ag(II) can react directly with water in the anode compartment to

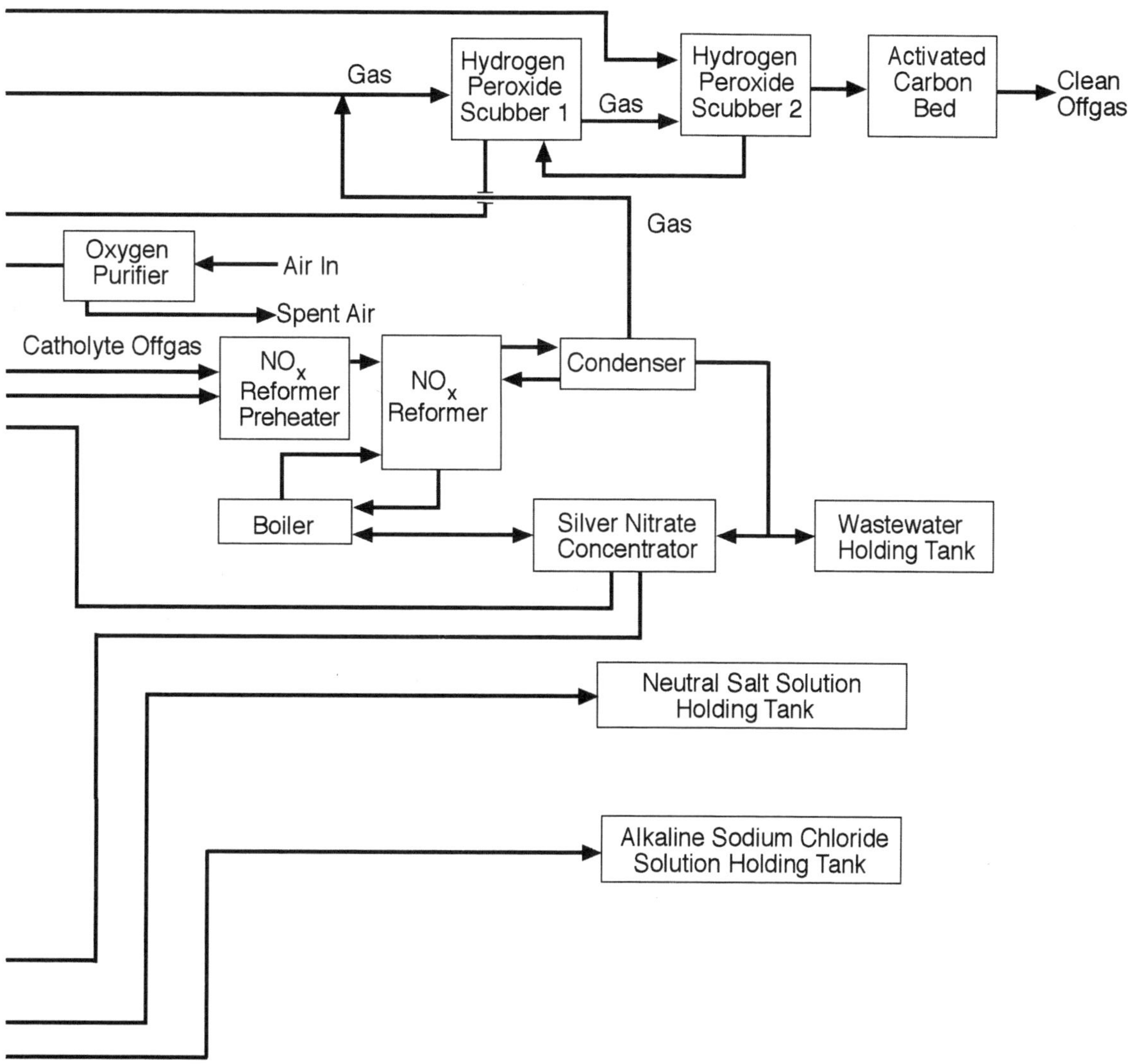

form oxygen gas (O_2). Second, the Ag(I) can migrate across the membrane to the cathode compartment. Third, cationic impurities in the agent can migrate across the membrane to the cathode compartment. Analyses of the HD stored at Aberdeen show that such impurities are likely to include iron, copper, and possibly mercury. Organic impurities in the agent will be oxidized in the anode compartment by reactions analogous to the reactions with agent.

The process reactions involving agent cannot be reversed. Therefore, once agent is destroyed, it cannot reform. However, agent destruction is likely to proceed in several steps, some of which may produce volatile organic intermediates that will enter the gas phase and require further treatment. In laboratory tests, for example, the TPC identified varying levels of alkyl nitrates in the anolyte offgas, which was mainly CO_2. Nonvolatile organic intermediates that may also form will remain in the anode compartment and will ultimately undergo complete conversion to simpler inorganic products, such as sulfate, phosphate, chloride, and CO_2/CO.

In common with virtually all commercial electrochemical processes, Silver II requires continuous feed systems to both the anolyte and catholyte chambers and treatment systems for anolyte and catholyte products. Figure 5-3 is a block flow diagram of a total system, which comprises the following components:

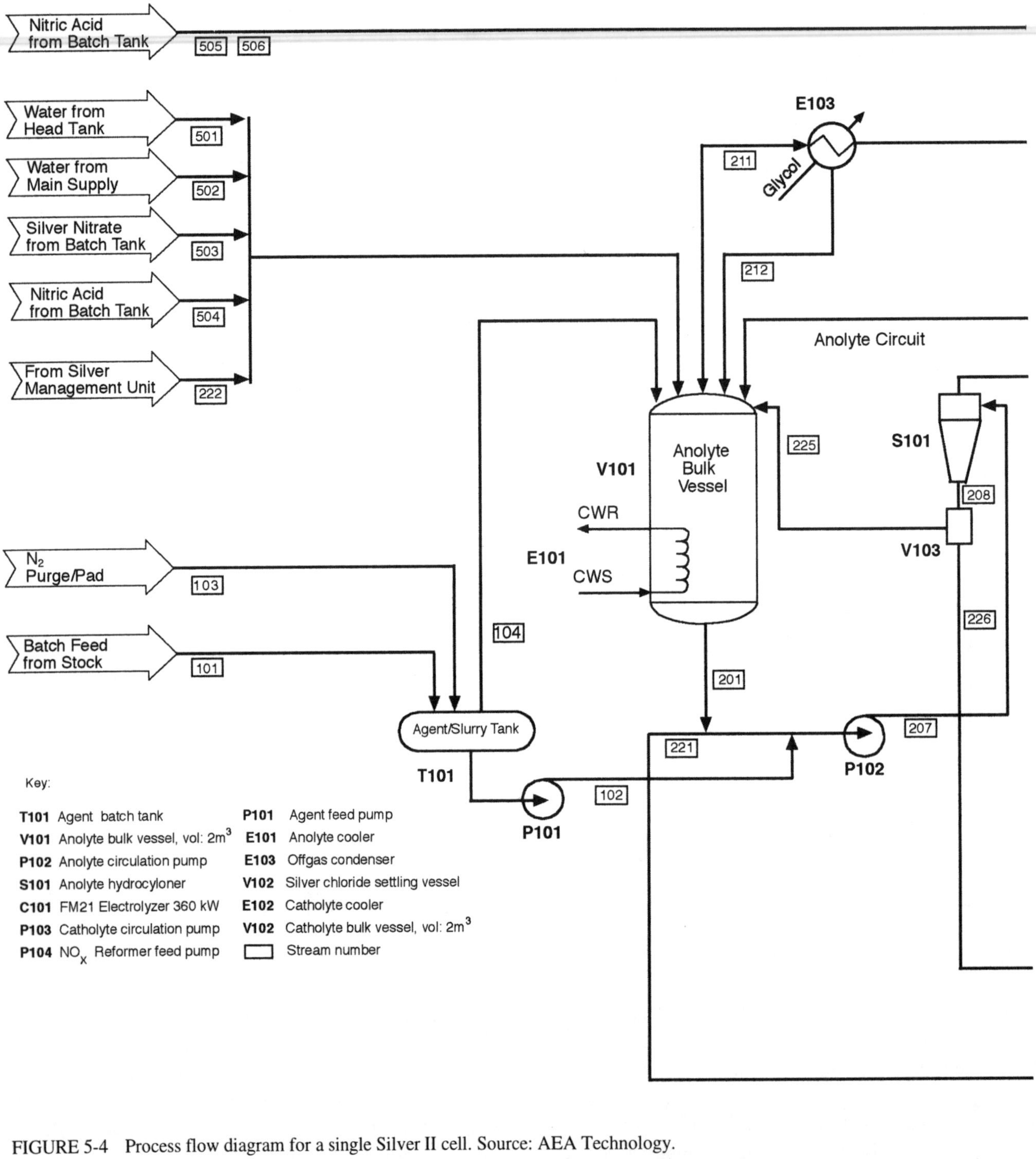

FIGURE 5-4 Process flow diagram for a single Silver II cell. Source: AEA Technology.

- agent receipt and supply
- anolyte feed circuit
- catholyte feed circuit
- electrochemical cell
- anolyte offgas condenser
- NO_x reformer system
- catholyte silver nitrate recovery circuit
- combined offgas treatment circuit
- silver management system
- utilities infrastructure

Figure 5-4 is a process flow diagram. Each of the key system components is discussed below.

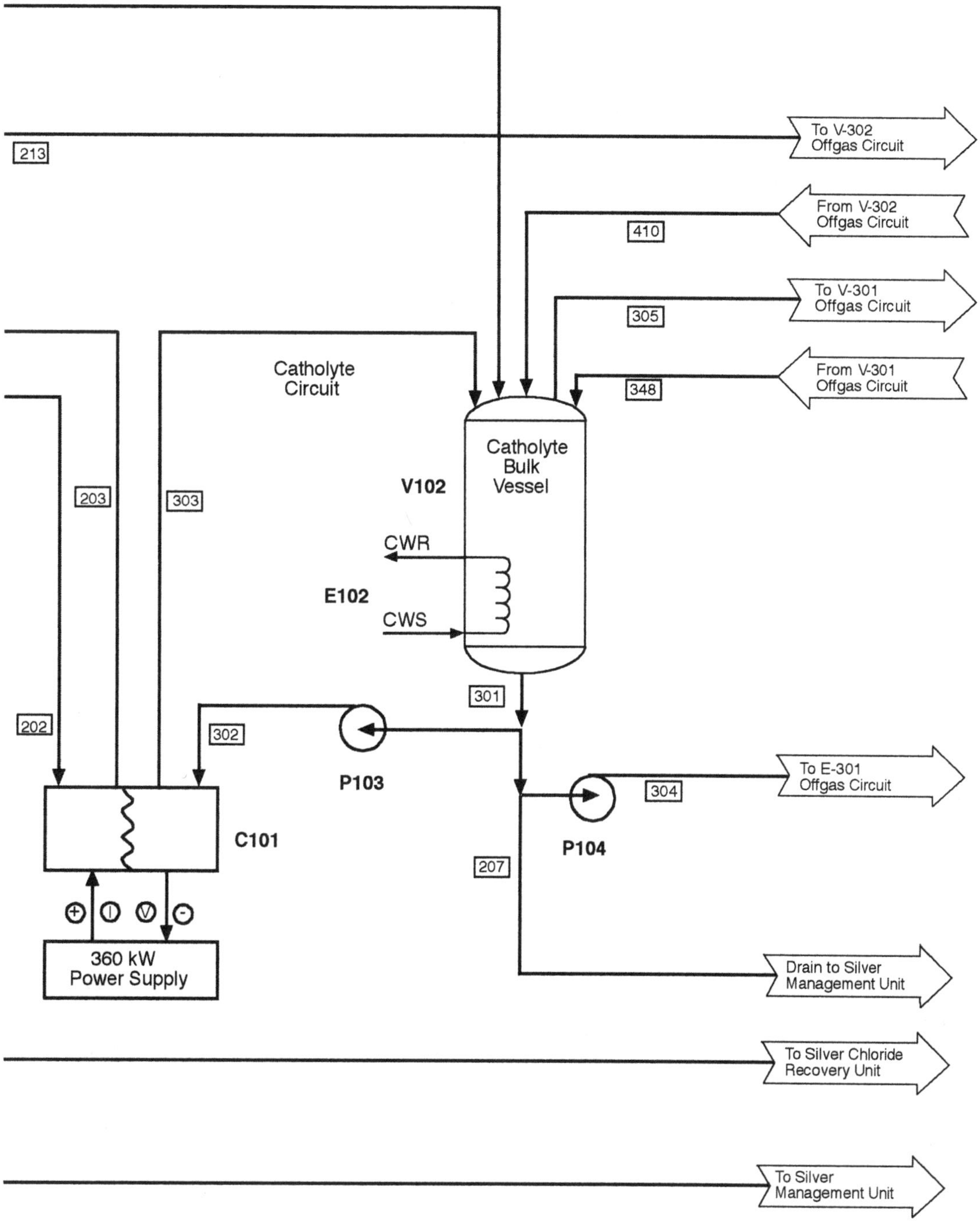

Agent Receipt and Supply. The TPC plans to use the same systems developed and tested by the Army for the baseline system.

Anolyte Feed Circuit. The anolyte feed circuit includes a 2-m^3 anolyte vessel, the anolyte compartment of the electrochemical cell, a circulation pump, and connecting pipework. For HD processing, a hydrocyclone is added to remove some of the silver chloride precipitate. The anolyte vessel is fed from batch tanks of silver nitrate and nitric acid, a head tank of water, the catholyte silver nitrate recovery circuit, and an agent-slurry tank.

Catholyte Feed Circuit. The catholyte feed circuit consists of a single loop by which 4.0 molar nitric acid is pumped from a 2-m^3 bulk vessel through the cathode compartment of the electrochemical cell and back to the bulk vessel. The nitric acid concentration in the bulk vessel is maintained by additions from the NO_x reformer, which reclaims nitric acid from spent catholyte and NO_x separated from the catholyte.

Electrochemical Cell. Anolyte and catholyte solutions circulate through the cell at flow rates up to 45 m^3/h and temperatures up to 90°C.

These four components make up the basic agent-destruction system. This system runs in a semibatch, or campaign, mode. Each of the FM21 electrochemical cells has an associated agent receipt and supply unit to process a ton container of agent, as well as its own anolyte and catholyte feed circuits. A campaign consists of processing a ton container of agent through this system. A campaign for the standard 360-kW module (two FM21 cells) therefore involves handling and processing two ton containers of agent simultaneously. The TPC expects each campaign to last 7 to 10 days, during which time the system will be run continuously. The 360-kW module is the basic unit of facility scale. Increased throughput, or facility scale-up, consists of adding additional 360-kW modules and the infrastructure to support them. The silver management system is operated in batch mode at the end of a campaign. It operates totally apart from the agent destruction process and does not affect the time for destroying agent (throughput rate).

Anolyte Offgas Condenser, NO_x Reformer, Catholyte Silver Nitrate Recovery Circuit, and Combined Offgas Treatment Circuit. These four components, which are shown in Figure 5-5, operate continuously throughout a campaign. They constitute the auxiliary and downstream processing and recycling components of a fully functioning agent destruction system. The anolyte offgas condenser removes water vapor, nitric acid vapor, and condensable organics from the offgas. The NO_x reformer reconstitutes nitric acid from the products of the cathode reaction. The catholyte silver nitrate recovery circuit captures silver that has migrated across the cell membrane from the anolyte. The offgases from the cell and the noncondensable overheads from the distillation circuits are processed through the combined offgas treatment circuit before being released to the atmosphere.

Silver Management System. The silver management system, shown in Figure 5-6, operates independently of the agent-destruction system. At the end of a campaign, it is used to treat residual chemicals that have accumulated in the anolyte and catholyte circuits and to recover silver. Residuals in the anolyte circuit can include phosphate, sulfate, nitrate, and chloride anions in acid solutions. The specific anionic mix depends on whether HD or VX has been treated. The anode compartment of an FM21 cell, at 2.5 m^3, is large enough to keep the phosphate from VX and the sulfate from VX or HD in solution throughout a campaign. After a campaign, the silver management system removes the phosphates and sulfates from the cell electrolytes and recovers any silver remaining in the catholyte and anolyte circuits. Not shown in Figure 5-6 is the auxiliary system that will be needed to recover silver from the solid silver chloride formed when HD is processed.

Utilities Infrastructure. The Silver II process is energy-intensive. The electrical energy required is 72,600 kW•h per metric ton of HD destroyed and 134,900 kW•h per metric ton of VX destroyed.

SCIENTIFIC PRINCIPLES

Ag(II) in an acidic medium is one of the most powerful oxidizing agents known (Lehmani et al., 1996). The standard reduction potential of the Ag(II)/Ag(I) couple is 1.98 volts, whereas the standard reduction potential of the O_2/H_2O couple is only 1.23 volts in nitric acid. Several published studies report on the use of anodically generated Ag(II) to oxidize organics in an acid solution (e.g., Lehmani et al., 1996; Farmer et al., 1992; Steele, 1990; Mentasti et al., 1984).

The basic half cell reactions for the Silver II process are as follows:

$$\text{Anode:} \quad 2Ag^+ \rightarrow 2Ag^{++} + 2e \qquad E^o = -1.98\ V$$

$$\text{Cathode:} \quad HNO_3 + 2H^+ + 2e \rightarrow HNO_2 + H_2O \qquad E^o = +0.94\ V$$

The net reaction is therefore:

$$2Ag^+ + HNO_3 + 2H^+ \rightarrow 2Ag^{++} + HNO_2 + H_2O \qquad E^o = -1.04\ V$$

In these equations, E^o is the standard equilibrium potential at zero current flow when all reactants and products are at unit activity. In practice, the required potential is larger than the standard equilibrium potential because of ohmic heating and other effects. The TPC uses an applied potential of 2 V.

Oxidation of Ag(I) to Ag(II) at the surface of a platinum anode is rapid, and the required overpotential is low: 120 mV at 5kA/m^2. The principal Ag(II) species formed is $AgNO_3$+, which has a dark brown color. The color disappears almost instantaneously in the presence of organics due to several complex reaction steps that result in the complete oxidation of the organics and the reduction of Ag(II) back to Ag(I). Silver is not consumed in the process but functions as a mediator between the electric power fed into the cell and the organic compounds being destroyed.

The reaction mechanisms in silver-mediated electrochemical oxidation are not well understood but are believed to involve highly reactive, short-lived species, including hydroxyl and other radicals. In a study of the electrochemical oxidation of ethylene glycol and benzene by Ag(II), several relatively long-lived reaction intermediates were identified, but with sufficient time complete oxidation was achieved as evidenced by measurement of stoichiometric quantities of CO_2 in the final product. (Farmer et al., 1992)

TECHNOLOGY STATUS

The Silver II process has yet to be operated on a commercial scale. The largest-scale pilot tests have been conducted with 4-kW cells consisting of a single anode-cathode pair. The most extensive tests have been conducted with spent tributyl phosphate dissolved in kerosene, from the Purex process, as the feed material. These tests, which were run continuously, 24 hours per day for up to 14 days, destroyed a total of 150 liters of the feed material. The TPC has successfully completed laboratory tests on 10-g batches of agent and has constructed a pilot plant at Porton Down, United Kingdom, that is suitable for tests on 15-liter batches of agent. All of the tests prior to startup of the Porton Down plant had been conducted with only the electrochemical cell component of the agent-destruction system. The Porton Down facility also includes anolyte and catholyte feed circuits, an anolyte offgas condenser, an NO_x reformer system, and a modified version of the combined offgas treatment circuit, which culminates in a sodium hydroxide scrubber. The silver management system will be tested at Dounreay on the effluent generated at Porton Down.

A preliminary draft report received by the panel on May 31, 1996, summarizes the results of a test conducted by the TPC at Porton Down on 14.62 kg of "as supplied VX," which contained 12.7 kg of agent. The test consisted of a single continuous run of 6.5 days. At the end of the run, no agent was detected in the catholyte or in the process residuals. The lower detection limits for VX were 7.6 mg/m^3 in the anolyte, 9.2 mg/m^3 in the catholyte, and 1.7 mg/m^3 in the residuals discharged during the trial. The corresponding volumes were 0.0724 m^3 of anolyte, 0.0854 m^3 of catholyte, and 0.0929 m^3 of process residuals. The total residual VX was therefore less than 1.5 mg out of an input of 12.7 kg of VX, corresponding to an agent destruction efficiency of greater than 99.99998 percent.

The TPC calculated that the 14.62 kg of "as supplied VX" contained 7.21 kg of organic carbon. At the end of the run, the total organic carbon remaining in the anolyte and catholyte circuits was 0.816 kg. Therefore, the destruction and removal efficiency for conversion of organic carbon to CO_2 and CO was 88.7 percent. The TPC suggests that further removal might have been possible by continuing the operation of the cell after the organic feed was ended.

The TPC operated the test cell at Porton Down at currents between 600 and 1,400 A. The test was not able to operate at the design current of 2,000 A because of pressure increases in the anolyte compartment when VX was added. The TPC traced the problem to lower than expected efficiency of the NO_x reformer, which resulted in the passage of more than expected unreacted O_2 and NO_x gas through the condenser and into the scrubber. This increased the pressure drop across the scrubber, causing an increase in pressure in the anolyte gas stream.

OPERATIONAL REQUIREMENTS AND CONSIDERATIONS

Process Operations

In concept, the Silver II process as a complete system will operate as follows. Prior to the introduction of agent to the system, all other constituents are present in the anolyte and catholyte solutions, the feed circuits are operating, and all systems are at their set-point temperatures. Once flows and temperatures are stable, the current is turned on and agent is pumped into the circulating anolyte solution from the 1-m^3 agent-slurry tank. The flow rate of this agent feed is about 0.01 m^3/hr, which should maintain the agent concentration in the anolyte

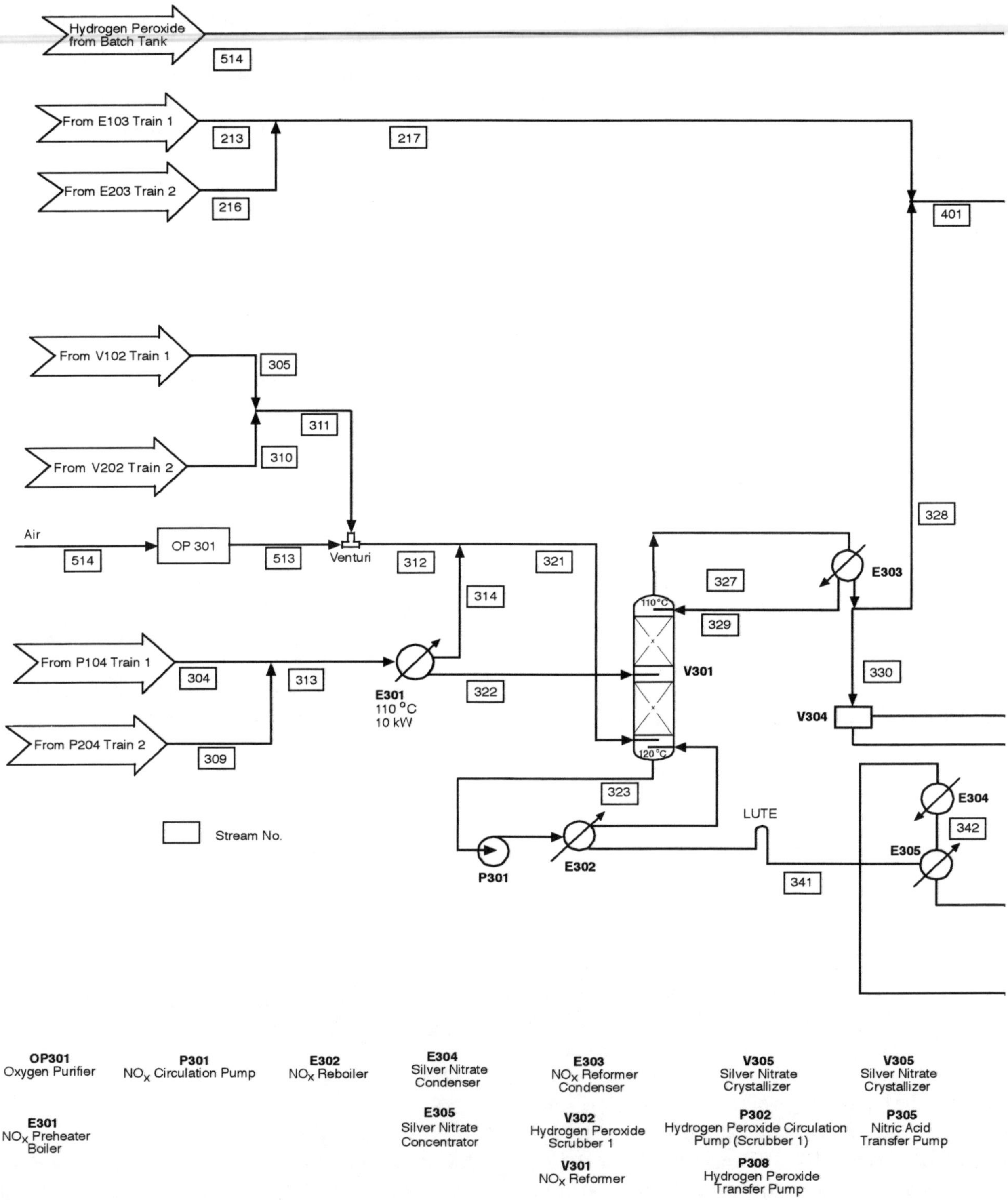

FIGURE 5-5 Anolyte offgas condenser, NO_x reformer, silver nitrate recovery circuit, and combined offgas treatment circuit. Source: AEA Technology.

at about 5,000 ppm. To ensure good mixing of the agent with the anolyte feed, the agent is added at the inlet to the circulating pump (see Figure 5-4).

The TPC has proposed several options for transfer-

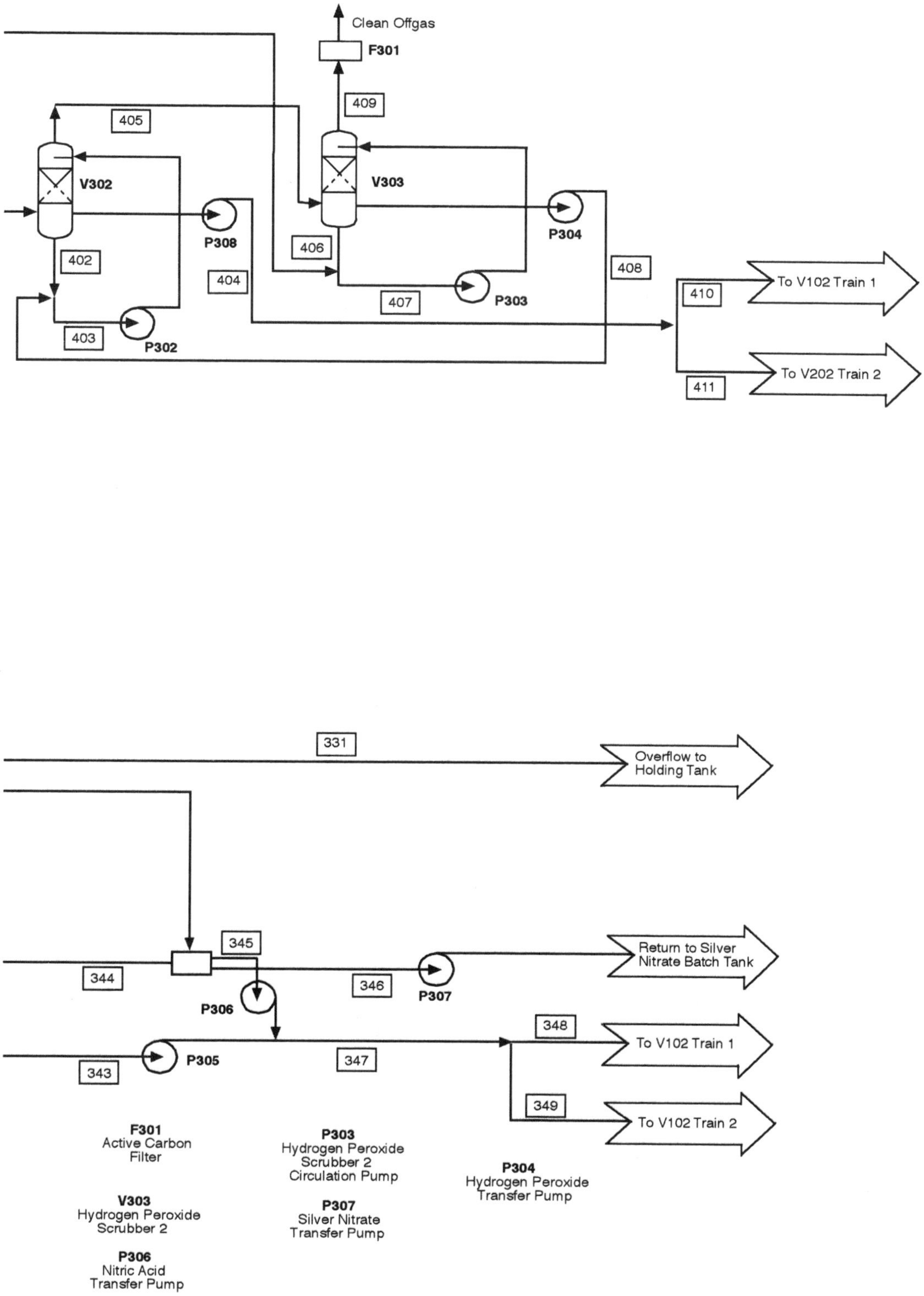

ring agent from a ton container to the agent-slurry tank. The agent-transfer system that the Army has proposed for use in the neutralization process (see Chapter 7) is equally well suited to Silver II.

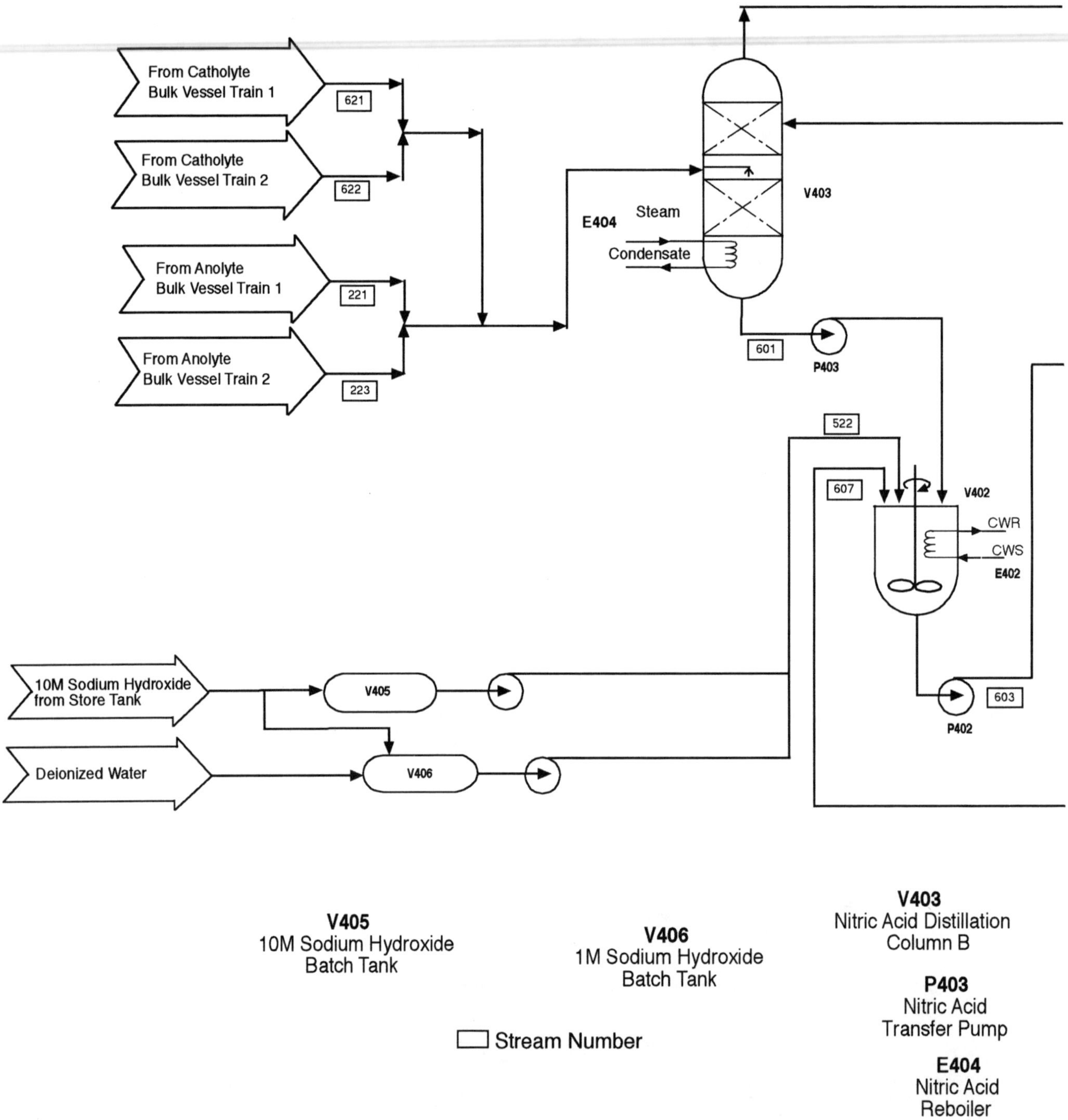

FIGURE 5-6 Silver management system. Source: AEA Technology.

Compositional Changes during Normal Operation

Normal cell operation depletes certain constituents of both the anolyte and catholyte, so continuous addition of makeup chemicals is required. Silver nitrate must be added to the anolyte circuit; nitric acid must be added to the catholyte circuit.

The loss of silver nitrate has two causes: the transport of Ag(I) from the anode to the cathode compartment, which occurs with any organic feed material, and the precipitation of silver chloride, which happens when a feed material contains chlorine, as does HD. The TPC reports that transport of Ag(I) accounts for about 1 percent of the total charge transferred. The total theoretical charge

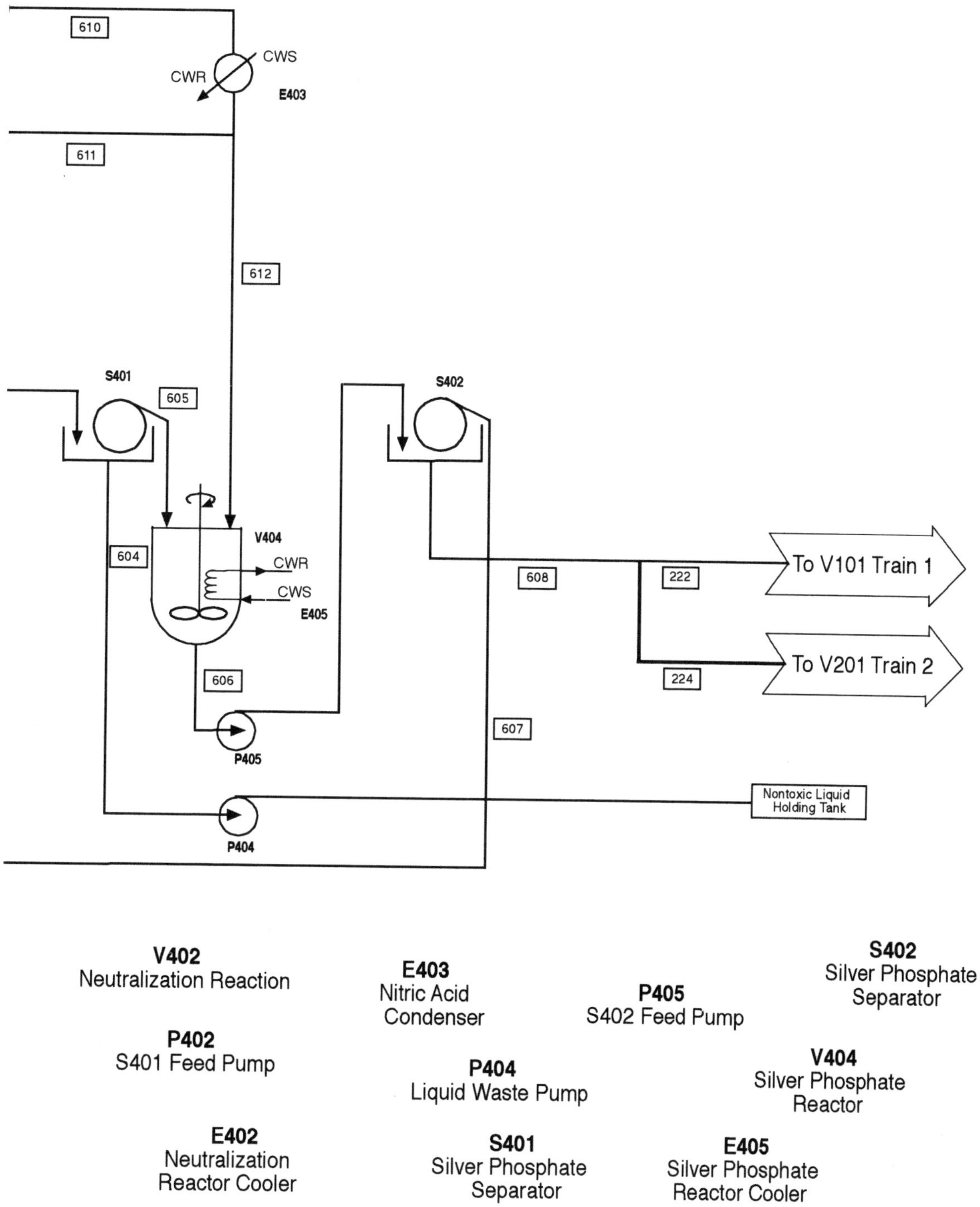

transfer per metric ton of agent destroyed is 17 x 10^9 coulombs for HD and 29.6 x 10^9 coulombs for VX. The anolyte circuit starts out with 2.5 m^3 of solution that is 0.5 molar in silver nitrate, which represents an initial inventory of 1.25 kg-mols of silver nitrate or 134 kg of silver. During the course of an HD campaign (one ton container), 190 kg of silver will transfer from the anode to the cathode compartment; during a VX campaign, 332 kg will transfer. In both cases, therefore, the total quantity of silver transferred to the cathode compartment during a campaign exceeds the initial amount of silver in the anolyte circuit. The catholyte silver nitrate recovery circuit, which is discussed below, recovers the silver from the catholyte by crystallizing silver nitrate from the concentrated solution and dissolving it in nitric acid for return to the anolyte circuit.

During the destruction of HD, major losses of silver from the anolyte occur from precipitation of insoluble silver chloride. By the end of the campaign, 12.58 kg-mols of silver chloride, containing 1,357 kg of silver, has precipitated. Therefore, the silver nitrate additions during an HD campaign must make up for silver losses of 1,547 kg from both Ag(I) transport and AgCl precipitation. This means that 1.5 metric tons of silver must be added to the anolyte circuit for each metric ton of HD destroyed.

The makeup silver nitrate is added to the anolyte feed circuit through a manifold in the top of the anolyte vessel and mixes into the bulk anolyte as the solution circulates. Silver concentration must be monitored during a campaign, and feedback systems must be designed to automate the addition of proper quantities of silver nitrate to the anolyte circuit.

The acidity of the anolyte solution increases substantially during a campaign. Sulfur from the feed becomes sulfuric acid; phosphorus becomes phosphoric acid; and in HD processing, chlorine precipitates with Ag(I) as AgCl, leaving nitric acid. It appears that the resulting increases in acidity will not be corrected during a campaign.

The catholyte solution loses nitric acid continuously because the nitric acid is reduced to nitrous acid as the principal cathode reaction. The nitrous acid subsequently decomposes to NO_x gases. To compensate for this loss, a bleed stream from the catholyte circuit is pumped continuously to the NO_x reformer system, where some of the excess water is boiled off and the nitrous acid is oxidized to nitric acid for return to the catholyte circuit. (The NO_x reformer system is discussed in detail below.)

Water Management System

A water management system is needed to control the water level in both the anode and cathode compartments. The water balance is complex, involving two countervailing forces. Water flows from the anode compartment across the membrane to the cathode compartment in the form of hydrated protons (hydronium ions, H_3O^+) generated as a product of the anode reaction. Water flows in the opposite direction, from the cathode compartment to the anode compartment, because of the osmotic pressure maintained by the lower acidity (i.e., higher water concentration) in the cathode compartment.

The transport of hydrated protons from the anode compartment to the cathode compartment can be calculated readily from the basic electrochemistry of the cell (Appendix E). The compensating effect of osmotic diffusion must be determined empirically. In pilot-plant commissioning tests observed by the panel at Porton Down, in which triethyl phosphate was the organic feed, the level of the anolyte visibly rose within a few hours of operation, while the level of the catholyte fell. Thus, under those conditions, the rate of osmotic diffusion was clearly exceeding the rate of water transport via hydrated protons. Further tests with agent as the organic feed will be required to engineer the system for proper water balance. The osmotic flow will also vary during a campaign, as the acidity of the anolyte increases.

NO_x Reformer

The principal reaction at the cathode is the reduction of nitric acid to nitrous acid. A bleed stream (flow rate of 0.168 m^3/h) from the bottom of each of the two catholyte bulk vessels used in a standard 360-kW module is pumped to a boiler, where the nitrous acid undergoes thermal decomposition to NO gas and nitric acid. The NO gas is mixed with 90 percent pure oxygen, heated to 110°C, and fed at a rate of 196.7 m^3/h to the base of a distillation column. This column forms the heart of the NO_x reformer (see Figure 5-5). The aqueous phase from the boiler, containing nitric acid and silver nitrate, is fed into the midsection of the distillation column at a rate of 0.377 m^3/h. The overhead stream from the distillation column passes through a condenser. The condensate stream, a dilute solution of nitric acid, is split; one part returns to the top of the distillation column, and the rest goes to a holding tank for reuse or eventual discharge (after being neutralized to a salt such as sodium nitrate). Noncondensables enter the combined offgas treatment circuit (discussed below).

Catholyte Silver Nitrate Recovery Circuit

The bottom stream from the NO_x reformer column passes to a boiler. Nitric acid and water vapor from the boiler return to the bottom of the distillation column, and the remaining, more concentrated solution of silver nitrate in nitric acid passes to a concentrator. Approximately every 6 hours, the liquid accumulated in the concentrator is transferred to a crystallizer, where the solution is cooled. Silver nitrate crystallizes out, and the supernatant nitric acid is drained off and returned to the catholyte circuit. The silver nitrate crystals are then redissolved in the dilute nitric acid from the overhead of the NO_x reformer. This solution returns to the batch tank for silver nitrate solution, to be used as makeup for the anolyte circuit.

Anolyte Offgas Condenser

Reactions in the anode compartment produce several gaseous products including CO_2, O_2, and possibly CO. Volatile organic products of incomplete oxidation may also form from the stepwise oxidation of agent. These gaseous reaction products form an offgas saturated with water and nitric acid vapors. The offgas is released from the anolyte vessel to a condenser chilled by a mixture of water and glycol at 0°C. The gases are cooled to 10°C, causing any nitric acid, water, chemical agent, or condensable organic products to condense and drain back to the anolyte bulk tank. The noncondensable gases enter the combined offgas treatment circuit.

Combined Offgas Treatment Circuit

The noncondensable gases from the anolyte offgas condenser and the NO_x reformer are combined for further treatment (Figure 5-5). The combined gases pass through two hydrogen peroxide scrubbers that are 30 to 35 feet tall. The scrubbers reduce the concentration of NO_x to less than the permitted discharge limit. The scrubbed gas passes through an activated-carbon filter bed and is released to the atmosphere. The process flow diagrams do not show a condenser and reheater that will be required upstream from the carbon filter bed to remove water from the scrubbed gas. Gas from the hydrogen peroxide scrubbers will be saturated in water vapor, which, if not removed, would impair the capacity of the carbon bed to adsorb trace organics.

Silver Management System

At the end of a campaign, solutions from both the anolyte and catholyte circuits are transferred to the silver management system (Figure 5-6). The combined solutions are distilled through two columns in series (columns A and B in Figure 5-6). Still bottoms from the first column are drained to a mixing tank, where they are neutralized by sodium hydroxide added from a batch tank. These highly acidic still bottoms contain a solution of silver nitrate in nitric, sulfuric, and phosphoric acids; as nitric acid is removed by distillation, silver sulfate and silver phosphate may precipitate. The exact composition depends on which agent was treated. Addition of sodium hydroxide converts the acids to their sodium salts in solution, which becomes a process residual. Any precipitated silver salts (silver sulfate, phosphate, or oxide) are filtered out and reacidified to recover silver.

Figure 5-7 shows the adjunct to the silver management system that will be required after an HD campaign. As previously discussed, the residuals in the anolyte circuit will contain more than a metric ton of precipitated silver chloride, which must be filtered out. This filtration could be difficult because precipitated silver chloride tends to form very small particles. The supernatant acid mixture is double-distilled as described above, and silver nitrate is ultimately recovered from the still bottoms.

The precipitated silver chloride is transferred to a separate mixing vessel to which excess sodium hydroxide is added. Any sulfuric and nitric acids accompanying the silver chloride are converted to dissolved sodium salts. The silver chloride is partially converted to silver oxide (Ag_2O) via a solid-state, diffusion-controlled reaction. This conversion therefore proceeds from the outside of the particle in, so that each particle has a core of silver chloride and a coating of silver oxide. The liquid, containing sodium salts, is filtered off and becomes a process residual. The precipitate is reacidified with nitric acid, which dissolves the silver oxide as silver nitrate. The silver nitrate solution is filtered off for reuse as anolyte feed. Any remaining silver chloride solids are recycled to repeat the treatment with sodium hydroxide for conversion to silver oxide. This sequence is repeated until all the silver chloride from the campaign has been converted back to silver nitrate solution in nitric acid.

The TPC has not described this post-campaign neutralization and silver recovery system in detail. It appears that neither part of the system has been tested. Actual quantities and compositions of feed and product streams were not reported to the panel. The silver management system will operate as a batch process totally separate from the agent-destruction campaign. The proposed process, which appears to be scientifically

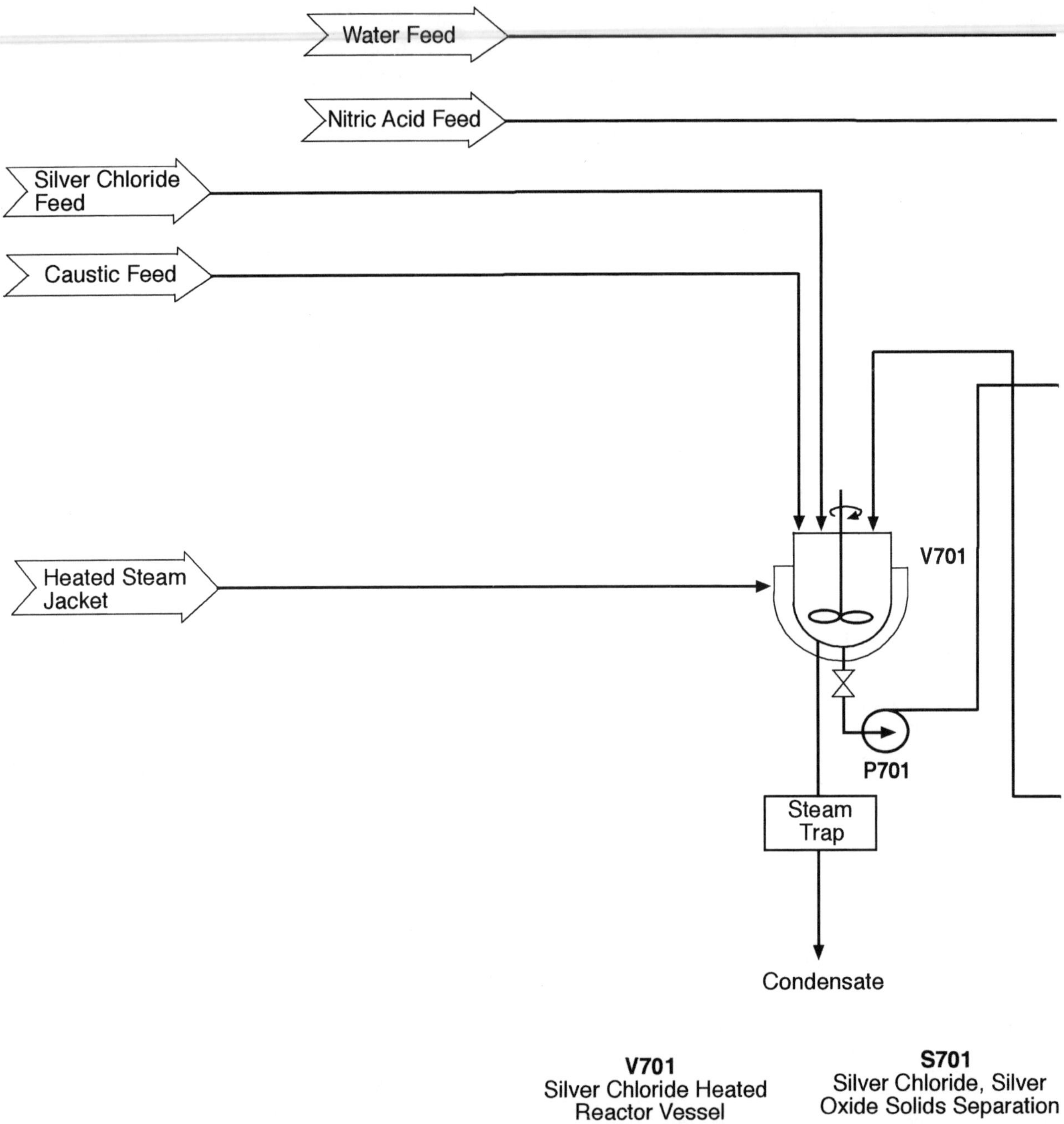

FIGURE 5-7 Silver chloride treatment system. Source: AEA Technology.

sound, will be tested by the TPC on the post-campaign electrolyte solutions from the pilot tests at Porton Down.

Energy Requirements

The Silver II process consumes a great deal of electrical energy for cell operation and for auxiliary heating, refrigeration, and pumping. The theoretical energy for a 2-volt cell is about 9,400 kW•h per metric ton of HD and 16,440 kW•h per metric ton of VX. The TPC assumes a 60 percent electrochemical efficiency, which raises the energy requirements to 15,700 and 27,400 kW•h per metric ton of agent destroyed for HD and VX, respectively.

The TPC estimates the total electric power consumption for operation of a basic two-cell module and auxiliary equipment at 1.7 MW, consisting of:

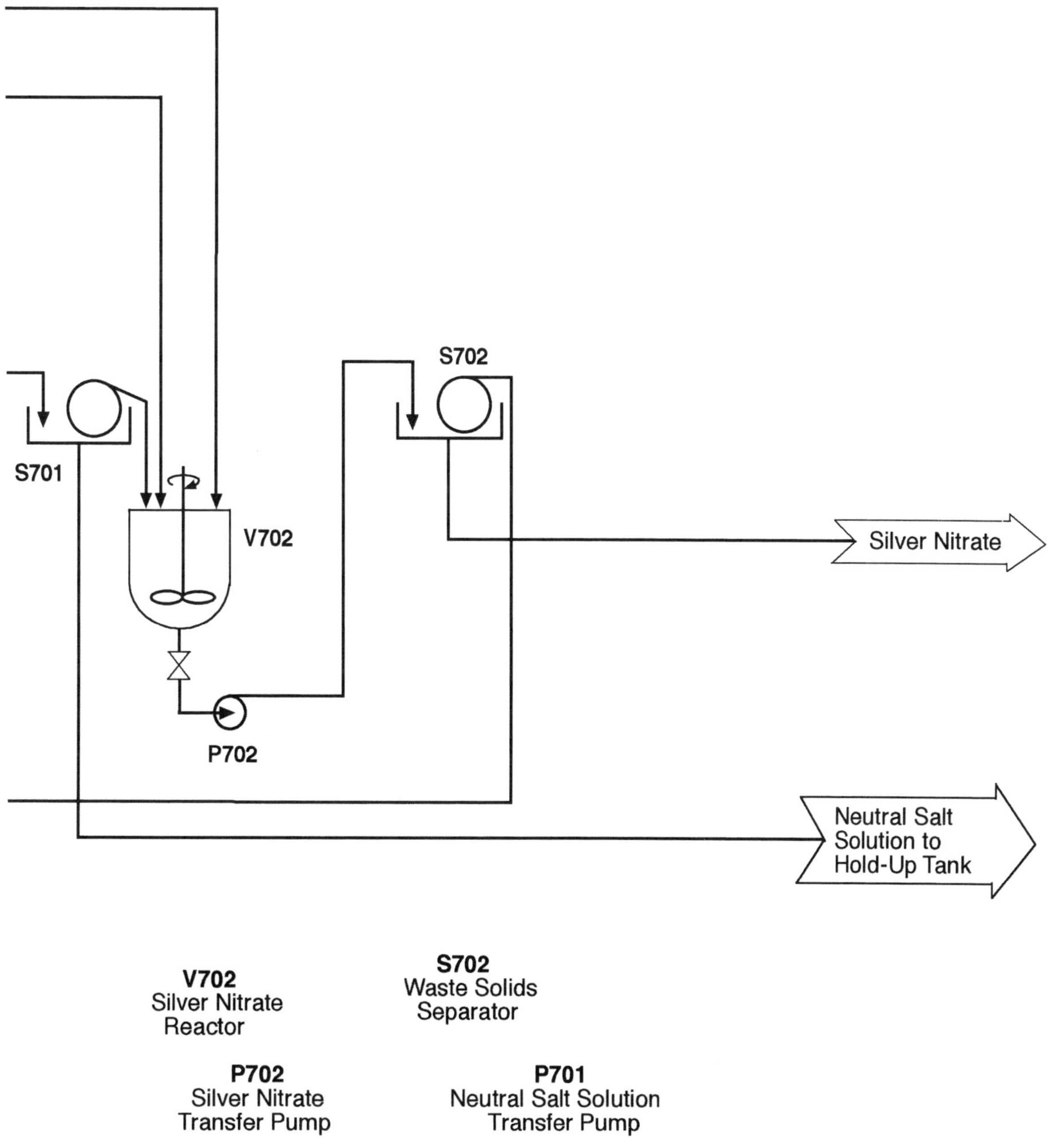

Cell requirement	360 kW
DC power supply losses	360 kW
Refrigeration	2 kW
Steam	622 kW
Compressor for plant air	10 kW
Instrumentation and control	10 kW
Blast air coolers	360 kW

Based on the TPC's estimates that a single 360-kW module, operated 24 hours per day, could destroy 137.6 metric tons of mustard or 74.1 metric tons of VX in 245 days, the total electric energy consumption is 72,600 kW•h per metric ton of HD destroyed and 134,900 kW•h per metric ton of VX destroyed.

The silver management system, which requires additional electric power of 507 kW, is expected to operate for about 6 hours following completion of each campaign. The electrical energy consumption for silver management after a two-cell (two ton containers) campaign is therefore about 3,000 kW•h.

The power requirement shown above is for one 360-kW module. The TPC's design for processing HD calls for two modules; the plan for VX calls for three modules. This scaling of facilities would provide sufficient capacity to destroy the agent inventories at Aberdeen and Newport in 6 years. It would require 3.4 MW of power for the HD facility and 5.1 MW for the VX facility. About 40 percent of this power must be transformed to 2 volts and then rectified to DC (direct current) to supply the electrochemical cells. The remainder is needed for motors and resistance-heating to produce steam. A power system of this scale will have to be carefully designed, although it is well within the state of practice. The power requirement is large enough that either facility will require its own power substation, where power will probably be drawn directly from a high voltage grid (around 13,800 volts) and transformed down to the voltages needed. There will probably be a requirement for phase correction. These requirements do not appear to pose any unusual problems for a local utility. Destruction of the agent inventories in a shorter time period would require additional modules and, of course, additional power.

All of this electrical energy input becomes heat. Additional heat is generated by the reactions (effectively the same as the heat of combustion of the agent being destroyed), which amounts to another 10 percent on top of the total electrical energy input. The heat from both sources must be removed, primarily by cooling water. The location of heat transfer equipment is shown on the various flow diagrams, and Figure 5-8 summarizes the various heating and cooling requirements. More than 1,000 square feet (93 m^2) of heat exchanger surface is required for each module. The heat exchanger materials must be suitable for service in contact with concentrated nitric acid.

Startup and Shutdown

It is preferable to run the agent-destruction system continuously during a campaign. Although agent oxidation can be stopped and restarted with a touch of the switch that controls current to the electrochemical cell, procedures need to be defined for shutting off electrolyte flows and downstream systems, if necessary. Before resuming cell operations after a shutdown, flows and temperatures of many process streams would have to be re-established. There is no time pressure in restarting the system because no reaction occurs until the cell current is turned on.

Emergency shutdown procedures have not been fully worked out, but if conditions do not require immediate shutdown of the cell, the sequence of steps would probably be as follows:

1. Shut off agent injection.
2. Shut off feedstock chemical injection.
3. When the total organic content in the anolyte circuit has been reduced to a predetermined level, shut off the current to the cell.
4. Shut off the circulation pumps in the anolyte and catholyte circuits.
5. Continue operating all scrubber, stripping, gas stream, and ancillary circuits until the system is purged, and then shut them down.

The same procedure would be followed for planned maintenance and at the end of each campaign.

Feed Streams

Table 5-1 summarizes the data submitted by the TPC on feed stream compositions and mass requirements per metric ton of agent destroyed. Tables 5-2 and 5-3 show the overall mass balances, also supplied by the TPC, for the destruction of 2 metric tons of HD and VX, respectively, in a single-module campaign. The panel assumes that the obvious discrepancies between the quantities in Table 5-1 and the mass balance quantities in Tables 5-2 and 5-3 will be resolved as the TPC continues to develop the technology toward a detailed engineering basis. (Appendix E contains the elemental balances corresponding to Tables 5-2 and 5-3.) Silver nitrate is not included as an input stream in the mass balance on the assumption that there is no significant net loss of silver. The mass balances are presented to the nearest tenth of a ton. They therefore do not address trace quantities of organics (i.e., concentrations of 1 percent or less) that might be present in the offgas. Nor do they include trace quantities of silver that might be present in the neutral salt solution. Material balances showing the flow of all fluids into and out of each component subsystem of the Silver II process are not available.

TABLE 5-1 Feed Stream Compositions and Quantities

Feed Streams	Composition	Tons Per Ton of Agent Destroyed VX	HD
Nitric acid	69 wt% (16 M)	4.4	5.0
Silver nitrate	200 g Ag/liter (1.2 M)	4.2	5.9
Water		47.1	41.3
Hydrogen peroxide	35 wt%	19.0	10.9
Sodium hydroxide	10 M NaOH	0.2	0.1
Oxygen	90 vol% (10% nitrogen)	2.6	0.7

Process Effluent Streams

The thermodynamics and kinetics of the electrochemistry underlying the Silver II process, coupled with the TPC's design conditions, such as a low concentration of agent in a highly acidic anolyte, clearly indicate that, *in principle*, the required DRE of six 9's or higher should be technically feasible. In laboratory-scale tests on both surrogates and agents (10 g per test), no agent was detected in the residuals. However, because of the small quantities involved in these tests and the limits of detectability of the analytical methods used, the computed DREs are only four 9's (99.99 percent). As was noted above (under Technology Status), preliminary results for VX from the pilot testing under way at Porton Down indicate a destruction efficiency of at least six 9's (actually, 99.99998 percent or almost seven 9's), with detectability again being the limiting factor. These results show that the technology can destroy agent. However, even the more sensitive analyses being run at the current Porton Down facility do not demonstrate that a full-scale cell (an FM21 cell), configured for the operating conditions of a fully functioning basic agent-destruction system over the course of a campaign, will in fact achieve or exceed the required DRE. In addition, the destruction efficiency for agent does not address issues of the composition and concentration of process products in the residual streams, including trace quantities of toxic residuals or the environmental burden of residuals. (For further discussion, see Scale-Up Requirements below.)

Under normal operating conditions, the submitted design for Silver II anticipates that the following process residuals will be produced:

- End-of-pipe gaseous emissions from the combined offgas treatment circuit will be a mixture primarily of carbon dioxide, oxygen, and nitrogen.
- Aqueous effluent from the silver management system will be a solution of sodium nitrate, sodium sulfate, sodium phosphate, and sodium chloride. The exact composition will depend on the agent that was treated.
- Sodium nitrate solution is the residual from neutralization of the effluent (0.6 percent nitric acid, pH 1, 13.2 m^3 per ton of agent) generated from the NO_x reformer. This salt solution is likely to be combined for discharge with the aqueous effluent from the silver management system.

No residuals have been tested for toxicity, but the principal constituents are common materials that are not considered hazardous to health or the environment.

The gases are released to the atmosphere after passing through two hydrogen peroxide scrubbers in series and a filter bed of activated carbon. This treatment should reduce any organics in the offgas to nondetectable levels, but the final emissions will not be retained for analysis prior to release. The panel considers it highly improbable that any agent will escape from the anolyte to the offgas. In any case, the severe treatment

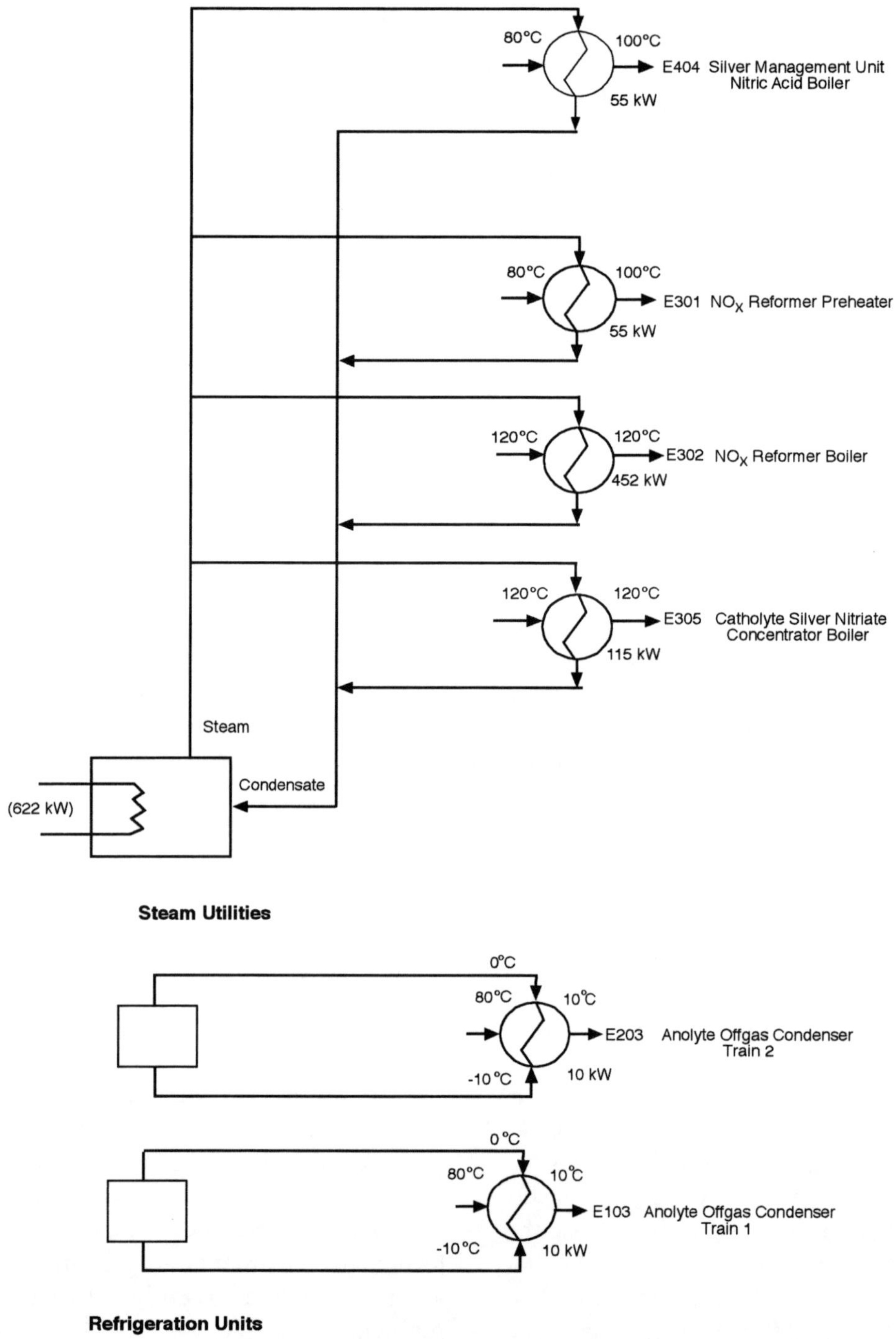

FIGURE 5-8 Process flow diagram for services and utilities. Source: AEA Technology.

of the offgas with hydrogen peroxide, followed by carbon filtering, will remove both agent and volatile organics from the offgas.

The TPC reports that the aqueous effluent from the silver management system is slightly acidic (pH 6). Although this effluent is primarily a solution of

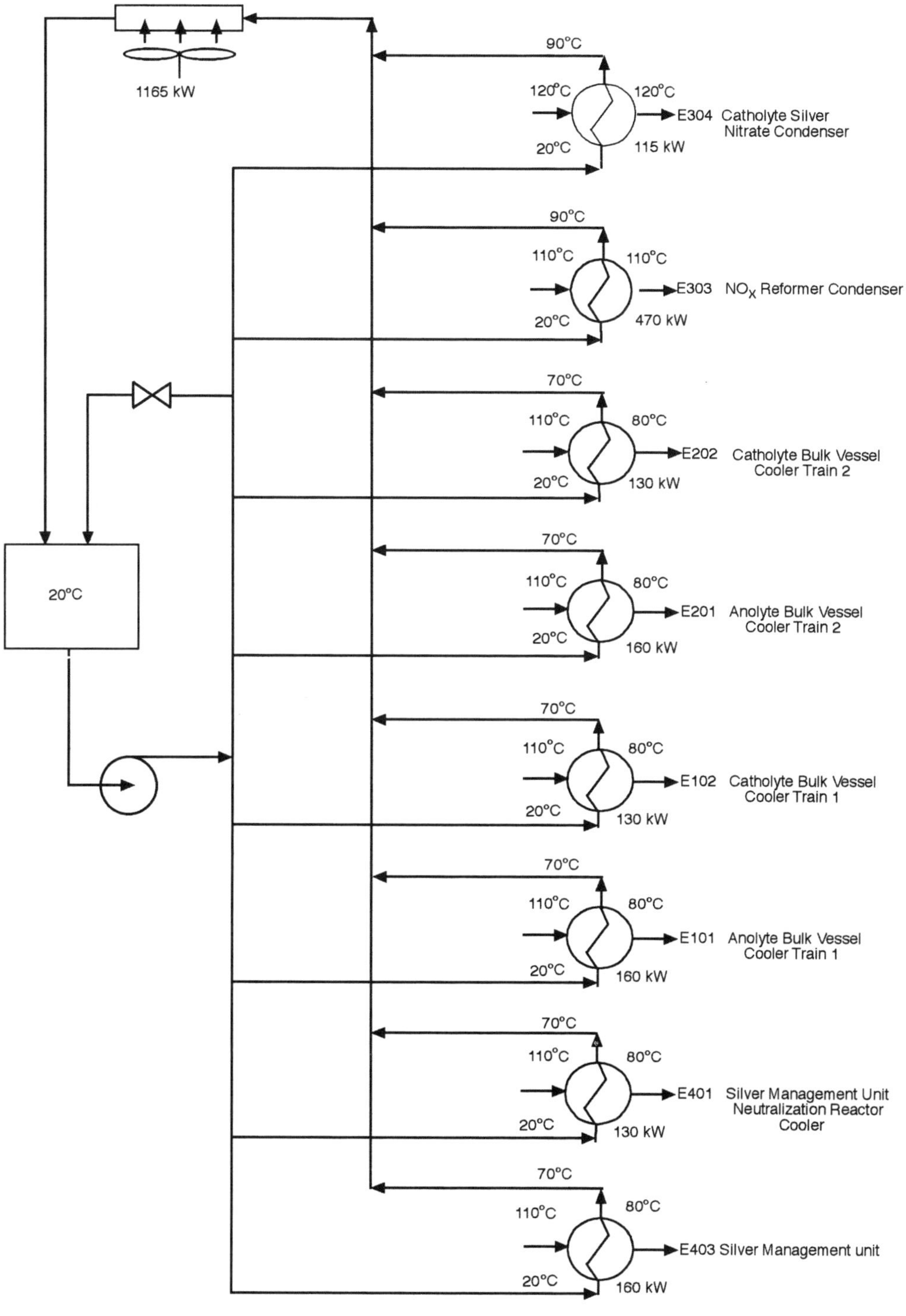

Cooling Water System

sodium salts, it could contain trace quantities of silver salts as well. The TPC also reports that laboratory experiments show that the silver concentrations in the effluent will be on the order of 50 $\mu g/m^3$ (about 50 parts per trillion); the panel did not receive details of these experiments. The maximum allowed concentration in

TABLE 5-2 Mass Balance for HD Destruction (all figures in metric tons)

Inputs	Agent	Nitric Acid	Hydrogen Peroxide	Sodium Hydroxide	Oxygen	Total
HD (mustard)	2.0					
HNO_3		0.4				
H_2O		0.2				
H_2O_2			1.1			
H_2O			2.0			
NaOH				2.0		
H_2O				0.1		
O_2					2.8	
N_2					0.3	
Total Input	2.0	0.6	3.1	2.1	3.1	10.9

Outputs	Offgas	Waste Acid	Neutral Salt Solution	Total
CO_2	2.2			
O_2	0.1			
N_2	0.3			
NO_x	0.002			
HNO_3		0.6	1.8	
H_2O		2.2	1.5 2.2	
Total Output	2.6	2.8	5.5	10.9

the United States is 50 ppb (parts per billion). The expected volume of aqueous discharge per metric ton of agent treated is 11.2 m^3 when treating HD and 4.7 m^3 when treating VX.

The aqueous residuals from the silver management system and the NO_x reformer are retained in a holding tank for analysis. After that, disposal may be by one of three routes: (1) direct discharge to the environment in accordance with an National Pollutant Discharge Elimination System(NPDES) permit; (2) indirect discharge to a publicly owned treatment works (POTW); or (3) transport to an off-site facility for recovery of the salts. The third option will have to be preceded by evaporating the solution to dryness, if the Army does not allow transport of liquid residuals.

Ton container cleanout will follow the protocol established and tested by the Army. (This protocol is described in Chapter 7.)

TABLE 5-3 Mass Balance for VX Destruction (all figures in metric tons)

Inputs	Agent	Nitric Acid	Hydrogen Peroxide	Sodium Hydroxide	Oxygen	Total
VX	2.0					
HNO_3		0.7				
H_2O		0.3				
H_2O_2			1.9			
H_2O			3.6			
NaOH				1.8		
H_2O				0.1		
O_2					4.9	
N_2					0.5	
Total Input	2.0	1.0	5.5	1.9	5.4	15.8

Outputs	Offgas	Waste Acid	Neutral Salt Solution	Total
CO_2	3.8			
O_2	0.1			
N_2	0.5			
NO_x	0.004			
HNO_3		1.1		
H_2O		3.9		
$NaNO_3$			0.6	
Na_2SO_4			1.1	
Na_3PO_4			1.2	
H_2O			3.6	
Total Output	4.4	5.0	6.5	15.9

PROCESS INSTRUMENTATION AND CONTROL

The heart of the proposed system of process instrumentation and control is a computer-based system for supervisory control and data acquisition (SCADA). This system allows the operators to monitor and control facility operations from a dedicated control room or cabin. To protect cabin personnel from on-site gases, the control cabin would have its own filtered air supply and be ventilated at positive pressure relative to the rest of the facility.

The control parameters to be monitored by a suitable SCADA software package are listed in Table 5-4, as are basic requirements and features. Key elements of this integrated system that are particularly relevant to Silver II are discussed below.

The *current and voltage measurements* indicate whether the cell is operating properly. They provide warning of cell malfunctions such as membrane failures.

Electrolyte flow rates must be monitored because high flow rates through the cells are necessary for good mixing. Each cell compartment is 10 mm wide by 240 mm high. Electrodes occupy about half the volume of a cell. The volumetric flow through a full cell is 45 m^3/h. The TPC estimates the hydraulic radius of each electrode compartment to be 4.8 mm, giving a Reynolds number of around 4,600 (density = 1,000 kg/m^3; viscosity = 1 centipoise), which is at the lower end of the turbulent range.

Gases released from the anolyte and catholyte circuits will be monitored for CO_2, O_2, NO_x, CO, volatile organics, and chemical agent as indicators of proper cell operation. For instance, an abrupt elevation of oxygen concentration indicates that direct oxidation of water by Ag(II) has become the predominant anode reaction. The same gaseous components are monitored in the offgas before and after carbon filtration to ensure safety and to confirm proper operation of the hydrogen peroxide scrubber train.

Liquid composition must be monitored to obtain the feedback necessary for the controlled addition of key constituents in the electrolytes. For satisfactory cell operation throughout a campaign, the addition of chemical agent is controlled to maintain about 5,000 ppm in the anolyte circuit. Monitoring data are also needed to control the addition of silver nitrate to the anolyte circuit and the addition of nitric acid to the catholyte circuit. Composition monitoring also follows the progressive buildup of sulfate or phosphate in the anolyte and indicates whether agent and organic intermediates are being oxidized.

Monitoring temperatures and pressures is important for confirming proper operation of the cooling system, particularly because of the large heat-transfer requirements for sustained operation of the Silver II process. (See preceding discussion of electrical energy and heat of reaction as sources of heat to be removed.)

During an HD campaign, another important parameter to monitor is the amount and location of precipitated silver chloride. By the end of a campaign a large amount of silver chloride will have precipitated in the anolyte circuit. The hydrocyclone in this circuit is intended to deposit most of the precipitate in a collection vessel (shown in Figure 5-4). The efficiency of the hydrocyclone is critical to proper functioning of the anolyte circuit. Some sampling at various points in this circuit will be needed to determine the solids content, with particular attention to the anolyte flowing into the electrochemical cell and the possible retention of precipitate in the cell.

All the parameters listed in Table 5-4 must be monitored without human intervention and the results fed into the SCADA system for control of operations. Analogous monitoring and control systems are used for industrial processes but will have to be adapted specifically for the Silver II process.

One of the commercially available SCADA-type software packages that operate on a personal computer and are used in the chemical industry may prove suitable for use in Silver II. Higher-integrity packages based on the UNIX operating system are also available. The SCADA system that the TPC is testing at Porton Down uses Paragon TNT software with Allen Bradley controls. The system was not yet fully operational at the time of the panel's visit. In any case, final SCADA system selection and integration will not be part of the piloting program under way at Porton Down. These actions are being deferred to an early stage of detailed design for a full-scale operating facility.

For agent monitoring, which will be required throughout the plant, the standard equipment approved by the U.S. Army will be used. All agent sensors must interface with the SCADA system to ensure automatic alarm and response capability.

PROCESS STABILITY, RELIABILITY, AND ROBUSTNESS

Stability

The Silver II process as presented in the submitted designs is composed of two systems that operate independently of one another. One is the agent destruction system, which is composed of the electrochemical cell and its supporting circuits; the offgas treatment circuits; and all supporting unit operations, processes, and plumbing. The other is the silver management system, which operates separately at the end of a campaign. Separation of the two systems contributes to stability and ease of operation.

TABLE 5-4 Elements of a Supervisory Control and Data System for Silver II

Control Parameters to be Monitored	
	DC current and voltage, particularly to the cells
	Electrolyte flow rates
	Gas flow rates
	Gas composition (volume percent O_2, CO_2, CO, NO_X, volatile organics, and chemical agent vapor)
	Liquid composition (pH; dissolved silver, sulfate, phosphate, total organic carbon, and chemical agent; suspended silver as AgCl)
	Temperature
	Pressure
Additional Required Software Features	
	Validation of operation inputs
	Interlocks to prevent inappropriate operator commands
	Mimic diagrams of plant subsystems
	Alarms that are triggered from process or facility sensors and that can initiate plant responses
	Software control and display of data from subsystems
	Operator control of plant actuators and processes based on graphical interface display of piping and instrumentation diagrams
	Automatic data logging
	Trend display of logged data
	Plant data (i.e., SCADA system data) accessible from remote sites
	Automatic report generation
	Multiple SCADA displays around the plant
	Automatic responses to fault conditions
	Detection of rate of change alarms

The agent destruction system operates in a semibatch mode. Catastrophic failure from uncontrolled reactions is highly unlikely because of the nature of the process and the conditions under which the various modules operate. Agent is fed slowly to the anolyte to maintain a constant, low concentration and therefore will not accumulate in the anolyte circuit. The agent feed rate is controlled by monitoring the CO_2 concentration in the anolyte offgas. If the CO_2 level drops below a set-point determined by the agent feed rate (i.e., by the carbon feed to the process), a fault condition exists and the agent feed will shut down automatically.

For a runaway condition to occur, the cell reactions must release enough heat to raise the electrolyte temperature from the normal 90°C at which it is controlled to 105°C, the boiling point of nitric acid. For this to happen, three independent trip or interlock systems must malfunction: the cooling circuit controls, the anolyte high-temperature trip, and the agent addition inhibition interlock. Simultaneous failure of these three control systems is highly improbable. Minor process fluctuations under normal operating conditions might vary the temperature between 87°C and 93°C.

During the course of a campaign, some process conditions will change substantially, particularly in the anolyte circuit, but the rate of change is slow under normal operating conditions. Therefore, the response time for most control instrumentation is not very demanding. For example, as stated previously, the total outflow of silver from the anolyte circuit during an HD campaign of 5 days at 24 hours per day is about 1,550 kg. The required makeup is therefore 12.9 kg per hour, which is less than 10 percent of the initial 134 kg inventory of silver in the anolyte compartment. The required silver makeup in a VX campaign is about 2.8 kg per hour, which is 2 percent of the initial silver inventory. Silver makeup is in the form of 1.2 molar silver nitrate, which contains 12.84 kg/ m^3 of silver.

None of the processes in the system modules is particularly sensitive to small excursions in composition or temperature. However, compositions of some constituents will change substantially during the course of a campaign, and a test program is needed to verify that the planned control systems are adequate to ensure stable operation over the full range of operating compositions.

Both the agent destruction and silver management systems operate at low temperatures and close to atmospheric pressure, which substantially reduces the requirements for sensitivity and response time of control systems, compared with high-temperature systems. Even though the system can tolerate small temperature excursions and a runaway reaction is unlikely, there are large heat loads produced in a system with relatively small volumes. Therefore, temperature control in each of the modules and in the system as a whole must be tested and validated.

A large loading of silver chloride precipitate during an HD campaign can cause many problems, including malfunction of the electrochemical cells, inadequate heat transfer in the heat exchangers, and pump malfunctions. The pilot demonstration is critical not only to determining the effectiveness of the hydrocyclone in removing the very fine precipitate expected but also to assessing the effect of suspended particles on cell operation. The pilot plant at Porton Down is testing only a single anode-cathode pair. In a full-scale cell, if one compartment should become plugged, the flow will increase through the remaining anode compartments and further precipitation will occur in the plugged compartment. Plugging would lower cell efficiency and, in the plugged anode compartments, increase the alternative reaction of Ag(II) with water to produce O_2. The TPC has identified the further potential consequences of plugging as overheating and failure of the Nafion membranes in the blocked compartment. To reduce the risk of solids settling in the anode compartments, the TPC has designed the system for turbulent flow. In addition, the temperature of the anode compartment will be monitored to detect overheating in time to exercise process controls, if plugging does occur.

Reliability

With respect to the *reliability of equipment*, the electrochemical cell to be used in Silver II is identical in design to commercial cells that have been used reliably for decades to manufacture chlorine gas and caustic (NaOH) by electrolysis of brine (NaCl solution). However, the two applications are totally different from a *process* perspective. Cells that produce chlorine and caustic operate in a pH-neutral to alkaline environment. The Silver II process requires a highly acidic environment. Furthermore, the anode and cathode reactions in the two processes are completely different.

Laboratory and pilot tests conducted by the TPC for reprocessing radioactive waste and for destroying many other organic materials have demonstrated that the general Silver II cell technology and conceptual framework are sound. There have been no commercial applications to date.

The other components of the agent destruction system are standard unit processes and operations to be conducted with readily available, off-the-shelf equipment. Tests conducted as of May 1996 have not included these other components. The key components are included in the scheduled pilot testing at Porton Down, but the facility itself and the planned tests will not provide an end-to-end proof of design sufficient for scaling to full operation. A higher level of pilot testing will be required to verify materials of construction (the Porton Down plant is constructed largely of glass), operational reliability for the full-scale FM21 cell under varying conditions, and integration of all system components that must operate simultaneously and in concert for the duration of a campaign.

In addition to the reliability of equipment and the reliability of the basic processes, there are several additional aspects of reliability relevant to an assessment of the Silver II process. With respect to *reliability of agent detoxification*, the agent is hydrolyzed, and therefore detoxified, upon contact with nitric acid in the anolyte circuit. The agent feed to the anolyte circuit is maintained at a level low enough that this hydrolysis occurs immediately.

With respect to *reliability as backup operability*, the standard 360-kW module for the basic agent-destruction system consists of two identical, separately fed 180-kW cells. If one cell fails, it can be removed for cleaning and replacement, while the other continues to operate.

The design includes a standby generator to provide electrical backup in the event of a power failure. This backup power must be adequate to continue operating scrubbers and pumps in the event of an emergency shutdown.

With respect to *reliability against unplanned downtime*, an individual 180-kW cell can be removed to a remote area for repair or maintenance while a replacement module in good working order is substituted and processing continues. Thus, the modular design of the system reduces the risk of unplanned downtime.

Robustness

In the panel's judgment, the Silver II system is capable of operating satisfactorily over a wide and varying range of temperature, pressure, energy input, and feed composition. Anionic and cationic impurities in the agent could reduce cell current efficiency but would not otherwise interfere with the basic process operations.

With a well-designed SCADA system, upsets in feed, in key reaction conditions (temperature, pressure, agent concentration, and reactant concentrations), or in energy input or heat removal should be readily detectable in time to take appropriate corrective action. However, repeated upsets, although not a major threat to human health or the environment, would be highly undesirable from an operational standpoint. Current test data are insufficient to estimate the probable frequency of events that could lead to upsets.

MATERIALS OF CONSTRUCTION

Systems and Materials

In the design submitted for Silver II, the core agent-destruction process is carried out in aqueous concentrated nitric acid at close to atmospheric pressure and at temperatures below 90°C. Temperatures at points in the secondary circuits where nitric acid solutions are distilled will reach the boiling point of the still bottoms. (The boiling point of concentrated nitric acid is 105°C; additional salts in the still bottoms may further elevate the boiling point.) The NO_x reformer heats the NO gas stream to 110°C.

The technology design, including the selection of materials of construction, is based mainly on the TPC's experience with nitric acid for reprocessing radioactive wastes. The materials selected, which are well known to be compatible with concentrated nitric acid, include titanium, low-carbon stainless steels, platinum, zirconium, and polytetrafluoroethylene (used for Nafion 324 cell membranes and for gaskets).

In the submitted design, anodes are made of platinum or platinized titanium. Cathodes are made of low-carbon stainless steel. The piping and vessels in the anolyte feed circuit are made from titanium to ensure integrity. Boilers are constructed from zirconium because their conditions of operation were judged to be too close to a corrosion band for titanium.

Although the materials and design are conventional for applications involving concentrated nitric acid, the panel believes the following issues require further consideration:

- The primary metals of construction (stainless steel, titanium, and zirconium) all sustain stress corrosion cracking in nitric acid solutions at various concentrations and potentials. The possibility of stress corrosion cracking must be carefully investigated, particularly given the presence of a high concentration of dissolved silver.
- The possibility of intergranular corrosion should be addressed because nitric acid is highly oxidizing, and the chemistry of oxidation at grain boundaries is not well defined for any of the metals being considered for Silver II. Of particular concern are changes in chemical potentials at grain boundaries, as a result of adsorption.
- Plugging of the anode compartments, particularly in HD campaigns, may significantly affect reliability. The conditions under which plugging occurs are not known at present. Also, a simple and reliable technique for replacing an FM21 cell when the system is on line is highly desirable. A means of detecting plugging and conditions that could lead to a short circuit or hot spots should be pilot-tested and incorporated into the final design.
- The electrochemical oxidation of agent in nitric acid will produce species containing carbon, sulfur, and phosphorus (VX only) in the anolyte. This environment is substantially different from the environments in previous industrial experience with nitric acid baths. In addition, the concentrations of species containing sulfur and phosphorus

increase throughout the duration of a campaign. The effects on corrosion resulting from this wide and cyclical variation in electrolyte composition should be examined.

- The pilot plant at Porton Down is constructed of glass and therefore will not test the construction materials to be used in a full-scale installation.
- The plant will be designed for a 20-year lifetime, but membranes will have to be replaced every two years at a minimum and possibly more frequently when processing HD.

Environmental Conditions and Chemistry

The principal issues for the internal environment of materials of construction derive from exposure of materials to concentrated nitric acid and have been addressed above. The SCADA system will be able to detect changes in temperature from a loss of circulation or cooling in time for appropriate actions to be taken. Local hot spots at a Nafion membrane, caused by plugging or some other loss of electrolyte circulation, may damage the membrane. Methods of monitoring for hot spots and plugging in the cell are necessary. Although the system operates at close to atmospheric pressure, the equipment is designed to withstand internal pressures of up to 4 atmospheres.

Startup and Shutdown

The procedures for startup and shutdown are described under Process Operations. Neither normal nor emergency procedures will cause significant thermal stress on the materials of construction.

Failure Definition

The TPC assembled a multidisciplinary team for two days in September 1995 to conduct a first-phase hazard and operability study for the design of a Silver II facility for chemical agent destruction. The team assessed the consequences of the hazard challenges listed in Table 5-5 to each of the key system components individually and to the facility as a whole, including the interfaces between components.

For each challenge and each component, the team identified causes, consequences, and safeguards. The team then recommended additional safety measures or additional information required to assess whether further controls were needed. Fifty recommendations were made. Most of the cases leading to an accidental release to the atmosphere were generated for the challenges of missiles, terrorism and sabotage, and other external events (seismic events, aircraft crashes, or fire affecting the agent receipt and supply system). The fact that atmospheric releases were identified for these external challenges does not reveal any particular vulnerability of the Silver II technology or the TPC's design because these challenges were not specific to the agent-destruction process at the facility. Other consequences worth noting were release of nitric acid as the result of corrosion or maintenance problems, in-plant fires, and releases of agent inside the secondary containment. The majority of consequences from these internal events affected the operability of the plant but not the safety of the public.

TABLE 5-5 Hazard and Operability Challenges

Fire	Human error
Explosion/implosion	Corrosion
Maintenance	Erosion
Containment	Effluents
Contamination	Missiles
Toxicity	Terrorism and sabotage
Loss of services	Other external events
Extreme weather	Industrial hazards

The TPC assembled a team to review this initial hazard and operability study for two days in May 1996. Taking into account the likelihood and severity of potential failures, the team identified only one possible occurrence of concern: the possibility that chemical contamination of the electrical system might degrade cable insulation or seals, leading to potential failures.

OPERATIONS AND MAINTENANCE

See Process Operations above for the operational details of each system component. This section describes operational experience of the TPC relevant to operating an agent-destruction facility and maintenance planning for such a facility.

Operational Experience

Operational experience with the Silver II process has been limited to the electrochemical cell. However, the pilot testing under way at Porton Down will combine the electrochemical cell with the auxiliary fluid systems (anolyte and catholyte feed circuits, anolyte offgas condenser, NO_x reformer, and a modified version of the combined offgas treatment circuit). This pilot system will include all the key components of the agent-destruction system except the agent feed and supply and the catholyte silver nitrate recovery circuit.

The TPC has conducted 12 laboratory tests to demonstrate the destruction of organophosphorous and mustard agents, including three nerve agents (GA, GB, and VX) and three mustard agents (HD, HT, and THD). The tests were performed with an FM01 electrochemical cell, which is a 1/35th scale model of the FM21 cell that would be used in full-scale operations. Figure 5-9 is a schematic flow diagram of the test rig for the FM01 cell.

In each test, 10 g of agent was injected into the anolyte vessel of the test rig. The anolyte vessel contained a 0.5 molar silver nitrate solution in 8 molar nitric acid. The catholyte vessel contained 4-molar nitric acid. Anolyte temperature was maintained at 50°C. Tests lasted for up to six hours. In all cases, final agent concentration was below detectable limits for the analytical methods used, but the limits of detectability were not specified. The anolyte offgas, which was measured throughout each experiment, contained varying levels of nitrous oxide and volatile alkyl nitrates.

The preliminary results from the Porton Down pilot testing of VX are discussed in the section above on Technology Status. Longer duration tests of a Silver II cell on a scale similar to the scale of the Porton Down facility have been undertaken with mixtures of tributyl phosphate and kerosene. In these tests, an FM01 cell was operated continuously, 24 hours per day, for up to 14 days.

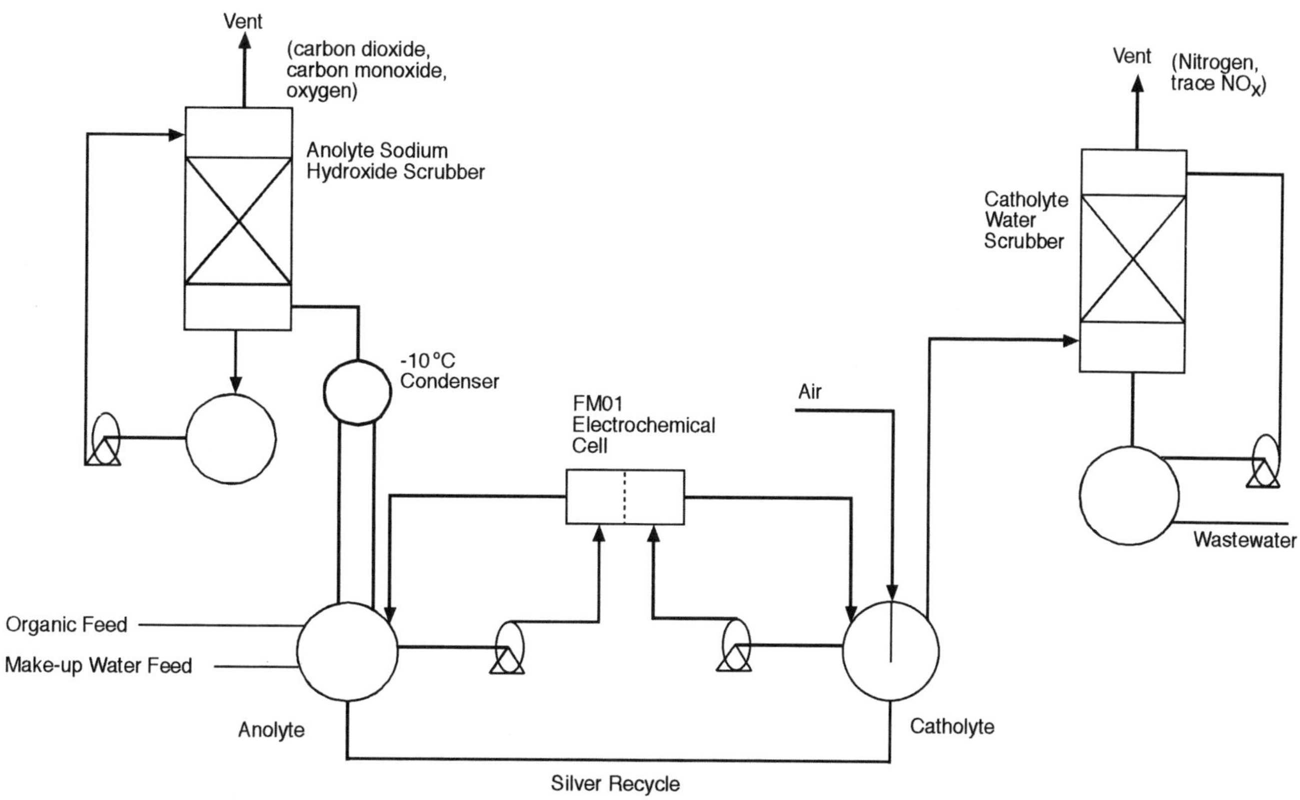

FIGURE 5-9 Schematic flow diagram of the FM01 test rig. Source: AEA Technology.

Maintenance

No maintenance schedule has been established at this stage of technology development for Silver II. Because the plant would operate under highly corrosive conditions with a hazardous working fluid (nitric acid), continuous inspection and maintenance must be a priority.

The electrolysis cell is the same as the cells used for chlorine production. Membrane cells have revolutionized that industry and have a good record for durability. The TPC states that a normal maintenance schedule for replacing membranes in chlorine production is 27 months. The maintenance required during agent destruction will have to be developed; process conditions for Silver II are quite different from the conditions for chlorine production.

SCALE-UP REQUIREMENTS

Plant scale-up in the submitted design is based on adding 360-kW modules (two 180-kW FM21 cells per module) to the facility. However, neither that module nor its 180-kW cell unit has been piloted for Silver II, and the FM21 cell represents a large scale-up from the 4-kW pilot test at Porton Down.

The TPC has stated that scale-up from the Porton Down pilot plant to a 180-kW cell with 45 electrode pairs and 45 parallel flow paths for circulating fluid will not be a problem because the FM21 cell has been used successfully in industry. However, the reagents and reaction chemistry for Silver II are very different from those in industrial production of chlorine and caustic from brine.

A technical issue of concern to the panel is the precipitation of silver chloride in HD campaigns. The TPC expects the hydrocyclone to be highly effective in removing silver chloride, with a solids concentration in the underflow of about 0.9 percent by volume. The TPC states that blockage of the hydrocyclone discharge line is unlikely below 30 volume percent solids. The TPC also states, based on information from the vendor of the FM21 cell, that heavy solids loadings will not adversely affect cell operation. However, in chlorine production the brine is treated with soda ash (crude Na_2CO_3) or caustic (NaOH) to precipitate out oxides and hydroxides of calcium, iron, and magnesium prior to electrolysis, because precipitation within the cell has been found to foul the membrane. Therefore, the effect of the anticipated loading of silver chloride solids on cell operation in the Silver II process clearly must be pilot-tested.

The NO_x reforming process to regenerate nitric acid is conventional; it is very similar to the process used commercially to treat offgases from the manufacture of nitric acid. Nonetheless, inefficiency of the NO_x reformer in the first pilot test at Porton Down indicates that the design must be improved and more tests must be done. The hydrogen peroxide scrubbing is also conventional, although not commonly used at the scale proposed. The silver management system is not conventional but appears to be based on sound chemistry.

There are certainly significant heat transfer requirements, although none seems unconventional. As an example, in the silver management system, which operates independently from the agent-destruction system, a concentrated acid solution (well over 8 molar) of a mixture of nitric, sulfuric, and phosphoric acids, plus silver nitrate, silver chloride, and various impurities, is neutralized with sodium hydroxide. This reaction has a high heat release and is prone to spattering, but the operation is well within the current state of practice.

PROCESS SAFETY

Plant Safety and Health Risks

Based on the first-level hazard and operability study performed by the TPC and on the panel's preliminary, qualitative evaluation, the possibility of a catastrophic accident with a cause internal to the Silver II technology is extremely low. However, anode and cathode reactions are carried out in concentrated nitric acid, which has been described as the common chemical most frequently involved in reactive incidents because of its exceptional ability to function as an effective oxidant even when fairly dilute or at ambient pressure (Bretherick, 1985). Many reported incidents have involved closed or nearly closed vessels that have failed from internal gas pressure created either by oxidation of organic compounds to CO_2 or auto-decomposition of nitric acid to NO_x fumes and oxygen.

Such incidents are unlikely in the Silver II process because the system is essentially open and the concentration of organics in contact with nitric acid is low. As was already noted, three independent controls or interlocks would have to fail simultaneously for a sufficiently high concentration of agent and derived organics to build up in the nitric acid and create potentially explosive conditions.

Community Safety, Health, and Environmental Risks

The planned containment system will reduce the risk of a release of either agent or other hazardous chemicals to negligible levels during normal operations. Abnormal events that might threaten the health or safety of the community or the surrounding environment are unlikely because the system is operated at low temperature and atmospheric pressure, the chemical reactions are slow and easily controllable, and the agent is processed at low total amounts at any one time.

SCHEDULE

The panel anticipates that pilot testing of a 360-kW module at Newport will require 12 months for design, 12 months for construction and commissioning tests, and an additional 12 months for agent testing. Installation of additional modules and associated infrastructure will require 12 months; commissioning tests, 6 months; and agent processing, 36 months. Pilot testing at Aberdeen is to take longer because of the added complication of silver chloride precipitation.

The duration of operation to complete destruction of agent at the Aberdeen or Newport sites depends on the number of basic modules installed for simultaneous operation. If full-scale operations start on January 1, 2001, and agent destruction must be completed by December 31, 2004, then the facility for destruction of VX at Newport will require five 360-kW modules with a total footprint of 33 m by 61 m. Under the same schedule requirements, the facility for HD at Aberdeen will require three 360-kW modules with a footprint of 33 m by 37 m. The footprint is only for the operating plant and does not include agent handling buildings, administrative offices, workshops, electrical substation, and tank farms. Agent destruction could be completed in a shorter time by adding modules. As noted in the section on Utility Requirements, the electrical power requirement correlates with the number of modules.

6

Gas-Phase Chemical Reduction Technology

PROCESS DESCRIPTION

The gas-phase chemical reduction process reviewed by the AltTech Panel was submitted to the Army by ECO LOGIC, Inc., of Rockwood, Ontario. ECO LOGIC is the developer and TPC for this technology and will be referred to as the TPC. The acronym GPCR will be used in the remainder of this report to refer to the particular process design submitted by this TPC for a gas-phase chemical reduction technology to destroy chemical agents. The process uses hydrogen and steam at elevated temperatures (up to 850°C) and nominally atmospheric pressure to transform organic wastes into simpler substances that are either less toxic or convertible to less toxic materials; these substances are also easier and safer to reuse or to release to the environment. The overall process requires a high-temperature reaction vessel, where the chemical reduction occurs, followed by a gas scrubbing train to remove inorganic by-products. The process also includes provisions for removing other byproducts and regenerating hydrogen gas through steam reforming. Figure 6-1 is a schematic illustration of the process.

Chlorinated hydrocarbons, such as polychlorinated biphenyls (PCBs), are chemically broken down and reduced to methane (CH_4)and HCl with CO and CO_2 as by-products. Nonchlorinated aromatic hydrocarbons such as toluene are reduced primarily to methane, with minor amounts of other light hydrocarbons. Carbon and presumably some heavier hydrocarbons are also produced.

The flow-through stainless steel reactor has nozzles to accelerate the vaporization or dispersion of liquid wastes, which are injected directly into the reactor mix of hot gases consisting of H_2, H_2O, CO, and CO_2. Within the reactor, radiant-tube heaters heat the mixture to 850°C. The residence time in the reactor is 2 to 6 seconds, although the TPC has stated that reactions occur in less than one second.

The gases exiting the reactor are scrubbed to remove by-products. Water is used as a quench to decrease the gas temperature and absorb water-soluble products, including HCl. These and other acidic products are further scrubbed by caustic scrubbers. A heavy-oil scrubber can be used in the scrubber train to remove some hydrocarbons. A standard monoethanolamine (MEA) scrubbing system removes most of the H_2S (produced from sulfur-containing feeds) and CO_2 from the gas train. The separated H_2S requires further treatment to convert it to elemental sulfur and water.

The TPC has also developed and employed a sequencing batch vaporizer (SBV), which is a high- temperature chamber (up to about 550°C) in which hot gases from the recirculating process stream, including H_2, H_2O, CO, and possibly CH_4, desorb organic contaminants reactively and thermally from drums and bulk inorganic solids. The SBV consists of two autoclave-like chambers that are operated independently in batch mode. The chambers can be fairly large—large enough to hold a ton container. A high temperature thermal reduction mill (primarily a bath of molten tin) can also be used to separate contaminants from soil or solids; the tin is a heat transfer medium to drive off volatile material, leaving inert solids behind. The gases from the thermal reduction mill and SBV are swept into the reactor for treatment. GPCR incorporates equipment for catalytically reforming most of the methane from the reactor to H_2, CO, and CO_2; the reformed gas is recirculated to the reactor to provide part of the necessary hydrogen.

The TPC has also developed mechanisms for holding gaseous process residuals for analysis prior to release or storage in containers. The overall process is monitored at a number of points using several methods: on-line gas chromatography, chemical ionization mass spectrometry, a NOVA® oxygen analyzer, and a NOVA® gas analyzer to monitor H_2, CO, CO_2, and CH_4.

The reactor (Figure 6-2) is constructed of stainless steel with a ceramic lining. The feed stream and hot reactant gases are injected through several ports mounted on the reactor. Special nozzles disperse liquid wastes into the hot gas. The gas mixture is heated further by 18 vertical radiant-tube heaters, which are isolated from the reaction mixture by an atmosphere of CO_2.

Effluent gases leave the reactor through a stainless steel central tube that leads to the scrubber system.

GPCR has been under development since 1986 and has progressed from bench-scale testing through commercial-scale operation. A number of organic feed materials, particularly chlorinated wastes, have been tested at bench scale. Several kinds of feed materials are currently being treated at commercial scale (tons per day), including pesticides, chlorinated hydrocarbons, and PCBs. A full-scale facility to treat mixtures of toluene and the pesticide dichloro diphenyl trichloroethane (DDT) is operating in Australia. A plant in Canada for PCB destruction, which was visited by an AltTech panel team in January 1996, went on line in the spring of 1996.

SCIENTIFIC PRINCIPLES

Feed-Destruction Chemistry

The chemistry by which GPCR destroys organic feed material is much more complex than a simple high-temperature reduction with hydrogen of organic compounds to produce methane. The complexity results from the reduction with hydrogen being accompanied by reactions of carbonaceous intermediates, including elemental carbon, with steam to yield the final products. Although the thermodynamic principles of reducing organics with hydrogen to carbon and the resulting reactions of carbon with steam (carbon–steam chemistry) have been thoroughly studied and are well understood, the interplay of kinetics and thermodynamics in the GPCR reactor are more difficult to ascertain.

The chemical agents HD and VX contain a high proportion of heteroatoms (atoms other than carbon, hydrogen or oxygen, such as chlorine, phosphorus, sulfur, or nitrogen). The reaction products containing these heteroatoms will generate a large volume of inorganic process residuals. HD is 45 percent chlorine, 20 percent sulfur, and 30 percent carbon by weight; VX is 12 percent phosphorus, 5 percent nitrogen, 12 percent sulfur, and 49 percent carbon by weight (hydrogen and oxygen make up the rest of each compound). This heteroatom content raises two unanswered questions. First, what are the final heteroatom products from the reactor? Second, how are they scrubbed or otherwise removed? The acid gases and other inorganic products must first be scrubbed from the reactor effluent gas and then converted to a form suitable for disposal or recycling in commerce. The reactions of organic compounds containing heteroatoms are even more difficult to predict without the same kind of detailed experimental work the TPC has carried out on the feed materials it currently treats successfully.

The TPC, which has considerable operational experience treating a number of highly halogenated wastes such as PCBs, hexachlorobenzene, and DDT, has found empirically that a fine balance of hydrogen and steam is necessary to avoid generating substantial amounts of carbon and polyaromatics in the reactor. The TPC has developed empirical models to predict operating parameters that yield optimal product composition: primarily methane, with CO and CO_2. Nonetheless, the TPC allows in the design for some production of carbon (as soot), and the panel believes that some high-molecular-weight aromatics are produced. Therefore, carbon and other solids must be managed downstream in addition to the gaseous products (see Appendix F).

For simple hydrocarbons, the TPC describes GPCR as a high-temperature reduction by hydrogen to produce methane. Simple thermodynamic calculations reveal, however, that considerable amounts of carbon would be expected from the initial reaction with hydrogen. Therefore, carbon must react subsequently with H_2O to generate CO, CO_2, and, ideally, more hydrogen. Some high-molecular-weight carbon residue is also generated. This postulated pathway is supported by results reported by the TPC. Steam is added to the hot feed gas to react with the carbon to form CO_2 and CO; the H_2 content of the reactant gas is maintained above 55 percent, a level at which experience indicates the major product will be methane.

Feed materials that contain heteroatoms must yield products that contain these elements, products such as acid gases (e.g., HCl) and reduced inorganics (e.g., H_2S). The TPC has found that chlorinated wastes yield HCl as a primary product. The clean formation of HCl under the reaction conditions can be understood in terms of simple thermodynamics, given that chlorine probably cannot speciate to many other products under the reaction conditions. For chlorinated hydrocarbons, the overall reaction can then be visualized as:[1]

$$\begin{aligned} C_xH_yCl_z + H_2 &= CH_4 + HCl + C + \text{other products} \\ 2C + 3H_2O &= CO + CO_2 + 3H_2 \\ CH_4 + 2H_2O &= CO_2 + 4H_2 \\ CH_4 + H_2O &= CO + 3H_2 \\ CO + H_2O &= CO_2 + H_2 \\ \hline C_xH_yCl_z + H_2 + H_2O &= CH_4 + CO + CO_2 + C + HCl + \text{other products} \end{aligned}$$

[1]Not all of the equations shown here and below are balanced.

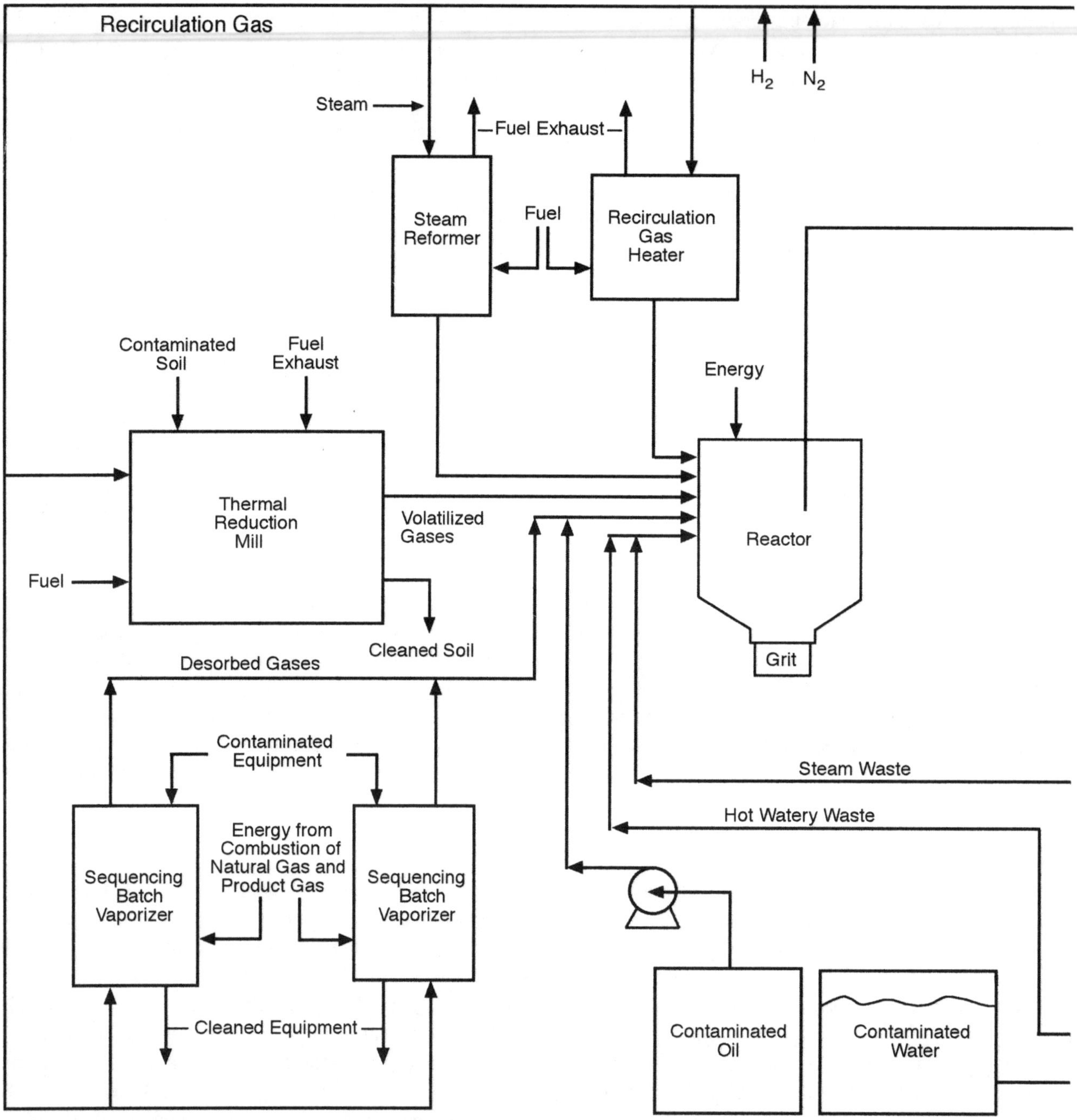

FIGURE 6-1 Schematic diagram of commercial-scale process. Source: ECO LOGIC, 1996a.

In principle, the third, fourth, and fifth reactions could occur in the waste destruction reactor to produce all of the H_2 needed for the hydrogenation reaction (first reaction). In practice, however, methane remains a major product from the reactor. The methane is converted to H_2, CO, and CO_2 in the steam reformer to provide enough H_2 for the reactor.

Another significant factor is that the rate of reaction of carbon with steam (second reaction) is slow, even at 850°C. For example, at 850°C, the time to react 99 percent of the carbon would be 23 days; in the SBV at 550°C, the same completion would require 500 years. Although reactive carbon-containing intermediates might react much faster, it is likely that some carbon will

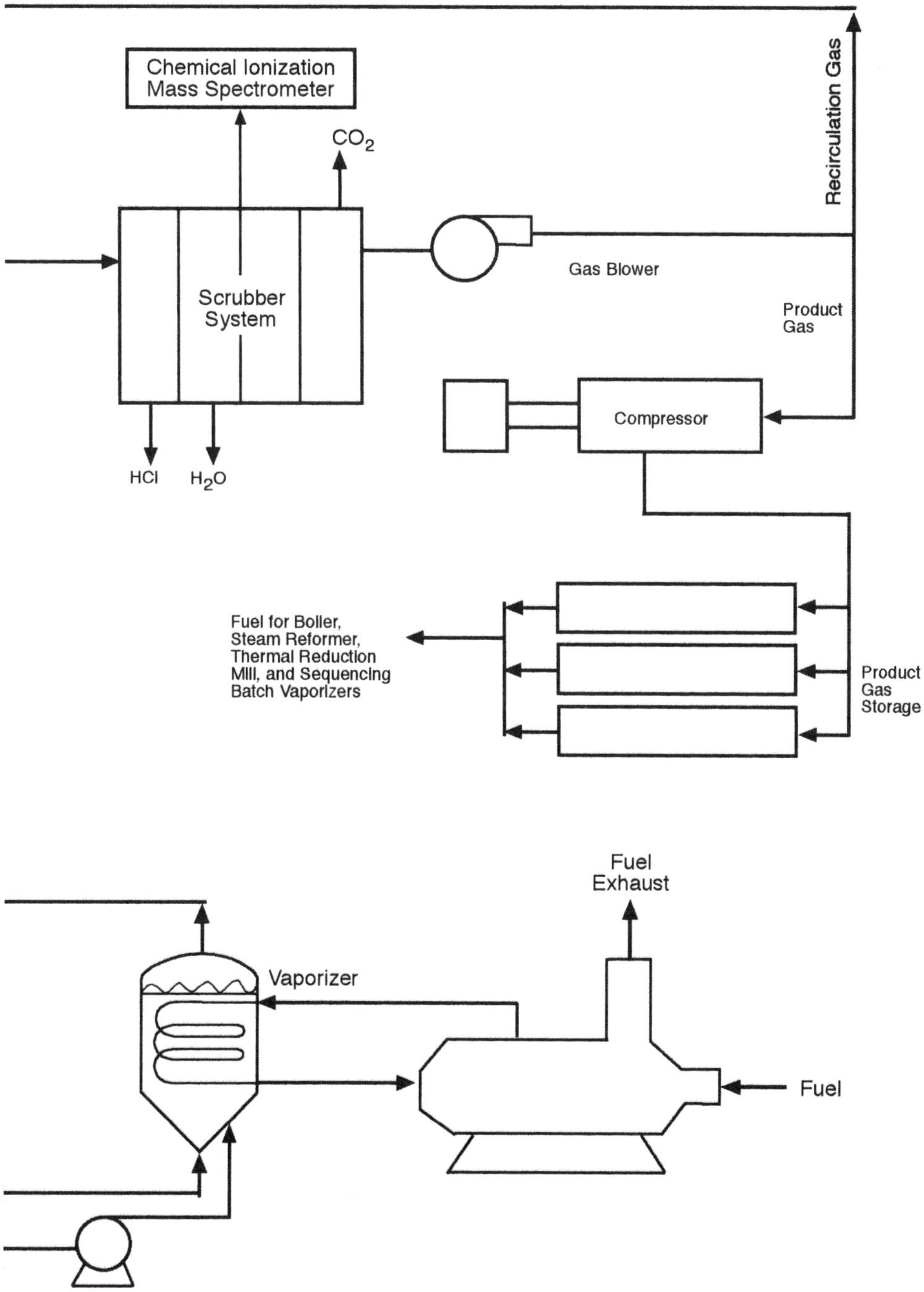

be formed and must be managed by the downstream treatment of the reactor effluent. For simple chlorinated hydrocarbons, the TPC has sufficient practical experience to operate the process at conditions that generate the least amount of carbon. Even so, some carbon is produced and must be managed, and additional hydrogen must be regenerated or added.

Far less is understood, fundamentally or empirically, about the fate of other heteroatoms—such as sulfur, nitrogen, and phosphorus that are present in the chemical agents HD and VX—in feed streams entering the GPCR reactor. The reactions of these heteroatoms have not been investigated extensively, and the interplay of kinetics and thermodynamics is difficult to predict a

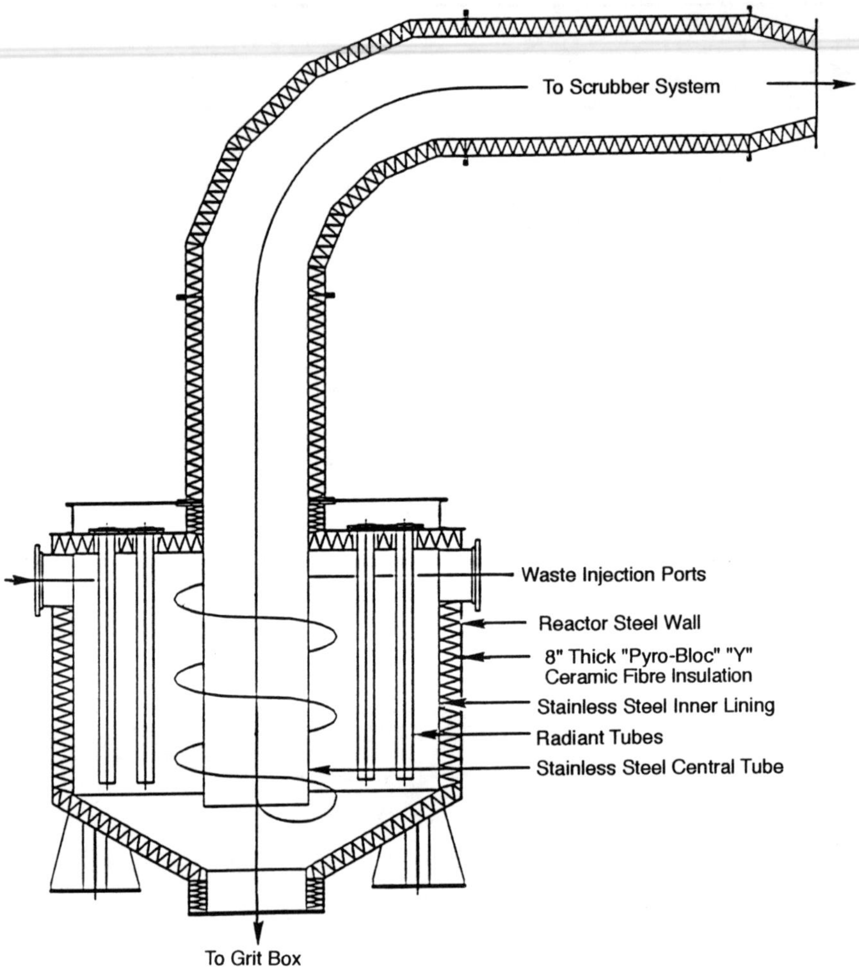

FIGURE 6-2 Main reactor in the gas-phase chemical reduction process. Source: ECO LOGIC, 1996a

priori. Predictions are necessary both for developing appropriate scrubber systems and for identifying and managing toxic residuals.

Predicting the residuals from HD appears to be much more straightforward than predicting the residuals from VX. One can reasonably expect H_2S to be the principal sulfur-containing product exiting the reactor from HD destruction. The TPC reports that this expectation is borne out by its experimental and full-scale work on wastes containing small amounts of sulfur. Moreover, the hydrogenolysis of organosulfur compounds to H_2S is well known from commercial hydrodesulfurization processes. For HD destruction, the overall reaction can be summarized as:

$$SC_4H_8Cl_2 + H_2 = CH_4 + C + CO + CO_2 + H_2S + HCl + \text{other products}$$

The TPC's empirical knowledge and operational experience with other feed materials should be sufficient to develop the appropriate conditions for HD destruction. However, provisions will be needed for handling the large sulfur (as H_2S) residual stream. Although the TPC has some experience with small amounts of sulfur in feed materials, it will have to scale up the MEA scrubber to handle the much larger quantities of H_2S that would be generated by HD. Adding a new, scaled-up scrubber unit to the flow plan will bring the usual complement of potential problems in both startup and continuing operations. The TPC's plan to use commercially available technologies to convert H_2S to elemental sulfur for ultimate disposal seems sound but considerably increases the complexity of the overall process.

The reduction of VX is much more complex, and the products are more difficult to predict. The speciation of the phosphorus and nitrogen present in VX is considerably more difficult to predict without laboratory bench work. The overall reaction for VX can be summarized by:

$$C_{11}H_{26}SNPO_{26} + H_2 + H_2O = CH_4 + CO_2 + CO + C + H_2S + \text{P-products(?)} + \text{N-products(?)}$$

In contrast to hydrodesulfurization chemistry, the removal of phosphorus from organophosphorus compounds by hydrogenolysis has not been studied extensively. A more thorough understanding or at least empirical knowledge of the fate of nitrogen and phosphorus is clearly necessary for destruction of VX. Identifying the phosphorus and nitrogen products is also necessary for developing appropriate scrubbing systems and delineating the ultimate form and disposal of process residuals. The TPC believes that nitrogen-containing feed materials will yield both N_2 and NH_3 in the reactor; some HCN is another possibility though not favored thermodynamically. The analogy for phosphorus is tenuous, however.

The main issue in heteroatom speciation can be illustrated with phosphorus-containing materials. Phosphorus–steam chemistry is not well understood, nor is reduction of the pentavalent phosphorus [P(V)] compounds found in the environment of a GPCR reactor. [P(V) is the form of phosphorus present in VX.] Although the TPC initially suggested that phosphine (PH_3) would be the main phosphorus-containing material exiting the reactor (by analogy to the TPC's experience with methane production from carbonaceous material in the highly reducing steam environment of the reactor), the TPC has not reported detecting or characterizing any phosphorus-containing products from the laboratory-scale tests of VX surrogates. The panel's own thermodynamic calculations suggest that reduction of oxides of P(V) to phosphine is unlikely. From thermodynamic considerations, more likely products are oxyphosphorus acids (e.g., HPO_2) and perhaps elemental phosphorus. (Appendix F describes thermochemical calculations made by the panel to understand potential speciation for phosphorus in the reactor.)

A cautionary note is that oxyphosphorus materials are probably much less volatile than their carbon analogues, CO and CO_2, and therefore might remain in the reactor or foul the exit tube or downstream piping. (A metric ton of VX would yield 375 kg of phosphoric acid.) Experimental work will be necessary to define the phosphorus end-products in the reactor and explore these possibilities, particularly because the models used by the TPC are empirical and derived from experimental data for carbon speciation.

These speciation issues are serious and will require substantial laboratory testing to resolve them prior to pilot-scale work. The TPC understands these issues and has stated that work is being done on them. The TPC has developed a plan to determine the speciation of phosphorus and design a method of scrubbing phosphorus-containing residuals from the reactor effluent. This aspect of the underlying chemistry is difficult for the panel to assess further without these empirical studies.

Reactor Effluent Scrubbing

The principal inorganic products of the gas-phase reduction of HD would be HCl and H2S. Both can be managed by conventional scrubbing systems that the TPC has previously employed for other feed materials. Although the large volume of H2S from HD will require scaling up the caustic and MEA scrubbers that the TPC currently uses with other feed materials, doing so should not be arduous. The plan by the TPC to convert H2S on site to elemental sulfur using conventional commercial technology is preferable to storing and transporting large volumes of H2S, which is highly toxic. However, the conversion will increase the complexity of the overall system.

There are two main problems associated with scrubbing inorganic products and acid gases from VX destruction: (1) determining the primary phosphorus-containing products exiting the reactor and (2) developing or implementing the scrubbing systems needed to handle these products in the effluent stream. The TPC has presented proprietary chemistry for scrubbing phosphine, but the technique requires further demonstration and may be inappropriate if phosphine is in fact not produced (see discussion above).

For both VX and HD, the scrubbers must also remove not only the elemental carbon formed in the reactor but also any high-carbon-content precursors that may be present, such as aromatics and polycyclic organic compounds. The elemental carbon will probably be present as very finely divided particulates (soot) and will wash out with the initial water quench. The TPC should have experience with this process from its current operations. The TPC has stated that it expects that any polycyclic organic compounds will be present in very small amounts and can be recovered by heavy-oil scrubbing.

Test work on agents will show whether or not this scrubber will be needed. The TPC has not described the ultimate disposition of this process residual, which conceivably could be managed by recovering the high-carbon material from the oil, by burning it along with the recovery oil, or by other means.

Regardless of the status of the scrubbing technologies, the recovery of process residuals containing the speciated products of chlorine, sulfur, phosphorus, and carbon and their conversion for ultimate disposal is clearly a complex process that will require a number of unit operations.

TECHNOLOGY STATUS

The TPC has experience treating organic wastes, including PCBs, other chlorocarbons, and hydrocarbons such as toluene, and has been developing GPCR for more than 10 years. It has conducted a significant amount of work at laboratory, pilot, and commercial scales.

Pilot-scale work has been performed since 1991 at several sites in the United States and Canada. The TPC has a laboratory-scale system available for waste treatability studies and has use this system for preliminary tests on agent surrogates, such as the organophosphorus pesticide malathion. Pilot-scale demonstrations have been performed on several materials, including polyaromatic hydrocarbons at Hamilton Harbor, Ontario, in 1991 and PCBs (PCB-contaminated oily water, highly concentrated PCBs in oil, and contaminated soil) at Bay City, Michigan, in 1992.

Commercial units are currently deployed in Australia and Canada, and others are in progress. The TPC has been treating a mixture of DDT and toluene on a commercial scale in Australia. Another system at St. Catherines, Ontario, processes PCBs both as a concentrated material and in PCB-contaminated concrete, for General Motors. For treating these feedstocks, the status of the technology is advanced, since there are commercial facilities in operation.

In the judgment of the AltTech Panel, the TPC has considerable experience with these feed streams in all aspects of facility operation, including operational requirements and considerations, mass balances, gas recycling, and management of residual HCl. Although the panel received detailed modeling data from the TPC, it did not receive detailed laboratory data from the agent-destruction tests, which were at laboratory scale. No bench-scale tests have been reported to the panel.

Full operational manuals, hazard and operability studies, process and instrument diagrams, and risk analyses have been developed and documented for processing DDT–toluene mixtures and PCBs.

The TPC's experience with organic wastes forms a basis for applying the technology to agent destruction, but further development specific to the chemical agents to be treated is still required. For instance, all operations to date have been outdoors. Agent-destruction facilities, however, will require containment of all unit processes where agent may be present. Containment of hydrogen gas within a building can be hazardous. Additional hazardous-operation procedures for handling these conflicting safety demands were not addressed in the TPC's submissions.

In past and current operations, the TPC has also tested the capability of the SBV to remove and destroy organic wastes from inorganic matrices. This experience qualifies, to some extent, as a demonstration of the SBV's efficacy for treating ton containers and dunnage.

The panel notes that the washing, cleaning, decontaminating, and shipping procedures for ton containers that the Army has proposed for neutralization (see Chapter 7) could also be used with GPCR.

In summary, the main uncertainty in applying this process to agent destruction centers on identifying and managing the inorganic by-products derived from the sulfur, phosphorus, and nitrogen in the agents. What are the primary inorganic products from the reactor, and how will they be removed and managed downstream?

The systems currently used by the TPC should work well for HD destruction after modification of the scrubber train to handle the large load of H_2S. The status of the technology for HD treatment is *near commercial* except for: (1) the lack of demonstrated handling of large volumes of H_2S, (2) the overall process demonstration on HD itself (small-scale tests to show the process can destroy agent have been successful), and (3) the resolution of secondary containment and safety issues specific to processing chemical warfare agents and hydrogen gas.

Because much less is known on both a fundamental and practical level about the identity and handling of phosphorus-containing residuals from GPCR, the technology for VX destruction is less mature than it is for HD. The TPC has little experience with phosphorus-containing materials, even at bench scale. Although the

TPC has developed a plan for addressing these issues, the time line for doing so is unclear.

OPERATIONAL REQUIREMENTS AND CONSIDERATIONS

Process Operations

GPCR consists of a number of sequential subsystems (e.g., feed system, SBV, reactor, scrubber train, methane reformer) that must be tightly controlled and integrated. Because the process is tightly integrated, provision must be made for safety cutoffs or mechanisms that recirculate excess materials back to earlier stages in the process.

At least one scrubber is needed to manage each heteroatom-containing product that exits the reactor. The recovery subsystem—that is, the scrubbers and the subsequent unit operations for generating final process residuals from the scrubber effluents, such as conversion of H_2S to elemental sulfur—consists of standard unit operations (with the possible exception of operations to scrub and handle the phosphorus-containing products). Nevertheless, this subsystem adds considerable complexity to the overall process for destroying agent. The conversion of H_2S to sulfur alone will require four or five unit operations and a compressor.

Mass Balance

The panel received a flow sheet for the processing operations and some material balance predictions. The TPC has developed an empirical model, based on past experience, from which it made the following predictions:

- Methane is the predominant hydrocarbon produced, as long as the H_2 content of the circulating gas stays above 55 volume percent (dry basis). The model uses 60 percent H_2.
- Steam is necessary to limit the production of elemental carbon or high-molecular-weight aromatic material. The steam content is not specified in the model but in practice has been 20 to 80 percent of the dry gas.
- The model assumes gaseous carbon species in the reactor effluent gas occur in the ratios of 42 percent CH_4, 34 percent CO_2, and 24 percent CO (see Appendix F for details). In addition, 10 percent of the carbon in the feed materials is assumed to exit as elemental carbon (soot). The hydrogen/methane ratio in the reactor effluent gas is assumed to be 2.0.

Energy Balance

The TPC did not provide a complete energy balance, but the electric energy required for the reactor heating elements was stated. For HD destruction, the reactor requires 5,019 kW•h/day, or 346 kW•h per ton of HD destroyed. The total heat input required is much larger, and the rest is supplied by burning fuel gas: either the gas produced in the operation or LPG (liquefied petroleum gas, primarily propane).

Process Residuals

Some of the process residual streams are reasonably well defined; some are not. They are discussed below in terms of the fate of the carbon and the heteroatoms in the feed material.

Carbon. Some carbon (as methane) and hydrogen are eventually burned in a steam boiler, which supplies the steam for the gas reformer that converts methane to CO_2 and H_2. The residuals from the combustion exit the process primarily as CO_2 and steam from a stack.

Some carbon (as CO_2) in the reactor effluent gas is scrubbed out with the H_2S in the MEA scrubber. The MEA will scrub out most of the CO_2 in the reaction gas; there is somewhat more CO_2 than H_2S expected in the gas. Some carbon (10 percent of the feedstock C) is estimated to exit the process as carbon soot. Some may also exit as hydrocarbon products that would be scrubbed by oil.

Heteroatoms. The heteroatoms are scrubbed from the reactor effluent gas in various ways. Chlorine exits the reactor as HCl, which is scrubbed out first with water, which eventually produces a residual of concentrated aqueous hydrochloric acid. After the water quench, a caustic scrubber completes the removal of HCl and other acidic components with sodium hydroxide solution. The residual stream from the caustic scrubber is a solution of sodium salts (sodium chloride and salts of other scrubbed acids).

The sulfur exits the reactor primarily as H_2S gas, which is scrubbed out with MEA, a standard commercial treatment. The H_2S is stripped from the MEA solution with hot steam. The TPC plans to convert the H_2S to elemental sulfur using a commercial process called the SulFerox process. The H_2S can be converted to sulfur even in the presence of the carbon dioxide that is stripped with it. The SulFerox process uses oxygen from intake air to oxidize H_2S to water and sulfur, with the oxidation mediated by an iron chelate that is regenerated with air. The sulfur is filtered from the SulFerox liquor. The gas from the subsequent sulfur plant, which is one of the effluents of GPCR, would have three components: (1) the CO_2 remaining after the SulFerox treatment removes the H_2S, (2) the water vapor from the water produced with the sulfur from oxidation of the H_2S, and (3) the spent air (lower in oxygen) from regenerating the iron chelate.

The flow sheet shows some sulfur, as sulfate and sulfite, appearing in the scrubber liquors from scrubbers preceding the MEA scrubber, although the major sulfur-containing product is expected to be H_2S. Some H_2S would be expected to dissolve in scrubber liquors (e.g., the sodium hydroxide solution in the caustic scrubber), which will make them unacceptable for discharge without further treatment.

The issues associated with inorganic by-products from VX have been addressed above and in Appendix F. The panel's calculations, based on thermodynamic equilibrium, suggest that the following products will form: chlorine will yield HCl; sulfur will form primarily H_2S with trace amounts of SO_2; phosphorus will yield higher-valent oxides such as P_4O_6 and perhaps elemental phosphorus; nitrogen will form N_2, NH_3, and possibly nitrogen oxides. If nitrogen reactions are kinetically controlled rather than reaching equilibrium, HCN may be a minor product.

Presumably the fate of phosphorus under the process conditions will be determined by the experimental work planned by the TPC. The exact nature of the scrubber effluent, in fact the type of scrubber to be used, will not be known until this work is done.

Ton Container Treatment

In the design as submitted, the emptied ton containers are cleaned in the SBV. The maximum residual agent in the SBV effluent gas stream, which is fed to the main reactor, cannot be predicted quantitatively prior to pilot testing, although the Army's experience with container cleanout at existing facilities suggests it will be very low under the SBV process conditions (described below). The SBV will also be used to heat the solid carbon residual that collects in the water quench effluent. The heating will drive off and react any volatile material associated with the carbon.

Ton containers will be placed in the SBV, two at a time, and treated with hot gas at 540°C (1000°F) for several hours (probably around 6 hours). The TPC has stated that this relatively long period at high temperature will be at least equivalent to the Army 5X cleaning conditions of heating for 15 minutes in an oxidizing atmosphere at about the same temperature. Evaporation and thermal cracking are stated to be the most important processes that occur, and they are independent of the composition of the circulating gas. Testing must be done to verify these statements. If regulatory approval can be obtained, the TPC has stated that it would prefer to recycle the ton containers. An alternative for treating the emptied ton containers in the SBV would be to use the Army process (hot water wash, followed by steam cleaning), described in Chapter 7, to clean them sufficiently to allow shipping them to Rock Island Arsenal, Illinois, for melting.

Materials and Energy Balance

The TPC provided material balance data (feed rates with product compositions) for the reactor, the scrubbers, the product-gas boiler, and the catalytic reformer when treating HD. Details on feed streams and products for these unit operations are given in the tables and discussion in Appendix F. The TPC provided a similarly detailed material balance for VX destruction, but that balance has not been reproduced for this report because the panel believes that the reaction chemistry is still too uncertain, as explained above.

Most of the chemical bonds in HD (or VX) should break at the temperature of the reactor. If some feed material were to exit the reactor without reaching reactor temperature, some agent might not be completely broken down into the simple residuals expected. According to the TPC, its experience with other feed materials suggests that this will not happen. However, based on the information available to the panel, tests have not yet been done for some process residuals that could result from partial breakdown of HD or VX. The least detectable concentration for most process substances and residuals from HD destruction will be very low (parts-per-billion range) because the compounds of analytical interest will be in liquids that can be stored for hours or days, allowing ample time for detailed analysis.

Feed Streams

There are six feed streams to the reactor: the liquid agent feed, the SBV effluent gas, steam, hot waste water recycled for processing, reformer gas, and recycled reactor effluent gas. The reformer gas is the largest of these streams (about 85 percent of total moles fed) and has the composition shown in Table 6-1. The reformer gas is already at high temperature when it reaches the reactor; the electric heaters in the reactor are needed only to provide enough incremental heat to reach the reaction temperature.

The two agent-derived feed streams are the liquid chemical agent from the ton containers and the effluent gas stream from the SBV. One ton container at a time is drained into a holding tank, and the liquid agent is pumped directly from this holding tank into the reactor. The ton container with its residual liquid or gel (the heel) is moved into the SBV, where remaining material is vaporized or reacted with the hot circulating gas (hydrogen, methane, steam, and CO). The effluent gas from the SBV is fed to the reactor. According to the TPC, the precise details of safety containment for holding tank, pumps, and lines have not been worked out.

The scale of the equipment the handles the agent-derived feed streams (pumps, reactor, and SBV) is the same as the equipment in operating plants in Australia and Canada. The exact mechanical layout and protective housing to be used when a ton container is punched and drained have not been designed. However, the system proven by the Army (see Chapter 7) for the punch-and-drain unit operation can be used. No pretreatment of gels or solids is needed if they are sent to the SBV inside the ton containers.

TABLE 6-1 Composition of Reformer Gas

Component	g-mols/m^3	Volume %
H_2	755	74.0
CH_4	15.3	1.5
CO	35.3	3.5
CO_2	55.3	5.4
H_2O	159.8	15.7

Process Residual Streams

Bulk Agent

The residual streams consist of the combustion gases from a steam boiler and a number of products from the scrubbing system. For HD, the overall reaction (agent in, residuals out) will be:[2]

$$SC_4H_8Cl_2 + 5O_2 \rightarrow 18.8\ N_2 + 4CO_2 + 2\ H_2O + 18.8\ N_2 + 2HCl + H_2S + H_2O + 1/2\ O_2 + HCl\ (\text{in solution}) + H_2O$$

The material balance for HD destruction in Appendix F is based on an agent feed rate of 5 metric tons per day and the TPC's model for assigning products of reaction. There appears to be some flexibility possible in the material flows through the process.

The product HCl is scrubbed out in the water quench, along with the solid carbon from the reactor. The carbon, which is filtered out, eventually goes to the SBV for drying and the removal of any adsorbed products of incomplete reduction. The residual carbon is a final process residual and will have to be disposed of.

Most of the H_2S is recovered as elemental sulfur in a multistep process. H_2S and CO_2 are scrubbed from the gas by MEA. Efficient removal of H_2S from the reactor effluent gas is required for two reasons. First, most of the gas will be steam-reformed, and the reforming catalyst may be sensitive to sulfur. Second, the remainder will be burned, and the combustion exhaust gases will have to meet regulatory limits on sulfur.

The MEA scrubber is a two-vessel system, with H_2S and CO_2 absorbed in one column and regenerated by steam-stripping in a second. The circulating MEA must be alternately cooled and heated as it flows from one vessel to the other. The effluent gas from the MEA stripper consists of H_2S and CO_2 (about 70 volume

[2]The caustic scrubber will also produce a small amount of NaCl. Some H_2S and NaSH/Na_2S may also be present in the quench and scrubber solutions, respectively.

percent CO_2). The TPC proposes to oxidize the H_2S to sulfur using the SulFerox process (which is also proposed for use with the CEP, see Chapter 4). The CO_2 is released to the atmosphere.

The form of the phosphorus-containing residual (or residuals) must be determined by further experimental work. Presumably the phosphorus would be converted to its most stable form as a phosphate salt. In a reducing atmosphere, P(III) (phosphorus with a valence of 3) is the stable form of phosphorus.

At this time, the TPC has not identified commercial facilities for receiving any of the liquid or solid process-residual streams described above.

Nonprocess Wastes

Nonprocess wastes will be treated like process wastes. Solids (such as personal protection suits) will be treated in the SBV. Solid residuals from the SBV will probably be classified as toxic waste. Liquids (such as used decontamination fluid) will be sprayed directly into the reactor, along with the liquid agent feed stream. Products from treating these nonprocess wastes will become part of the same residual streams as the process residuals described above.

PROCESS INSTRUMENTATION AND CONTROLS

The GPCR design states that the instrumentation and the control system for agent-destruction applications will be based on the ones used in the operating facilities in Australia and Canada, with appropriate modifications. The following instrumentation is being used.

Chemical Ionization Mass Spectrometer. This is a sensitive, soft-ionization mass spectrometer capable of monitoring, by visual display and recording, selected organic compounds at concentrations of parts per billion. Results are obtained in seconds. The gas outlet of the reactor will be sampled and monitored for compounds selected as indicators of the completion of the destruction process. Preliminary tests on the reaction chemistry will be used to select the compounds to be monitored. For HD, as an example, unreacted HD and simple sulfides or mercaptans might be the selected indicator compounds.

M200 Gas Chromatograph. This on-line gas chromatograph will also be used to analyze samples from the reactor effluent gas, for the same purpose as the mass spectrometer.

NOVA Process Gas Analyzer. A high-precision infrared detector will be used to monitor concentrations of methane, carbon monoxide, and carbon dioxide at various locations in the process gas flows. Hydrogen will be monitored by a thermal conductivity cell. As a safety measure, oxygen will be monitored continuously by a NOVA oxygen analyzer with an electrochemical cell. Gaseous residual streams coming out of the process will also be monitored, particularly for residual agent.

Pressure measurement and control are important for the process; to preserve a safe hydrogen atmosphere, negative pressure anywhere in the recirculating gas circuit must be avoided. The system pressure is maintained by continuous feed-gas inputs to the reactor and by adjusting the rate of removal of reactor effluent gas; the latter is controlled by a variable-speed blower with a gas bypass around the blower. As an added safety feature, gas from the high-pressure storage subsystem can be fed back to the reactor if the system pressure becomes negative.

An aspect of measuring and controlling pressure that the TPC did not specifically address in its written submissions or in dialogue with the panel is proper ventilation of the building that will serve as a secondary containment for the recirculating gas circuit. (Secondary containment is required as a backup line of control to prevent an accidental release of agent to the atmosphere.) The system circuit must be maintained at a slight positive pressure relative to the ambient building air, to prevent oxygen from leaking in. Any leakage will therefore be process gas leaking out. In current operations in Australia and Canada, there is no explosion hazard if hydrogen leaks *out* of the system because the entire system is outdoors (no secondary containment). A small hydrogen leak that causes a small, controllable flare in an unconfined system could lead to an explosion if hydrogen accumulates in an air-filled secondary- containment building. For example, the experience of the National Aeronautics and Space Administration in handling hydrogen, which it uses routinely and in large quantity, has been that all leaks in enclosed systems lead to fires. Therefore, any secondary containment for GPCR will require substantial ventilation to prevent the accumulation of hydrogen. At the same time, the ventilation system must prevent a release of agent from the secondary containment in case of a leak.

The most crucial temperature to control is the reactor temperature. The final reaction temperature is achieved by electric heaters, and small adjustments are made by controlling the electric power. (A major drop in temperature would trigger a cutoff of the feed.)

The overall process in the submitted design for GPCR is tightly integrated. Effluent gas from the reactor must be scrubbed in a series of scrubbers. The final, MEA scrubber must be regenerated continuously and the H_2S converted to elemental sulfur. A portion of the reactor-effluent gas stream must be fed continuously to the steam reformer to maintain hydrogen concentration. The effluent must be preheated for this process. Another part of the effluent gas stream is simply recycled to the reactor after preheating. Some of the effluent gas is compressed and stored and then used to even out flow and pressure requirements. The control program used in existing operations appears satisfactory and will have had many hours of operation on a commercial scale before a facility for treating chemical agent is ready for testing.

The TPC has assessed several failure modes and has developed control strategies for them. Process parameters have been identified that are critical for process control and safety. The monitoring and control operations, which are based on treating materials other than chemical agent, are well documented. The complete scrubbing system required for processing chemical agents has not yet been settled. (The TPC is considering incorporating a multistage sulfuric acid scrubber with other more conventional scrubbing towers, but the plan may change depending on the results of further laboratory work.) New controls may be needed for new scrubber equipment.

Certain control requirements will place stringent limits on the operation, particularly controls related to the safe handling of high-temperature hydrogen—namely, the control and monitoring of pressure and the control of oxygen. Close control of temperature will be required for the chemical reactor units: the main reactor, the steam reformer, the MEA stripper and regenerator, the sulfur recovery system, and possibly the HCl scrubber. In summary, GPCR technology involves a complex chemical plant that must be operated carefully.

PROCESS STABILITY, RELIABILITY, AND ROBUSTNESS

Stability

Of the reactions that are presumed to take place in the GPCR main reactor, hydrocarbon cracking followed by hydrogenation to CH_4 is highly exothermic. The steam-forming reactions to yield CO and H_2 are highly endothermic. For HD and VX, the net heat effect in the reactor appears to be very slightly exothermic. By monitoring the temperature and controlling the agent feed and the heat for bringing the feed streams to the reaction temperature, the process can easily be controlled.

The operation of the main reactor should be stable because the reactor can operate reasonably well over a wide range of temperature above the minimum needed for the reduction reactions to occur. Operation of the steam reformer will require closer temperature control to maintain catalyst activity. Scrubbers should operate satisfactorily within modest deviations from design conditions.

Reliability

Because GPCR is a continuous process rather than a batch operation, multiple reactions in multiple reactors must proceed in continuous balance. The main reactor has had considerable commercial operation and has proven reliable. The entire system, however, consists of a number of sequential unit operations that must be tightly integrated and controlled. Agent destruction in the reactor is followed first by a sequence of chemical scrubbers and a sulfur recovery reactor, then by a catalytic steam reformer whose catalyst would be poisoned by breakdown of the scrubbing system. Removal of the process residuals from the recirculating gas stream for treatment and disposal, as well as steam reforming of H_2 from methane, are carried out in a continuous loop. In a system operating in this continuous recirculating mode, a failure in one unit operation could significantly effect the others. For example, any reaction products containing sulfur or phosphorus that are carried past the scrubbers to the steam reformer are likely to poison the catalyst in the reformer. The system as a whole therefore must be carefully controlled at many points.

The information provided by the TPC indicates that the overall system has been stable and reliable in operations that treat simple chlorocarbons ("simple" meaning compounds containing just chlorine as a heteroatom). The reliability of the more complex system required to treat one or more chemical agents will require demonstration. Tighter controls will certainly have to be implemented.

The stored energy that might be of concern is the hot H_2, which obviously must be protected from air.

Scenarios can be envisioned in which air leaks into the process stream because of negative pressure. Multiple controls are built into the system to prevent negative pressure in the system.

The panel expects that the mechanical equipment will be highly reliable because it is generally standard in the chemical industry. However, existing plants have not been operating long enough to give a definitive answer on long-term operability and performance.

Robustness

The reactor can operate satisfactorily over a considerable range in temperature—±100°C from the design temperature. The catalytic steam reformer will require close temperature control and a "clean" feed; that is, no catalyst poisons in the input gas, such as sulfur or phosphorus. The scrubbers will require constant monitoring but should be able to handle a modest range of feed rates.

The system response to upsets will depend on the upset. A sudden drop in a feed stream will trigger the pressure control system, which will add recycle gas to maintain positive system pressure. A drop in energy input to the main reactor (electric power to the heaters) will presumably produce a temperature drop. The panel believes that the time constant for such a temperature change is on the order of several seconds. Drop in reactor temperature below a set point would shut down the agent feed as well as trigger the pressure control.

A change in the feed material will probably require major changes in the scrubbing system. A change in the feed material or feed rate may also require modifying the amount of gas circulating to the steam reformer, but the system should adjust for this by responding to the monitoring data provided by the gas analyzer.

MATERIALS OF CONSTRUCTION

Satisfactory performance of the overall GPCR system depends upon containing the reaction and the products of reaction. Designing an effective containment requires understanding how chemicals in the process environment interact with materials of construction to degrade them. In general, it appears that the unit processes of this technology are similar to or the same as unit processes for which substantial experience exists and for which there are conventional and satisfactory materials of construction.

To understand the possibility for premature degradation that would reduce the integrity of important materials in this system, it is necessary to consider specific environments to which they are exposed, the properties of the materials, the design features of design that affect degradation, and the possible failure modes.

Environmental Definition

Three operating environments are expected in the GPCR technology:

1. The reactor and the immediate downstream piping contain gas mixtures at temperatures as high as 900°C. This gas consists mostly of hydrogen and steam, but other corrosive species are present, such as HCl, H_2S, phosphorus oxides, and carbon. Particularly in the reactor itself, this gas may also contain some agent in the process of being destroyed.
2. In the scrubber systems, room-temperature acid solutions are produced that contain various acidic species, including chloride, sulfur acids, and phosphorus acids.
3. Products from the scrubber systems contain H_2S, CO_2, and steam.

Although these are the general steady-state operating environments in the GPCR process, other environmental conditions that will occur intermittently can be important to material degradation. First, shutdown conditions may lead to aerated acidic or other corrosive conditions on component surfaces. For example, in fossil-fueled power systems the chemicals formed by the reaction of humidity with deposits during shutdown are often more corrosive than the deposits. Second, environments on the outside of component surfaces may be corrosive during operation or shutdown because of humidity, dripping water, or industrial gases. Third, accidents or out-of-specification conditions may occur during normal operation.

Finally, the degradation of materials of construction usually involves subtle processes dependent on the relative amounts of chemical species. If a generally analogous system in terms of anticipated species present in the technology has proven satisfactory, a common presumption is that a similar system will also perform satisfactorily. Whether this argument by analogy is valid depends on subtle chemical differences, such as the following, which may apply to gas-phase reduction of chemical agents HD and VX.

Sulfur Valency. Sulfur species with valences less than +6 are generally corrosive, depending on the pH and the presence of other species. They are corrosive over a wide range of compositions for nickel and chromium alloys, although the susceptibility of an alloy to corrosion depends on how heat treatments have affected its local composition.

Sulfur-Chloride Ratio. Corrosiveness of the gas mixture changes greatly, depending on the relative ratios of sulfur and chloride species.

Hydrogenation. The presence of lower-valence sulfur species, as well as of phosphorus and cyanide species, influences various modes of hydrogen-related damage, such as cracking and blistering.

Materials to be Used

The TPC's submission did not define the materials of construction except to note that they will be materials that have performed satisfactorily in analogous industrial systems. The preliminary design suggests that a major structural material would be stainless steel. However, it is necessary to specify which of many stainless steels would be used, the fabricated condition of the material, the conditions of welding (heat-affected zone, weld metal, weld passes), and the residual stresses.

It is also necessary to define how the properties of materials change with exposure to processing conditions, especially to temperatures. For example, in the range of 800°C, the microstructures of certain stainless steels change in ways that increase the likelihood of corrosion-induced failure.

Design Features

Certain design features in any new system affect whether accelerated degradation occurs. Although the materials of construction may be typical of the materials used in analogous systems that perform satisfactorily, often a system-specific configuration of these materials promotes degradation. Among configurations and conditions that may lead to premature degradation in GPCR are the following:

- Operation at 800°C suggests that the system will be subject to high thermal stresses. Because this system has no prototypes, it will probably be subjected to thermal cycling because of the intermittent runs conducted during the pilot-testing and demonstration phases prior to long-term, continuous operation.
- Crevices in general, and heat-transfer crevices in particular, can accelerate degradation. The design as submitted does not specify whether crevices are prohibited or have been otherwise considered.
- The formation of deposits at the bottom of reaction vessels in a scrubbing system generally accelerates degradation processes. Other conditions that accelerate degradation include liquid-vapor interfaces, especially when the vapor contains oxygen.

Modes of Degradation

As a result of chemical and design conditions, a number of modes of degradation may occur either in the main reactor or in the scrubbing system. In *general degradation*, the material may "rust away" if surface corrosion increases as the environment becomes increasingly acidic or alkaline. General degradation can be minimized by using alloys that contain a larger percentage of nickel and chromium (high alloys). *Localized corrosion,* such as pitting and intergranular corrosion, may occur even in highly alloyed materials. In fact, localized corrosion is often more aggressive in highly alloyed materials and can be especially affected by operating temperatures. *Stress corrosion cracking* and *hydrogen embrittlement* can occur regardless of the alloy composition. These modes of degradation are particularly aggressive in alloys that are more resistant to general degradation. *Stress cycling*, such as thermal cycling, may interact with the process environments to produce fatigue cracks.

Failure Modes

In addition to the modes of degradation described above, there are more general modes of degradation and failure to consider. These are especially important given the toxicity of the agent and of some of the principal products of reaction, such as H_2S. Among the failure modes that need to be considered are (1) oxygen leakage mixing with hydrogen during startup, shutdown, or operation, (2) release of reactor effluent gases from piping defects, (3) leaks in the pregasifier, and (4) system problems caused by thermal variations.

OPERATIONS AND MAINTENANCE

Operations

According to the TPC, the system would be operated continuously (24 hours a day, 7 days a week) and would employ four separate shift-teams during treatment operations (12-hour shifts, 4-day rotations). Each shift-team of 10 to 11 people would consist of a shift supervisor (Professional Engineer [P.E.] certified or equivalent training), two process control engineers (P.E. or B.S.), two regular technicians for process maintenance and monitoring, one maintenance/boiler operator-engineer, one logistics coordinator, and three or four trained laborers, as required for handling material. In addition, an operational staff of five would be assigned to the project for a normal work week (5 days per week, 8 hours per day). This staff would consist of a project manager (P.E.), a project administrator, a quality assurance officer, a health and safety officer, and a senior member of the technical support staff (Ph.D.), as required. The TPC states that the normal staff (i.e., the shift-team and the operational staff as needed) would be capable of, and have the required training for, initiating shutdown and restart during normal operations or emergency situations.

The system operates on a 24-hour continuous basis. Although the system can be operated on an intermittent basis (e.g., an 8-hour day), cost-effectiveness would decrease because of increased startups, shutdowns, heating costs, etc. Standby mode is similar to operational mode but with reduced utility requirements. For long standby periods, the system would be purged with nitrogen. At ready mode, all system components are operating and up to temperature. Staffing for standby and ready modes is the same as for operational mode, with a provision for reduced staffing in the event of a long standby.

The TPC has complete operational manuals, hazardous-operations procedures, process and instrument diagrams, and risk analyses for its commercial operations. The process control system, which was described in the response to the panel's questionnaire, consists of a Moore Advanced Process Automation and Control System interfaced to a microcomputer. Upgrades to the control room (e.g., more screens and monitors) will probably be needed for an agent-destruction plant.

Although the TPC has considerable experience in pilot-scale and commercial-scale processing of dilute wastes for extended periods of time, it has stated that it is only in the early stages of commercial-scale operations and therefore does not have sufficient operational history to quantify the ratio of downtime to operational time. The TPC estimates about 20 percent downtime, based on its models, and will provide further information as it accumulates experience with current projects.

Startup and Shutdown Procedures

The panel had the following concerns about the startup and shutdown procedures provided by the TPC:

- Primary precautions are for keeping oxygen away from hydrogen. The TPC plans to purge the system with N_2 and monitor for O_2 during startup and shutdown to ensure that hydrogen and oxygen do not mix. The procedure for monitoring for hydrogen leakage out of the system during startup and shutdown was less clear.
- Gradual startup and shutdown appear to be necessary because of the thermal stresses.
- The possibility of surface deposits under moist shutdown conditions should be addressed.
- No criteria were provided for monitoring or assessing the stresses on materials or other damage to the system from an emergency shutdown.
- The reformer startup must follow a particular procedure to maintain an active catalyst.

Maintenance

The TPC appears to have considerable operational, and therefore maintenance, experience with full-scale treatment plants. For this technology, a full maintenance plan for an agent-treatment plant will require: (1) consistent implementation of routine maintenance or inspection; (2) objectives for maintaining barriers; (3) controls to prevent release of poisonous downstream gases; (4) attention in all procedures to conditions that might allow oxygen and hydrogen to mix; (5) process-specific maintenance manuals; and (6) understanding of failure modes and specification of appropriate inspections to prevent them.

Utility Requirements

The TPC stated that the electrical energy requirement for the reactor heaters is 5,019 kW-h/day, that is, average power of 209 kW. The general electrical requirement for pumps, other heaters, lighting, etc. was given as 20,000 kW-h/day, or 833 kW. Total average electrical power required is therefore 1,042 kW.

The fuel required during operation consists of propane plus part of the effluent gas from the reactor. In addition, the reaction is expected to be very mildly exothermic. The approximate energy requirements for treating 9 metric tons of HD per day are summarized in Table 6-2. below. The TPC's preferred rate of operation is somewhat lower, at 5 metric tons per day. Fuel will also be required for startup and for operation of the SBV. The amounts of fuel required have not been estimated.

The TPC stated that the water requirement is 100 gallons per minute of clean water. For HD, this appears to be on the high side. Some water will be used in the steam feed to the catalytic reformer, and some will be used in the caustic scrubber. However, these are small requirements (a few gallons per minute). A larger amount will be needed for cooling (perhaps 20 gallons per minute).

SCALE-UP REQUIREMENTS

The GPCR process can be broken down into six subsystems: the agent (or waste) feed system, the SBV, the reactor, the scrubber system, the catalytic reformer, and the evaporative-cooling/air-water treatment systems. Each of these subsystems consists of a number of unit operations. The various subsystems have not all had the same demonstrations of operability with scale-up. For example, the scrubber system required for HD will be quite different from the demonstrated system for chlorinated materials. The catalytic reformer, on the other hand, should be the same. As noted, the process has been demonstrated at pilot and commercial scales for processing aromatic hydrocarbons and chlorocarbons. The TPC has said that all system components pertinent to the treatment of chemical agents are demonstrated commercial technologies. The panel believes, however, that demonstration is lacking in the following areas:

- handling and disposition of high concentrations of sulfur-containing products (primarily H_2S) in reactor effluent gas, although commercial scrubbing technologies are available
- speciation and management of phosphorus-containing products in the reactor effluent gas, including scrubbing technologies or other methods for managing phosphorus-containing reaction products, as well as the final form and mode of disposal of high-volume process residuals containing phosphorus
- effects of reactor products containing sulfur and phosphorus on the catalytic reformer and mechanisms to avoid poisoning if these products are not fully recovered in the scrubber system (The TPC's experience has been with chlorine-containing materials, which do not present the same problem.)

With only preliminary information derived from VX surrogate tests using malathion, and without complete information from the initial agent tests conducted late in the study process, the panel is unaware of the fate of the phosphorus in VX. Moreover, the TPC provided little detail on the scrubbing system, and it was difficult for the panel to verify some of the necessary oxidation-reduction chemistry in the TPC's proposed technology for scrubbing phosphorus compounds from the reactor effluent gas. The TPC provided little detail the on treatment of process residuals for ultimate disposal.

Although the capability to clean out emptied ton containers has not been demonstrated, the SBV has been demonstrated on inorganic matrices. The panel believes that the SBV hold-up times, temperatures, and reactant

TABLE 6-2 Daily Energy Requirements to Process HD at 9 Metric Tons Per Day

Energy Source	Rate of Use (MJ/h)	Percentage of Total
Electric power (833 kW)	3,000	28.4
Burn product gas	3,876	36.7
Burn propane	3,517	33.3
Heat of reaction	160	1.5

gases are likely to suffice for this purpose. Laboratory-scale tests have demonstrated desorption efficiencies in excess of 99.9999 percent for organic residues in enclosed containers such as PCB-contaminated lamp ballasts.

PROCESS SAFETY

As noted in the section on Process Safety in Chapter 4, the risk factors for process safety in all the alternative technologies can be divided into two categories: factors related to handling agent prior to its introduction into the specific technology and factors related to the agent-destruction technology and associated system elements. The process-safety risk factors related to the handling of agent prior to entry into this technology, which are common to all the agent-destruction technologies, include storage risk, transportation risk, and the risk from the punch-and-drain operation. These factors can be exacerbated or ameliorated by unique aspects of a technology. For example, if the SBV treatment of ton containers is implemented, the risks in handling ton containers will differ somewhat from the risks for a technology that uses hot water and decontamination solution to bring the containers to a 3X condition.

The process-safety risk factors inherent in GPCR include safety issues associated with high-temperature hydrogen, hot water and corrosives in the scrubbers, and secondary containment. Many of the risk factors that are not specific to chemical agent have been addressed by the TPC in safety analyses and in hazard and operability reports.

The panel found no failure scenarios involving a loss of electrical power, loss of cooling, failures of pumps and valves, inadvertent overpressurization, or inadvertent temperature transients that would lead to off-site releases of agent or toxic process products. Based on the panel's preliminary and qualitative evaluation, the most significant off-site risk appears to be associated with handling agent prior to the agent-destruction process. The principal risk factors appear to involve mishaps in the punch-and-drain operation or damage from airplane crashes or other external events to holding tanks where agent is stored before being fed to the main reactor.

The following subsections on process safety address risk factors specific to the GPCR technology. The panel expects that the safety issues discussed below can be resolved through further design and demonstration prior to constructing a full-scale facility.

Off-Site Safety Issues

The following issues should be addressed fully and clearly in a final GPCR process design.

Hydrogen and Other Combustible Gases

The process uses hydrogen. In addition to the hydrogen circulating in the process gas stream, most of which is produced in the steam reformer from methane produced in the main reactor, compressed hydrogen is stored in tube trailers and in the product gas tank. Other combustible gases (carbon monoxide, propane) are also present and must be considered.

Hydrogen is commonly used in industry and can be used safely. Recent industrial accidents involving hydrogen are rare because of the care taken to handle it properly. The issue is mentioned here only because of the potential for a hydrogen explosion or fire to cause grave damage to personnel and structures if the hydrogen is not managed properly. Also, a hydrogen explosion could lead to a release of chemical agent. Although leaks of flammable gases are a risk factor for worker safety, they are not currently an off-site risk factor in either of the TPC's two current commercial operations. This was discussed above under Process Instrumentation and Controls.

When a GPCR system is housed in secondary containment (as is required for agent-destruction facilities), the potential increases for buildup of an explosive concentration of hydrogen. The potential increases for damage to agent-bearing structures from an explosion or fire. The containment of the system for an agent-destruction facility will need to be designed so that the hydrogen will neither stratify nor build up locally to a combustible concentration.

A large detonation or burn near the containers that store the agent could damage containment structures and cause a release of agent. This risk factor should be considered when designing component locations and shielding. The proximity of the hydrogen tube trailer, product gas tank, or any other combustible storage area to the agent-containing components (holding tank, reactor, SBV) is very important.

Another risk factor that must be considered is combustible mixtures of air and hydrogen inside the circulating-gas system, which could result from air leaking into the system combined with flow imbalances. The current system does have design features and controls in place that address this risk in appropriate ways.

Decontamination of Ton Containers in the SBV

Extremely large doors are needed on the SBV to insert and remove ton containers. These large doors must be sealed tight to prevent leakage of agent and hydrogen. Proper sealing of the SBV at high temperatures can be an engineering challenge. In past operations, two types of seals have been used: glass-fiber gaskets and U-shaped silicon rubber seals with nitrogen gas pumped into the seal. The TPC has stated that it will probably redesign the seals for additional reliability in applying the SBV technology to agent destruction.

Design of the Reactor Vessel

The design of the reactor vessel needs to consider thermal stresses, welding problems, crevices, and local design problems. These issues, however, are no different for an agent-processing facility than for the waste-treatment facilities that the TPC has already piloted and run commercially.

Worker Safety Issues

There are a number of worker safety issues associated with high temperature hydrogen, high temperature steam, hot water and corrosives in the scrubbers, and secondary containment (concerning both inadvertent leaks and maintenance activities). These risk factors need to be addressed in the final operational design. The status of the technology with respect to these risk factors and the nature of the risks have been discussed above.

Specific Characteristics that Reduce Risk Inherent in the Design

The system operates at low pressure, and it appears to be extremely difficult to overpressurize the system inadvertently. Upon slight overpressure, the reactor is relieved to the caustic scrubber through an 8-inch pipe. The SBV chambers are relieved in a similar manner. There are no apparent ways for the reactor or SBV to fail because of overpressure; there are no valves between either the reactor or the SBV and the pressure relief mechanisms.

Loss of electrical power, failure of cooling water to the heat exchanger, or failure of cooling to pumps will result in a "graceful" shutdown of the system. The integrity of the system does not appear to be threatened in any realistic failure scenarios.

SCHEDULE

The TPC has stated, "the schedules for design, construction, testing and evaluation of a pilot-scale system have been requested by the Army and will be provided according to their requirements." The TPC states that the time for facility construction is about 6 months, with systemization taking another three months. In its submission, the TPC assumed that the Army will provide the secondary-containment building and ancillary non-process facilities.

Although the reactor, feed systems, and steam reformer have been deployed commercially, the lack of details on scrubbers and on handling phosphorus-containing materials could mean further development is necessary. Other than the increases in monitoring requirements, the design of secondary containment, and the engineering necessary for managing the sulfur and phosphorus wastes, this technology is at the point where a unit like the existing commercial systems could serve as the pilot operation for agent-destruction. Still to be assessed are the effects on the schedule of designing the secondary containment and any associated reengineering. The effect on schedule is likely to be more severe for VX than for HD because the need for identifying and managing phosphorus-containing reaction products applies only to VX.

7

Neutralization Technology for Mustard Agent HD

The NRC Stockpile Committee recommended that the Army accelerate research and development on neutralization-based technologies for the destruction of chemical agents, particularly for use at sites where bulk agents are stored (NRC, 1994b). Neutralization[1] employs process conditions that are specific for each type of agent. Thus, a neutralization process for destroying a specific agent or class of agents would not be suitable for treating a wide range of other wastes (e.g., commercial hazardous wastes).

The virtue of neutralization is that it detoxifies HD agent rapidly at low temperature and low pressure. Batch or semibatch processing allows retention of the products from neutralization until testing can verify destruction of the chemical agent. Bench-scale testing indicates that most of the processing equipment for a neutralization process is commercially available. The ability to use equipment already being used in the chemical industry should minimize the time and cost of construction and process startup. The use of standard equipment should also enhance the reliability and ease of maintenance of the facility.

The U.S. Army, like the defense ministries of other nations, has evaluated many different approaches to the neutralization of HD (NRC, 1993; Yang et al., 1992). Intensive testing in the past two years has led to selection of direct hydrolysis with hot water followed by biodegradation of the hydrolysis product as the best candidate for scale-up to a pilot plant demonstration (U.S. Army, 1996b). Within the Army, the Alternative Technologies Program was established to pursue the testing and development of neutralization alternatives. With respect to the AltTech Panel's evaluation of alternative technologies, this Army program office, much like the companies have whose technologies are described in Chapters 4 through 6, have functioned as the technology proponent. For the remainder of this chapter, the Army Alternative Technologies Program will be referenced as the TPC (technology proponent company).

Proposed pilot-scale testing at the Aberdeen site would consist of a single process train, which subsequently would be replicated to scale up to the full-scale destruction facility for that site. Thus, successful pilot-scale testing would directly provide the technical basis for constructing and operating a full-scale facility at Aberdeen for disposal of the HD agent stored there.

BACKGROUND TO PROCESS CONFIGURATIONS

Although neutralization of HD detoxifies the agent, the resulting hydrolysate requires further treatment prior to final disposal. Treatment of the hydrolysate must destroy both thiodiglycol, which is the major residual in the hydrolysate, and chlorinated volatile organic compounds (VOCs), which originate as impurities in the HD (see Chapter 1). Management of hydrolysate from HD neutralization may be either on site, through additional treatment following the neutralization process, or off site, by shipping the hydrolysate to a permitted waste-management facility—a RCRA TSDF (treatment, storage or disposal facility). On-site treatment requires substantially more complex processing than does the neutralization process alone. The primary process considered for on-site treatment of hydrolysate is biodegradation. Aqueous effluent from an on-site biodegradation process potentially could be discharged to the existing federally owned treatment works (FOTW) at Aberdeen or recycled as process water.[2]

[1]In the context of this report, *neutralization* refers to the chemical hydrolysis of an agent to produce less toxic residues. *Hydrolysate* refers to the effluent from a neutralization process. *Biodegradation* refers to the use of microorganisms to further detoxify the products of neutralization. The biodegradation processes considered for use with HD hydrolysate would convert most organic carbon compounds in the hydrolysate to CO_2 and bacterial cell mass. Sulfur present in the hydrolysate is converted to sulfate.

[2]The FOTW at Aberdeen, Maryland, is a wastewater treatment facility that receives wastewater from several sources.

The processing options are further complicated by the possibilities for treating VOCs and recycling water. Separating and treating VOCs prior to on-site biological treatment of the hydrolysate is necessary because the process configuration selected for on-site biodegradation in sequencing batch bioreactors (SBRs) would cause the VOCs to be air-stripped from the hydrolysate and subsequently adsorbed onto the activated carbon filters, rather than being biodegraded. This outcome would result in an unacceptably high rate of use of activated carbon. The Army has proposed using photochemical oxidation to destroy VOCs during on-site treatment of hydrolysate. The primary process considered for off-site management of hydrolysate is shipping it to an off-site TSDF that includes biodegradation as a process step. VOCs in the hydrolysate would also be treated at the TSDF.

Selection of the specific process sequence for use at the Aberdeen site requires consideration of Army programmatic requirements, requirements for shipment to commercial wastewater management facilities or discharge to a FOTW, and regulatory constraints (i.e., permitting requirements). Currently, the policy of the program office for the CSDP (Chemical Stockpile Disposal Program) requires that no liquid effluents be discharged from an agent-destruction facility. This policy would have to be modified to make possible either off-site management of the hydrolysate or use of biodegradation followed by discharge of the effluent to a FOTW. In recognition of these policy limitations, the TPC has developed a process configuration of neutralization followed by biodegradation that requires neither shipment of hydrolysate for off-site treatment nor discharge of effluents to a FOTW. However, in the interests of process simplicity and cost-effectiveness, the TPC also has developed several simplified process configurations that may be implemented if CSDP policy is revised. The primary process options are (1) discharging liquids from the process or not and (2) on-site or off-site biodegradation of the hydrolysate.

There are four primary neutralization process configurations under consideration by the Army. Configuration 1 (Figure 7-1) is neutralization followed by biodegradation with on-site water recycling and photochemical

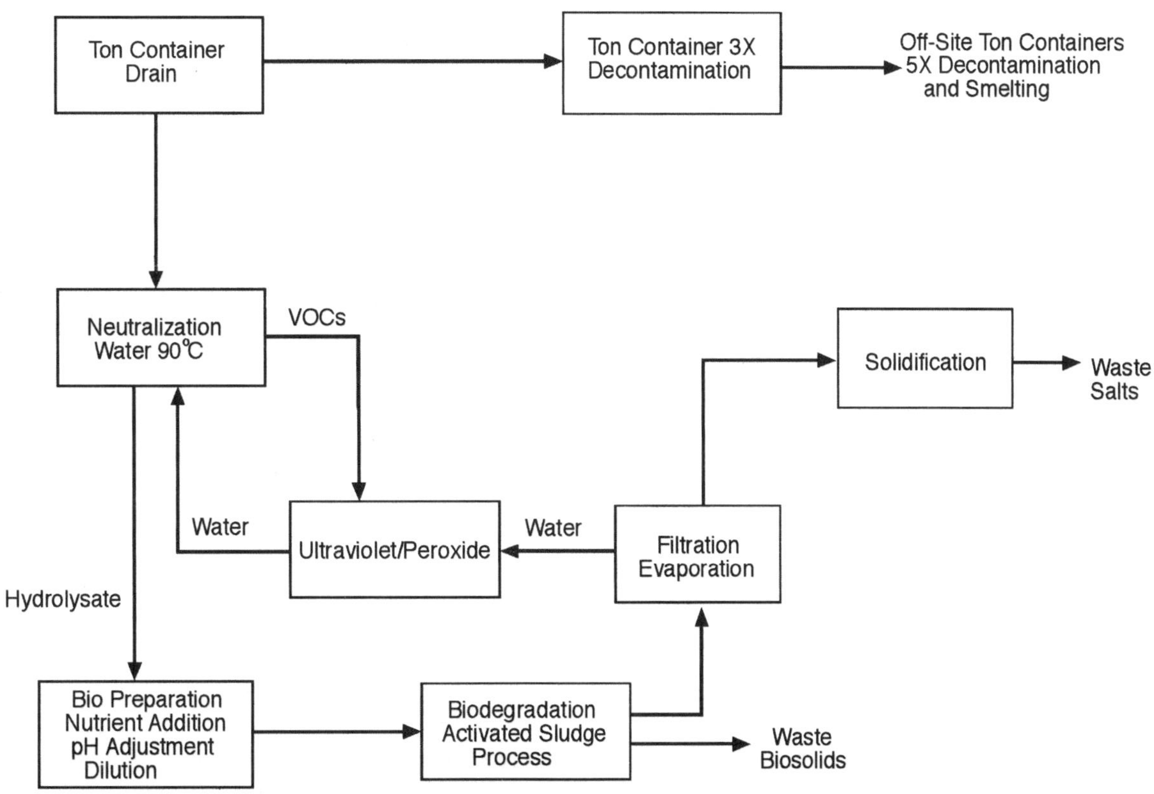

FIGURE 7-1 Process Configuration 1: Neutralization followed by on-site biodegradation, including water recycling and photochemical oxidation of VOCs.

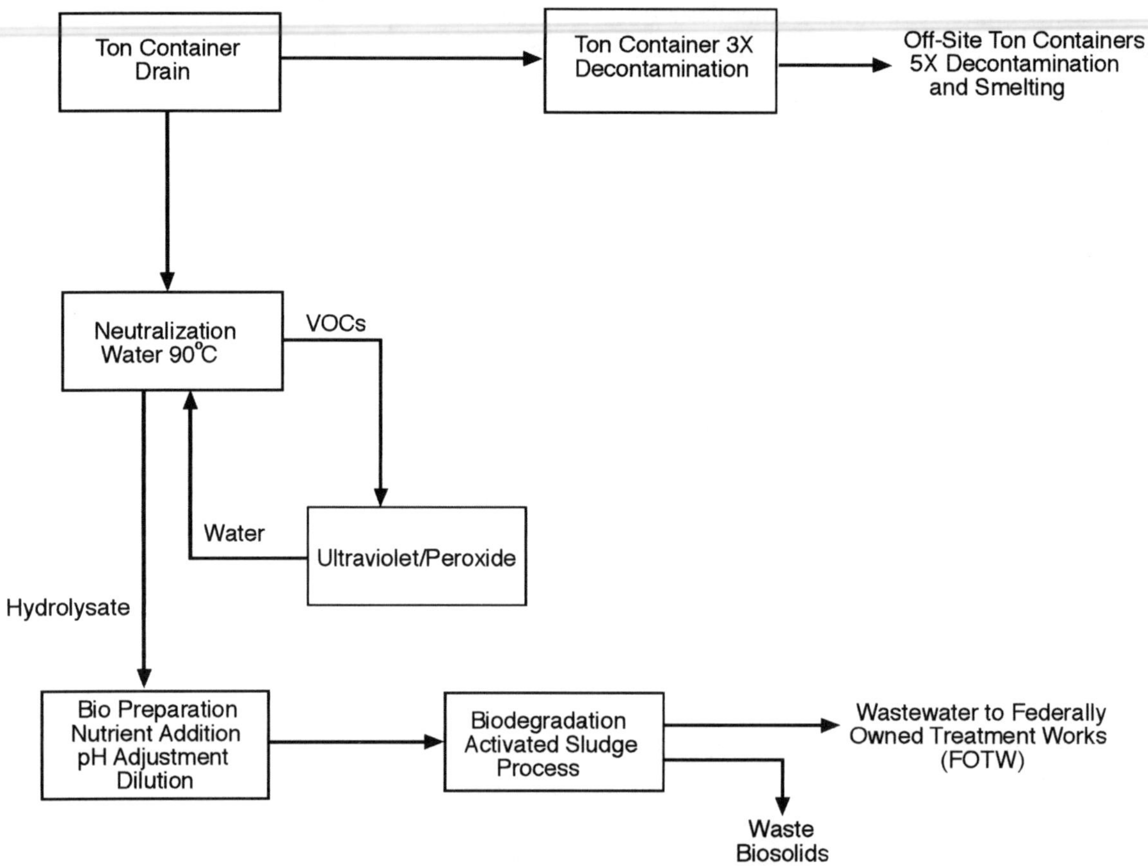

FIGURE 7-2 Process Configuration 2: Neutralization followed by on-site biodegradation. VOCs are treated by photochemical oxidation. Biodegradation process effluent is discharged to a FOTW.

oxidation to destroy VOCs. This configuration meets the current CSDP policy of discharging no liquid process effluents and fulfills treaty requirements under the CWC (Chemical Weapons Convention).

Configuration 2 (Figure 7-2) is neutralization followed by biodegradation, with process effluents discharged to a FOTW. Photochemical oxidation is used to destroy VOCs to FOTW standards, but water recycling is not used. This configuration fulfills CWC requirements (i.e., it destroys "scheduled precursors") while neutralization effluents are under Army control. The primary difference between configuration 1 and configuration 2 is how process water is managed. In configuration 1, process water is recycled; excess water is lost by evaporation in the cooling tower for water recycle and in the air discharged from biological treatment. In configuration 2, process water is used once (that is, it is not recycled) and then discharged in the aqueous process residual stream to a FOTW.

Configuration 3 (Figure 7-3) is neutralization followed by biodegradation, with process effluents discharged to a FOTW, but VOCs are separated from the hydrolysate and shipped to a TSDF for subsequent treatment and disposal. This configuration also meets CWC requirements by destroying scheduled precursors while neutralization effluents are under Army control, but the process is simplified by eliminating the photochemical oxidation step retained in configuration 2.

Configuration 4 (Figure 7-4) is neutralization followed by shipping the hydrolysate to a TSDF. VOCs in the hydrolysate would be treated at the receiving TSDF in accordance with permit requirements. This is the simplest configuration but requires acceptance of the hydrolysate by a TSDF. Accepting the hydrolysate would subject the TSDF to inspection under the verification requirements of the CWC because destruction of a scheduled precursor (thiodiglycol) would occur at the commercial facility. Configurations 2, 3, and 4 would all require modification of CSDP policy and could have different regulatory permitting requirements.

In a data submission to the panel late in the panel's review process, the TPC chose configuration 2 as its

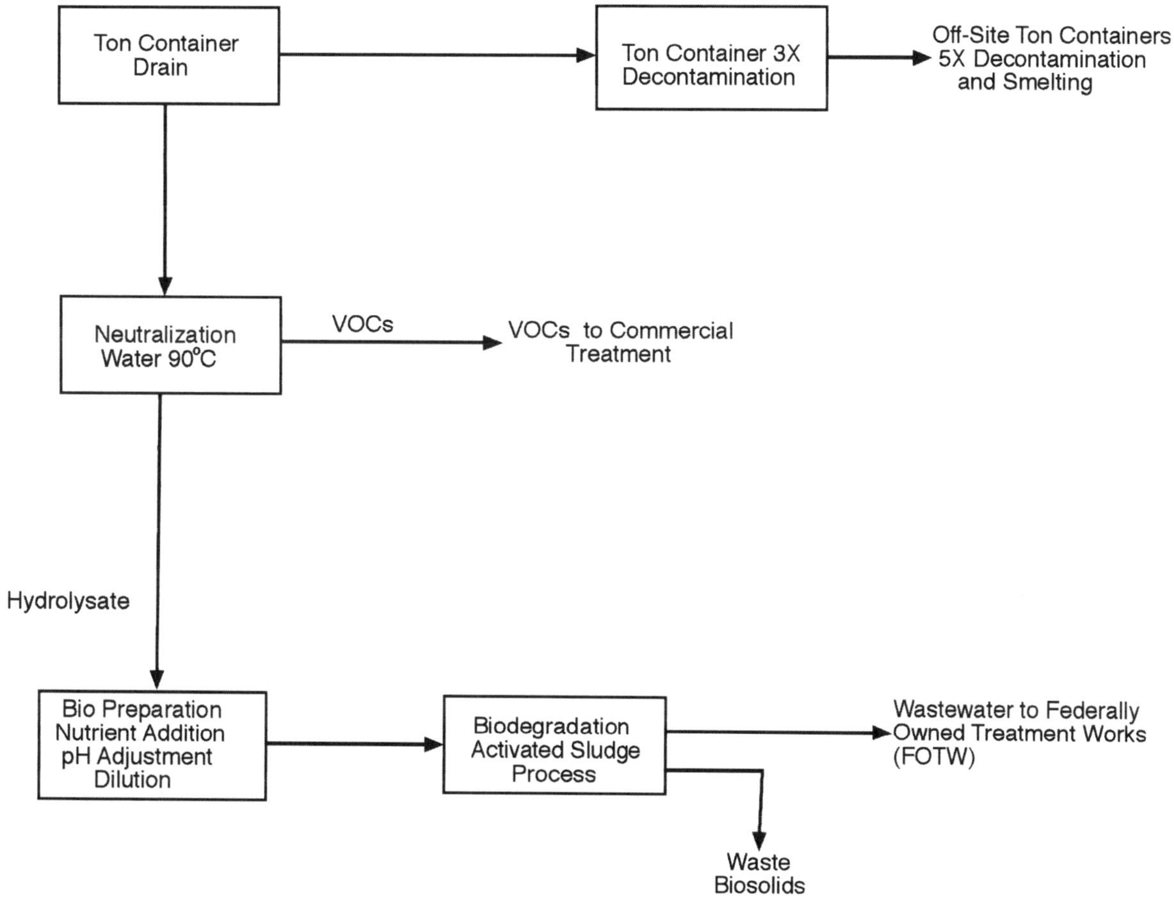

FIGURE 7-3 Process Configuration 3: Neutralization followed by on-site biodegradation. VOCs are shipped to an off-site TSDF. Biodegradation process effluent is discharged to a FOTW.

preferred candidate for development (U.S. Army, 1996b). However, to provide a complete evaluation of the TPCs technology in this chapter, configuration 1 of the proposed neutralization process is described and discussed in the most detail. Configurations 2, 3 and 4 are simplifications achieved by eliminating specific components of configuration 1 without requiring major modifications of other process steps. These simplified configurations are discussed by comparing by process flow diagrams and mass balances in the appropriate sections of this chapter. The neutralization technology submitted by the TPC for destroying VX nerve agent is discussed in Chapter 8.

PROCESS DESCRIPTION

Figure 7-1 shows the several steps involved in the configuration 1 process for neutralization of HD (U.S. Army, 1996b). The ton container, which contains the agent, is drained into an agent holding tank. The HD then is neutralized by vigorous mixing in water at 90°C (194°F) using a 4 wt pct feed of HD. At the beginning of a reaction batch, the reactor initially contains most of the required hot water. HD is added to the reactor over a period of one hour to minimize the quantity of unreacted agent in the reactor at any given time. The reaction completely destroys the HD and is 90 percent selective to formation of thiodiglycol. During the neutralization reaction, 2 wt pct of hydrochloric acid is produced, resulting in acidic reaction conditions.

Once the reaction is complete, sodium hydroxide (prepared as an 18 wt pct solution) is added to adjust the pH to 12. The dilute processing of HD and the addition of sodium hydroxide after completion of the neutralization reaction are designed to minimize the production of unwanted by-products during reaction. Laboratory testing has indicated that either increased loading of HD (up to about 10 wt pct) during neutralization or adding sodium hydroxide to the hot water prior to introduction

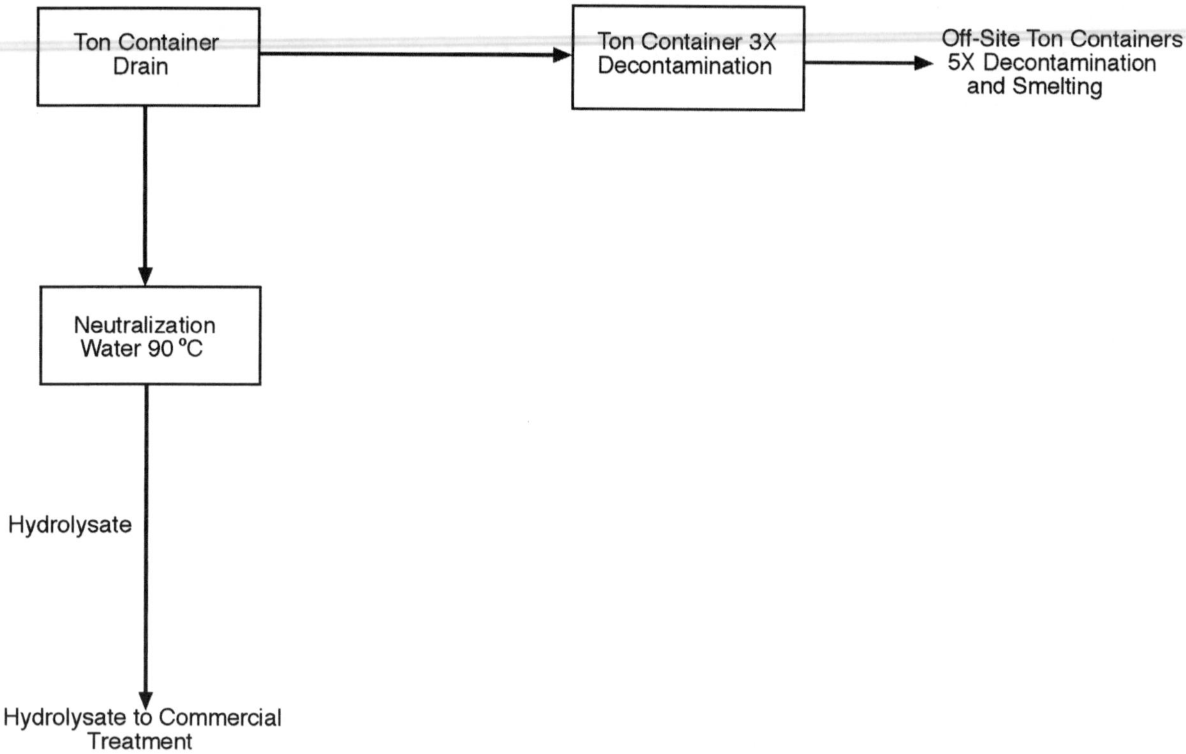

FIGURE 7-4 Process Configuration 4: Neutralization followed by off-site biodegradation of the hydrolysate at a TSDF. VOCs remain in the hydrolysate.

of HD (rather than after the neutralization reaction is completed) results in lower yields of thiodiglycol and increased concentrations of reaction by-products in the hydrolysate (U.S. Army, 1996b).

The solution resulting from the addition of sodium hydroxide also contains minor amounts of organic impurities that were present in the stored agent and metal salts from corrosion of the storage container or the processing equipment used in the manufacture of the HD (see Chapter 1 for details on impurities in HD). After testing to ensure agent destruction to less than 200 ppb HD,[3] the hydrolysate is transferred from the toxic control cubicle for further treatment.

In configuration 4, at this point, the hydrolysate would be shipped to a TSDF. In configurations 1, 2, and 3, the hydrolysate is partially evaporated to remove VOCs such as chlorinated ethylenes. The resulting aqueous condensate contains approximately 0.2 wt pct VOCs. The VOCs are then either shipped to a TSDF (configuration 3) or passed through a photochemical oxidation unit in which hydrogen peroxide is added and the solution is irradiated with ultraviolet light to destroy the organic compounds (configurations 1 and 2).

In configurations 1, 2, and 3, the aqueous solution from the evaporator bottom is adjusted to neutral pH and fed to SBRs which reduce the dissolved organic carbon content by 90 percent and destroy more than 99 percent of the thiodiglycol. This level of destruction meets the CWC definition for destruction of scheduled precursors. The vapor stream from the SBRs is filtered through activated carbon to control odors and remove traces of organic contaminants; it is then released to the atmosphere. The biomass (a thick slurry of solid organic material in water) from the SBRs is fed to an aerobic digester to reduce the volume, then dewatered in a filter press and disposed of as a hazardous waste under current Maryland requirements, even though the dewatered sludge does not contain hazardous concentrations of any constituents. Delisting of this material may allow it to

[3]The current detection limit for HD in hydrolysate is 10 ppb. Destruction of HD to less than 200 ppb represents a destruction efficiency of greater than 99.9995 percent; destruction to less than 10 ppb represents a DRE greater than 99.99997 percent. The calculated destruction efficiencies are independent of the 25-fold dilution of HD with water that takes place in the neutralization process.

be disposed of as a nonhazardous solid waste.[4] The liquid effluent from the SBRs is filtered and either recycled through the evaporator (configuration 1) or discharged to a FOTW (configurations 2 and 3).

The bottom stream from the evaporator consists of salt brines that are mixed with solidifiers (e.g., cement) and packaged for off-site disposal in a landfill. The evaporator distillate (overhead stream), which is predominantly water containing with low levels of organic impurities, is recycled to the neutralization operation for use in diluting the next batch of HD.

After HD is drained from a ton container, the empty container is flushed with hot water, cut in half, and cleaned with hot, high pressure water and steam. The cleaned container is then monitored to ensure adequate decontamination from agent and sent to Rock Island Arsenal, Illinois, to be melted.[5] The liquid effluent from this cleaning process is used in the neutralization process to replace part of the required process water.

The vapors from the ton container cleanout process, neutralization reactors, and hydrolysate holding tanks are all passed through a single caustic scrubber. Then they are reheated to reduce the relative humidity of the gas; filtered through activated-carbon beds, which serve as guard beds to ensure there is no release of toxic organic vapors; filtered through the plant-ventilation activated-carbon filter beds; and finally discharged to the atmosphere.

SCIENTIFIC PRINCIPLES

The neutralization process proposed for the disposal of HD is, in principle, a simple hydrolysis, that is, a reaction with water to form thiodiglycol (bis(2-hydroxyethyl) sulfide) and hydrochloric acid:

$$S(CH_2CH_2Cl)_2 + H_2O \leftrightarrow S(CH_2\,CH_2OH)_2 + 2HCl$$

Even though HD is only slightly soluble in water, the C-Cl bonds, which are essential to mustard's toxicity, react readily in hot water to produce the relatively innocuous thiodiglycol. Pure agent reacts with neutral or acidic water predominantly as shown in the equation above, although the detailed reaction mechanism, as presented in Figure 7-5, is complex (U.S. Army, 1996b; Yang, 1995). The reaction is carried out in hot water, with the final hydrolysate being a dilute aqueous solution (e.g., less than 10 wt pct hydrolyzed agent) to minimize the production of unwanted by-products such as sulfonium ions (R_3S^+ where R is an organic constituent). The hydrolysis reactions are exothermic, releasing about 15 kilocalories per mole of HD in the neutral-to-acidic hydrolysis (U.S. Army, 1996b).

Under alkaline conditions, much the same chemistry occurs, but it is accompanied by side reactions that give rise to many minor products, some of which are undesirable. Therefore, sodium hydroxide is not added until after the initial reaction, when it is used to neutralize the hydrochloric acid formed in the hydrolysate from the reaction of mustard with water and to react any remaining sulfonium ions. As implied by the equation, the hydrolysis is, in principle, reversible. But the reformation of mustard agent is prevented by adding sodium hydroxide to make the hydrolysate alkaline.

Munitions-grade mustard agent contains several impurities that are formed during manufacture. Even the distilled agent (HD) stored at the Aberdeen Proving Grounds is only 85 to 95 percent pure (U.S. Army, 1996b). Several significant impurities—dithiane and chlorinated ethanes—do not react extensively with water under standard hydrolysis conditions, and they remain in the hydrolysate.

On the basis of extensive laboratory and bench scale testing (Irvine et al., in press), biodegradation has been selected by the TPC as a preferred treatment for the hydrolysate (U.S. Army, 1996b). Microorganisms in sewage sludge can adapt to using thiodiglycol as a primary energy and carbon source. Biodegradation of thiodiglycol requires adjusting the pH of the hydrolysate to neutral by adding sodium bicarbonate buffer. Aqueous ammonia is added as a nitrogen source, phosphoric acid as a phosphorus source, and mineral salts as trace nutrients. The bacteria oxidize thiodiglycol efficiently to carbon dioxide, water, and sulfate with high efficiency, as expressed in the idealized equation:

$$S(CH_2CH_2OH)_2 + O_2 \rightarrow CO_2 + H_2O + H_2SO_4 + \text{biomass}$$

During actual operation, approximately 0.8 g of cell mass (dry weight basis) is produced for every 1 g of organic carbon removed from solution. Excess bacterial

[4]Delisting is a regulatory process by which a solid waste that has been classified as a hazardous waste based on its origin is demonstrated not to be hazardous. The delisted waste may then be disposed of at waste management facilities designated for nonhazardous wastes.

[5]The required level of decontamination is specified as a 3X condition (see Capacity to Detoxify Agent in Chapter 2).

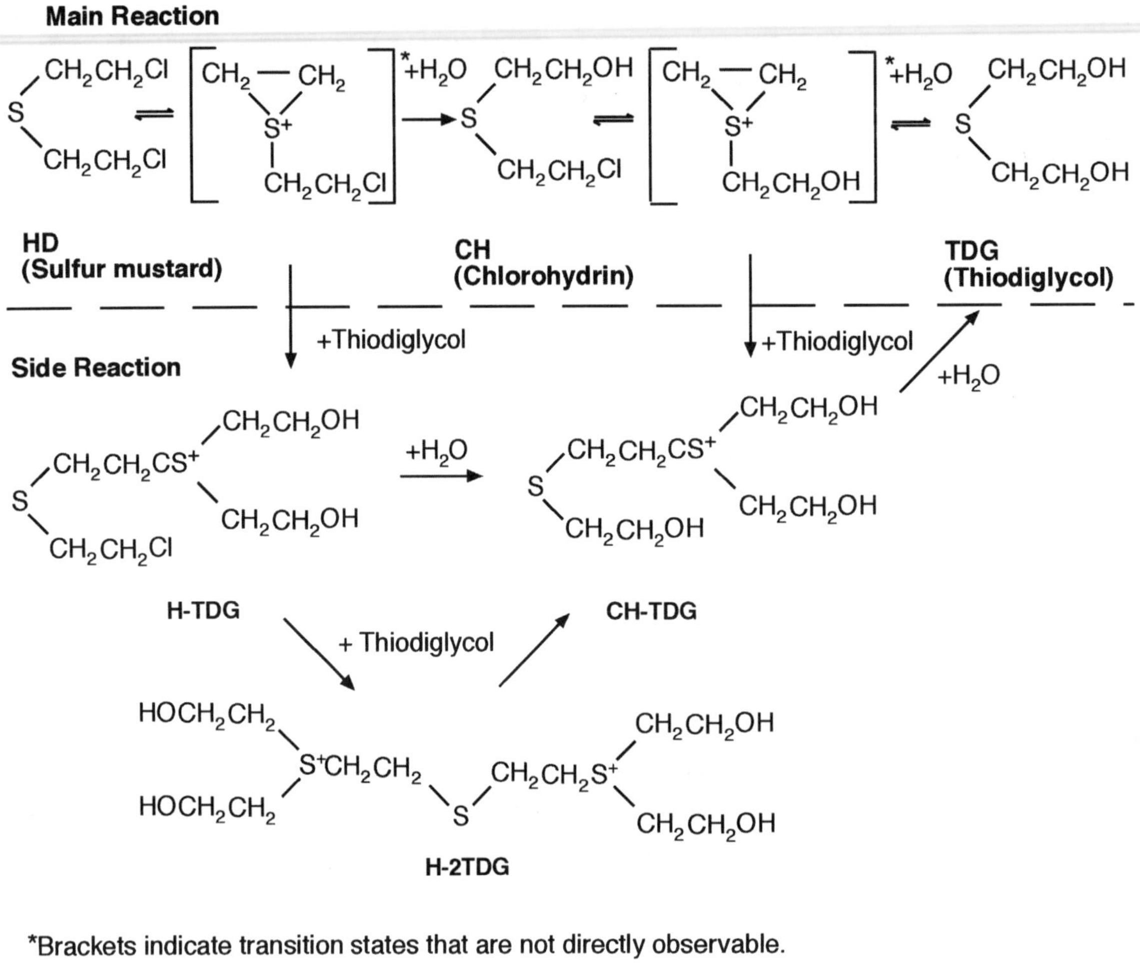

FIGURE 7-5 Chemical reactions during the hydrolysis of HD. Thiodiglycol is the primary product, and side reactions are minimized by maintaining HD as a dilute component in water during reaction. Sulfonium ions are indicated by S^+. Source: U.S. Army, 1996b.

cell mass is separated from the biodegradation process effluent. This biomass is further oxidized (degraded) through aerobic digestion, dried, and disposed of at a commercial TSDF. Fortunately, dithiane and the least volatile chlorinated ethane, hexachloroethane, are oxidized along with the thiodiglycol (U.S. Army, 1996b).

The TPC has proposed photochemical oxidation as a polishing step to destroy VOCs that were present in the HD and remained in the hydrolysate after neutralization. Because of their volatility relative to water, VOCs can be removed from the hydrolysate by a stripper unit. Volatilized VOCs and water vapor are then condensed, and the resulting condensate is photochemically oxidized by adding hydrogen peroxide in the presence of ultraviolet light. Under these conditions, the VOCs are both directly degraded by photolytic dissociation and oxidized by HO radicals, which are formed by the photochemical dissociation of hydrogen peroxide (Solarchem Environmental Systems, 1996). The products of the photochemical oxidation are simple organic compounds (aliphatic organic acids such as acetic acid), chlorides, and carbon dioxide. The panel's evaluation of the specific photochemical oxidation process proposed by the TPC was limited to reviewing the treatability study provided by the TPC and applying the prior experience of panel members with photochemical oxidation processes.

Overall, the neutralization, biodegradation, and photochemical oxidation operations yield a relatively simple set of final products: carbon dioxide, water, chloride and sulfate salts, metal salts (originally present in the HD), and biomass.

TECHNOLOGY STATUS

Hydrolysis of HD

Alkaline hydrolysis has been used extensively to detoxify mustard agents and the G family of nerve agents. (Application to nerve agents is discussed in Chapter 8.) Munitions-grade mustard agents have been hydrolyzed with methanolic NaOH on a pilot scale in Poland (Koch and Wertejuk, 1995). This procedure was effective for liquid agent and for solids that remained in the storage containers. The methanol solutions of hydrolyzed agent were incinerated. In Canada in the mid-1970s, mustard agent was hydrolyzed on a production scale (8-ton batches) with hot lime water prior to incineration (NRC, 1993, p. 63). Recent laboratory evaluations of the alkaline hydrolysis of HD by the U.S. Army have not shown any particular advantages in using lime instead of the more soluble sodium hydroxide (Harvey et al., 1994). Lime most likely was used in the Canadian hydrolysis process to avoid high concentrations of sodium in the hydrolysate during subsequent incineration, because sodium attacks common refractories.

Simple hydrolysis of mustard with hot water is not as well documented as alkaline hydrolysis, but it has apparently been used in France (Harvey, 1995) and is the basis for the long-used method of steam cleaning and decontaminating storage containers. Although hydrolysis with water under neutral or acidic conditions is slowed by the limited solubility of the agent, the reaction proceeds well at low concentrations (1 to 10 percent agent in water). With vigorous agitation and temperatures of 75°C to 90°C, the reaction is essentially complete in one hour, as was demonstrated in a laboratory-scale experiment witnessed by panel members.

The laboratory-scale tests were initially conducted in small glassware but were later scaled up to 1-liter flasks to permit hydrolysis of about 10 g of HD at a concentration of 1.3 wt pct or 67 g at 8.6 wt pct. Tests at this scale were used to optimize conditions for tests in bench-scale reactors. Experiments in 2-liter Mettler reactors (up to 150 g of HD per test at 9.3 wt pct) yielded precise thermodynamic data that guided larger tests and design studies (U.S. Army, 1996b).

Subsequent tests in a 114-liter reactor (U. S. Army, 1996b) provided valuable operating experience and basic engineering data on a scale that can be easily extrapolated for designing the pilot- and production-scale reactors. In addition, these experiments produced substantial volumes of hydrolysate for bench-scale biodegradation studies. In a typical run, 7.22 kg of munitions-grade HD was hydrolyzed to produce 88 liters of hydrolysate. When the 114-liter reactor was converted for VX hydrolysis studies (see Chapter 8), HD hydrolysate for ongoing research on biodegradation was supplied by a larger reactor fabricated from a 55-gallon (208-liter) drum. The stainless steel drum was lined with polypropylene and fitted with an efficient stirrer coated with Kynar™ resin. Operations with this reactor provided about 130 liters per run of hydrolysate, which derived from HD concentrations in the reaction mixture of either 5.7 kg HD at 3.8 wt pct or 13 kg at 8.6 wt pct. In addition to producing hydrolysate for the biodegradation studies, the runs in this reactor provided useful experimental data on the rate of HD disappearance under conditions similar to conditions expected in full-scale operations (Harvey et al., 1996).

These large scale tests showed that the reaction proceeds cleanly with thiodiglycol as the primary reaction product. The residual HD concentrations in the hydrolysate from these experiments dropped below 200 ppb in less than 20 minutes, and the toxicity was dramatically reduced (see Agent Detoxification) (Harvey et al., 1996). Processing at bench scale also has demonstrated the successful destruction of HD present in the heel removed from a ton container.[6] Thus, the neutralization process has been demonstrated to work well for the distilled HD stored at the Aberdeen site. Overall, 161 kg of HD were destroyed in the laboratory and bench scale studies at Aberdeen (Novad, 1996).

In the past, incineration has been the principal process used for disposing of wastes from the alkaline hydrolysis of mustard agents, as described above for the Polish and Canadian methods. To facilitate public acceptance of neutralization-based technology, the TPC has studied the biodegradation of thiodiglycol and other organic compounds obtained from the neutralization of HD. Previous attempts to use microorganisms to destroy HD itself failed because the agent is toxic to all life forms tested. However, various bacterial cultures readily oxidize thiodiglycol (Zulty et al., 1994).

---

[6]The heel is a sludge that does not freely drain from a ton container; it consists primarily of iron oxides and a cyclic sulfonium salt.

Biodegradation of Hydrolysate

Laboratory-Scale Tests

The biological treatment of hydrolysate has been extensively tested at laboratory scale using SBRs with a 1- to 12-liter working volume (Irvine et al., in press). The primary objectives of the laboratory testing were to determine treatment efficiency, the quality of effluent from the process, the optimum operating conditions, and the effects of HD impurities (e.g., ton container heel and iron floc). More than 500 days of continuous bioreactor operation were completed during laboratory-scale testing. Individual bioreactors successfully operated continuously for durations of 130 to 159 days. Unacclimated and acclimated mixed bacterial cultures were evaluated for treating hydrolysate produced from the neutralization of 1.27 to 9.5 wt pct HD. The feed to the SBRs was diluted to the equivalent of 1.27 wt pct HD for all tests. Good results were obtained with mixed cultures such as those obtained from the Back River municipal waste treatment plant in Baltimore, Maryland. Thiodiglycol removal was greater than 99 percent in almost all cases, and the mean TOC (total organic carbon) removal ranged from 86 percent to greater than 92 percent, depending on the specific operating conditions. The effect of the concentration at which HD was hydrolyzed was minimal for hydrolysis concentrations up to 8.6 wt pct HD. An operating regime of 10 days hydraulic residence time (HRT), 15 days solids residence time (SRT), and hydrolysate organic carbon loading of 0.08 to 0.1 g TOC/g MLSS-day[7] (mixed liquors suspended solids) was repeatably demonstrated to remove greater than 99 percent of the thiodiglycol and greater than 90 percent of the TOC. The SBRs were operated continuously for several months and demonstrated stable operation at temperatures between 8°C and 35°C. No significant detrimental effect was observed as a consequence of the iron floc in the hydrolysate feed. Hydrolysate toxicity was reduced by a factor of 50 based on Microtox assays.

Bench-Scale Tests

Based on the laboratory findings described above, the biodegradation of hydrolysate was demonstrated on an 80-liter scale (bench scale) in a SBR (U.S. Army, 1996e). Three 80-liter test cases for biodegradation of hydrolysate were conducted, with each case in continuous operation for more than 30 days. Unacclimated biomass from the Back River treatment plant in Baltimore was used as the seed population for each test. In each case, the feed was diluted to about 99 percent water for biotreatment regardless of the starting concentration of HD in the preceding hydrolysis step. The reactor was operated successfully with hydrolysate from 3.81 wt pct HD, a 10-day HRT, and 24-day SRT. Thiodiglycol removal was greater than 99 percent and mean chemical oxygen demand (COD) removal was greater then 90 percent. Operation with hydrolysate from 8.49 wt pct HD (13.5-day HRT, 56-day SRT) resulted in thiodiglycol removal of greater than 99 percent and mean COD removal of 88 percent. Operation with hydrolysate from 7.9 wt pct HD was unsuccessful in that poor biomass settling resulted in a gradual decline in the efficiency of biodegradation. However, the cause of the poor settling characteristics is not known. Startup operation of the SBR was not replicated in these tests, which were run after an initial startup period for the SBR.

Successful operation was defined as meeting removal targets consistently for at least 3 HRTs or 1 SRT, whichever was longer. The results suggest that the operations can succeed with hydrolysate from about 4 wt pct HD and, potentially, with hydrolysate from up to 8.6 wt pct HD. The panel suggests further bench-scale testing of SBRs with at least 3 SRTs at design HRT conditions and up to 8.6 wt pct HD hydrolysate to define the most beneficial operating regimes. If biodegradation of hydrolysate from 8.6 wt pct HD can be successfully demonstrated, schedule and cost may be reduced; for example, either smaller neutralization equipment or fewer HD neutralization batches would be needed.

Overall, the results of these tests are encouraging. Bioassay testing of the aqueous effluent from the biodegradation process indicated that the effluent had low toxicity (see section below on Agent Detoxification and Consistency of Standards). Final disposal must be decided upon for the products of the biodegradation, which consist of the biomass sludge from bacterial growth and an aqueous solution of sodium chloride, sodium sulfate, and low levels of organics. Full-scale operation of SBRs for treating industrial wastewater has been demonstrated extensively; they have been in commercial operation for several years (Irvine and Ketchum, 1988; Brenner et al., 1992). Reduction of the biomass sludge by aerobic digestion and dewatering are standard

[7] The unit of "g TOC/g MLSS-day" is a standard measure of the rate (per day) of substrate carbon loading (TOC) per unit amount of biomass in the bioreactor (g MLSS).

processes in municipal and industrial wastewater treatment facilities (Metcalf & Eddy, Inc., 1979).

Off-Site Biodegradation Options

Off-site biodegradation (configuration 4) would most reasonably occur at a commercial TSDF that receives wastewater from many sources and includes multiple processing steps. HD hydrolysate would therefore be a small contribution to the total loading at the TSDF. A survey of TSDFs carried out by the TPC indicated that several potentially could process HD hydrolysate (U.S. Army, 1996c).

The panel anticipates that thiodiglycol would be biodegraded at a TSDF, after microbial acclimation, as efficiently as in the SBR tests described above. Biological treatment at a TSDF that receives wastewater from a variety of sources also may be more tolerant than the SBRs of the side products that arise from neutralization at higher HD concentrations (e.g., 8 to 10 wt pct HD). Laboratory scale (5 to 12 liter) testing carried out by the Army (U.S. Army, 1996i) and a contractor (SBR Technologies, 1996) have successfully treated hydrolysate produced from HD concentrations of up to 9.5 wt pct. Unsuccessful operation of a bench-scale (80 liter) test of biodegradation of hydrolysate produced from an HD concentration of 7.9 wt pct was hypothesized to be the result of an increased concentration of sulfonium ions in the hydrolysate, produced by the higher HD concentration during neutralization. In a large TSDF that receives wastewater from multiple sources, the increased sulfonium ion concentrations would be diluted below the level that could upset treatment. In addition, the diversity of the microbial population would be greater at a TSDF, making the process more resilient.

Results from testing the fate of VOCs during normal process operations at one candidate TSDF indicate that approximately 40 percent of the VOCs would evaporate to the atmosphere, 5 percent would be adsorbed onto activated carbon and biomass, and 25 percent would be biodegraded (O'Brien and Teather, 1995; Douglass, 1996). The anticipated concentration of VOCs in the HD hydrolysate (approximately 250 mg/liter) at the design disposal rate (approximately 130,000 kg/day) also appears to be significantly less than the maximum allowable intake of candidate off-site treatment facilities under current permit restrictions. However, treatment effectiveness at a TSDF must be demonstrated through detailed treatability studies. Preliminary treatability studies based on respirometry with unacclimated biomass yielded an 84 percent biodegradation of the thiodiglycol present in the hydrolysate (U.S. Army 1996e). The panel anticipates that improved removal efficiency similar to that achieved with the laboratory SBRs (greater than 90 percent biodegradation of thiodiglycol) can be achieved with acclimated biomass. Once treatability has been demonstrated, off-site biodegradation has the potential to greatly simplify the process requirements, construction, operations, and decommissioning required at the Aberdeen site.

It is possible that off-site biodegradation at a TSDF may be successful at higher weight percentages of HD in the neutralization process (e.g., 8.6 wt pct), even if on-site biodegradation is not favorable under those conditions. If off-site treatment of the hydrolysate from higher HD loadings is successful, the neutralization step could proceed faster, accelerating the schedule and reducing costs.

Treatment of VOCs

For on-site biodegradation, VOCs present in the hydrolysate must first be separated and treated to prevent them from being air stripped during biodegradation. Air stripping would result in unacceptably high rates of use of activated carbon to remove the VOCs from the biodegradation offgas stream. In configurations 1 and 2, the TPC proposes to use a combination of ultraviolet light and hydrogen peroxide to destroy VOCs that would be subject to regulatory constraints if they were released to water or a landfill (land-ban chemicals). The VOCs would be distilled from the hydrolysate and diluted with a large volume of water, which in configuration 1 would be the main reuse for recycled water. The ultraviolet/peroxide technology can only be used on low concentrations (parts-per-million range) of organic compounds because higher concentrations decrease the photon efficiency and increase unwanted side reactions. Preliminary laboratory-scale tests using simulated VOC distillates have shown that this ultraviolet/peroxide treatment can destroy initial concentrations of 48 to 114 ppm of chlorinated hydrocarbons to less than 1 ppm (Solarchem Environmental Systems, 1996). However, two of the three tests were performed on samples lacking the other organic materials, such as dithiane, that might co-distill with the chlorinated hydrocarbons from the HD hydrolysate. If these organic compounds are present in the solution, the requirements for electrical power and hydrogen peroxide in the VOC treatment process will increase significantly. This possibility seems to have occurred in a third laboratory test, in which the VOC simulants were diluted with evaporated bioreactor effluent rather than with water. Further testing is required to validate the ultraviolet/peroxide process as a cost-effective means of destroying the VOCs from an HD neutralization process.

An attractive alternative to the photochemical oxidation process for destruction of VOCs may be to ship the VOC stripper condensate to a commercial TSDF for disposal, as in configuration 3. Off-site treatment of the VOCs would have the benefit of simplifying the process requirements and operations at the Aberdeen site. Similar mixtures of chlorinated VOCs are commonly treated at many TSDFs. Evaluation of shipping of the VOCs to a TSDF also warrants a reconsideration of the design of the VOC stripper unit. The current design incorporates a falling-film evaporator to achieve VOC separation. This design is fully compatible with the water recycling in configuration 1, but it may result in more water in the condensate than necessary and increased disposal costs. Alternative design configurations and operating temperatures to optimize stripper efficiency should be evaluated if configuration 3 is selected for pilot-testing.

OPERATIONAL REQUIREMENTS AND CONSIDERATIONS

Process Operations

Drainage, Clean-Out, Packaging, and Off-Site Shipment of Ton Containers

The ton container punch-and-drain system, which is common to all configurations, will be essentially identical to the well-proven JACADS system. There will be one ton-container cleanout-area, which is designed to operate 24 hours per day and drain 28 ton containers per week. The ton containers are moved into the toxic handling area, laid horizontally, and punched to create a 6-cm hole in the upper side. A tube is inserted into each container, and the liquid agent is pumped out into a 900-liter (240-gallon) holding tank. Residual agent and contaminated solids are removed by a high-pressure wash system with water at approximately 88°C. An abrasive is added if necessary to remove sludge. The containers are then cut in half by the equivalent of a large pipe cutter and washed with hot water (88°C), steam cleaned (177°C), air dried, and sent to a sampling area where an ACAMS (automatic continuous air monitoring system) tests for residual HD vapor over the metal.

Similar washing systems are commonly used in industry but have not been used with high levels of agent contamination. If decontamination to the 3X level is verified, the metal is packaged for shipment off site to the Rock Island Arsenal for melting operations previously demonstrated at that site (U.S. Army, 1996b). Melting ton containers after thermal treatment to a 5X condition is not necessary for disposal but is standard practice to facilitate recycling the metal.

Process gases from the ton container draining and decontamination area are passed sequentially through a condenser, a caustic scrubber, a reheater (to reduce the relative humidity of the gas), activated-carbon filters for the process, and the carbon filters for facility ventilation. They are then discharged to the atmosphere. Condensate from the condenser is recycled for use as rinse water for the ton container clean-out. The use of hot water for the ton container clean-out process should result directly in the destruction of a large fraction of the residual HD via the same hydrolysis reaction that occurs in the neutralization operation for the drained HD. Hydrolysis in the caustic scrubber should destroy any residual HD present in the process gases prior to filtering through activated carbon. Thus, the activated carbon filters serve as redundant safeguards to ensure the removal of all HD from the process gases prior to release to the atmosphere. The scrubber and process-specific carbon filters are designed to operate at a slight negative pressure (about -0.3 atm gauge) to avoid potential leaks from process piping. The scrubber and filter for treating process gas from the ton container cleanout area are the same units used to treat vent gases from the bulk HD neutralization process.

Although shields are used to contain splashes of decontamination fluid and metal cuttings, the spread of agent contamination will probably be somewhat greater than now occurs with the Army baseline system, where there is no water wash. The spray systems use recirculation as much as possible. All of the cleaning liquids are ultimately consolidated and transferred to the primary agent-neutralization reactor for complete destruction of the agent.

No major difficulties are foreseen with this operation, although somewhat greater contamination of the equipment and surrounding areas may be expected from the high-pressure spray systems. It will be difficult to predict the rates of use of water and decontamination solution, the requirements for their subsequent interim storage, and their dilution effect on the neutralization system. Filters provided to protect pumps must be cleaned out, decontaminated, and disposed of off site. In addition, the operating conditions for treatment of ton container cleanout and decontamination solutions, either separately or with agent, have not been defined, although similar materials from laboratory and bench-scale testing have been handled without difficulty.

The sodium content of the spent caustic solution used to decontaminate the work areas is not expected to be an issue for either the neutralization or biotreatment processes. It will simply replace some of the sodium hydroxide required by the agent neutralization process. The effects of the spent caustic solution on reaction pH can be balanced by controlling the rate of substituting spent solution for HD in the feed to the hydrolysis reactor.

Current analytical methods used in the Army's baseline technology should be adequate to monitor these front-end operations.

Agent Storage System

The drained agent is pumped to an interim holding and surge tank system, analogous to the proven Army system (NRC, 1994e). The capacity can easily be adjusted to account for local external-hazard risks. In the current design, gases vented from these tanks pass directly through an activated carbon filter bed located in the storage area and then are discharged to the room ventilation system. In the panel's judgment, this design does not provide the multiple layers of protection inherent in the scrubber and two-stage activated-carbon filtering used for vent gases from all other tanks that might contain agent. Processing vent gases from the agent holding tanks through the common scrubber and activated-carbon filtration system is an alternative that should be evaluated.

Neutralization Systems

In the TPC's design, there are three neutralization process lines in separate toxic containment areas; each process line includes two neutralization reactors designed to work in parallel. Neutralization is planned to continue 24 hours per day, 7 days per week. The liquid HD from the agent holding tank is pumped along with hot water through a static mixer, which is intended to disperse the agent in the water as droplets roughly 60 m in diameter. The aqueous dispersion of agent is pumped to a well-stirred 8.7-m^3 (2,300-gallon) Kynar-lined reactor partially filled with hot water. The agent concentration in the reactor is tentatively designed to be approximately 4 wt pct. The agent feed rate is controlled to maintain an excess of water which prevents the formation of sulfonium salts that slow the completion of neutralization and give rise to additional by-products. If neutralization is incomplete because of a process upset in the agent feed system, experience in the bench-scale reactors indicates that neutralization can be completed by extending the heating time in the neutralization reactor.

As the agent reacts with water, the neutralization reaction produces hydrochloric acid, which will lower the pH until the mixture is highly acidic (about pH 2). The reactors and related equipment must therefore be capable of withstanding highly acidic conditions.

Adding the agent and neutralizing it takes about one hour, during which time the exothermic reaction releases heat to the reaction mixture. The heat is removed by reactor cooling and a heat exchanger in the recirculation loop. Data have also been presented concerning the anticipated heating of the main reactor vessel by the slightly exothermic reactions.

During the first phase of a batch run, hot water is added to the reactor, and displaced gases are vented. While liquid agent is being added to the reactor, no venting occurs and the pressure increases by up to 1 atm from the compression of nitrogen in the headspace of the reactor. Because the reactor is operated at up to 1 atm gauge pressure, venting is only required during filling with hot water at the beginning of a reaction batch. Thus, only residual VOCs remaining in the empty reactor from the previous batch are vented during the filling process. Any residual agent in the gas stream is neutralized in the caustic scrubber.

After approximately 1 hour, 18 percent sodium hydroxide solution is added to bring the pH to about 12, which neutralizes the acid, completes the neutralization process, and prevents the re-formation of agent. Thus, the reactor and related equipment must be able to withstand alkaline as well as acidic conditions.

The neutralized hydrolysate is transferred to a 20-m^3 (5,300 gallon) storage tank and analyzed for residual agent prior to subsequent processing (e.g., biodegradation). Residual agent must be less than 200 ppb, which represents a destruction efficiency greater than 99.9995 percent. If agent is detected in the hydrolysate at greater than 200 ppb, the hydrolysate is returned to the reactor for further processing. The overhead gases from the hydrolysate tanks are sent to the scrubber and two-stage-activated carbon filter system prior to release to the atmosphere.[8] In configurations 1 and 2, the resulting hydrolysate is purged of VOCs and subjected to biodegradation. In configuration 1 only, VOCs are also solidified before final disposal.

VOC Stripping

Several VOCs remain in the neutralized hydrolysate. Because these are air stripped and not treated in the bioreactors, they are first removed in a waste VOC evaporator, then condensed in an overhead condenser and stored for subsequent treatment or disposal.

In configurations 1 and 2, the VOCs are destroyed by photochemical oxidation. Since this reaction must be carried out in dilute solution to allow adequate penetration of ultraviolet light through the mixture, the VOC condensates are added to the full water recycle stream and the entire stream is processed. This treatment oxidizes residual organic compounds in the recycled process water from biodegradation. Tests by a vendor have demonstrated at laboratory scale the technical feasibility of using a photochemical oxidation system in this way. Scale-up of the results has been extrapolated to a set of nine ultraviolet reactors and appropriate hydrogen peroxide feed systems (Solarchem Environmental Systems, 1996).

Biological Treatment of Hydrolysate

Biodegradation of the organic constituents of the hydrolysate (primarily thiodiglycol) can be carried out either on site in coordination with HD neutralization or off site at a commercial TSDF. Off-site biodegradation at the Aberdeen FOTW is not practical because the FOTW is not designed to treat organic constituents at the concentrations present in the hydrolysate to neutralize the sulfuric acid generated during the biodegradation of thiodiglycol.[9]

The TPC has selected biodegradation in SBRs (sequencing batch bioreactors) as the most robust design for on-site biological treatment (configurations 1, 2, and 3). SBRs have been used for full-scale treatment of a variety of industrial wastewater streams for several years (Irvine and Ketchum, 1988). An SBR is a large tank that contains piping for the injection of air, feedstock (the hydrolysate), and inorganic nutrients; a manifold for the withdrawal of settled sludge; a floating intake on an articulated arm, which is used to withdraw clear supernatant liquid; and a circulating pump to agitate the contents of the reactor.

SBRs are semibatch biological reactors that operate in several different states during a complete reaction cycle. During startup, a bacterial culture that has either been adapted to grow on thiodiglycol or that comes from a wastewater treatment facility is added to the reactor tank. During the first step in the reaction cycle, the hydrolysate diluted with additional water and supplemented with inorganic nutrients (Wolin salts) and sodium bicarbonate (for pH control) is pumped into the reactor tank. The filling process is carried out in the presence of air over a period of about five hours. Mixing and the addition of air initiates microbial oxidation of the thiodiglycol and other organic compounds in the hydrolysate. The major products are water, carbon dioxide, and bacterial cell mass; traces of methane may be evolved. During this aerobic phase, the sulfur in the thiodiglycol is oxidized to sulfate. The air injection is continued for about 17 hours, at which point greater than 99 percent of the thiodiglycol and 90 percent of the TOC in the hydrolysate have been oxidized. Next, the air injection and circulation are stopped, and the solids in the reactor are allowed to settle. The clear liquid at the top is decanted, and some of the settled sludge is pumped out through the manifold in the bottom of the tank. The residual liquid and sludge are left in the tank for a fresh cycle of reactions. The cycle of filling, reacting, and decanting is repeated every 24 hours. The residence time of the liquid contents in the SBR is about 10 days; the residence time of the solids is about 15 days.

The relatively clear liquid decanted from each SBR is sent to a water recycling facility. The sludge withdrawn from the bottom of the SBR is sent to a pair of aerobic digesters for further biotreatment. Polymers to facilitate dewatering are added to the digested sludge in a drum thickener, after which the sludge is dewatered in a filter press. Water exuded from the press is either sent for recycling (configuration 1) or discharged (configuration 2). The solid residue from the filter press (filter cake) is disposed of accordance with standard disposal practices for dewatered sludges from biological wastewater treatment. In this case, the filter cake may be disposed of at a

[8] Overhead gases from the hydrolysate tanks consist primarily of nitrogen displaced from the tank headspace during filling.

[9] The principal treatment mode employed by the FOTW at Aberdeen is biodegradation through use of at trickling filter. This process treatment is usually used for low loadings of organic constituents (Metcalf & Eddy, Inc., 1979)

TSDF (most likely in a landfill) or could be delisted and disposed of as a nonhazardous solid waste. In the past, the hydrolysate from bench-scale neutralization and the aqueous effluent from bench-scale bioreactors used to treat hydrolysate have been delisted.

Water Recycling

The process uses water for decontamination, the neutralization process, and dilution of the bioreactor feed. Much of this water leaves the process in the effluent from the bioreactors. If the effluents are discharged to a FOTW (configurations 2 and 3), there is no need for water recycling capability. However, if only the solids can be sent off site to a hazardous waste facility, water must be recycled to prevent an accumulation requiring off-site disposal. Water recycling is not required for configuration 4 because the hydrolysate is shipped to a TSDF immediately after the neutralization of agent.

To recycle water, configuration 1 uses a conventional thickener and filter press to separate the water from the biomass solids, a conventional sand filter with solids rejection by a clarifier to return residual solids to the bioreactors, and a conventional evaporator and mechanical vapor-recompression water-purification system. Salts recovered from the evaporator are to be solidified and stabilized prior to disposal at a TSDF (most likely in a landfill). Solidifying the salt stream is not an attractive option because it requires a large quantity of solidification agents. In addition, cement-based solidification processes have not been effective in reducing the long-term leaching of monovalent cations (e.g., Na^+) and halogen anions (e.g., Cl^-) (Kosson et al., 1995).

Agent Detoxification and Consistency of Standards

Laboratory and bench-scale tests have shown that the primary neutralization process can destroy the chemical agent to less than 200 ppb in the hydrolysate (see Technology Status section). The Army has proposed that residual HD in the hydrolysate must be less than 200 ppb before it can be transferred out of the toxic containment area. Preliminary analysis by the panel indicates that this standard appears to be inconsistent with the 3X standard for decontaminating solid materials and the airborne exposure limit for sulfur mustard.[10] Furthermore, no clear toxicologic or regulatory basis has been presented for the proposed release standard. Thus, the toxicologic and regulatory basis for the release of liquids that could contain agent or are derived from agent needs to be reevaluated. The consistency of standards for liquids with other related release standards such as the airborne exposure limit, the 8-hour time-weighted average, and the 3X standard should be considered in the reevaluation.

In conclusion, the Army needs to establish standards applicable to the transportation and disposal of the neutralization hydrolysate. This reevaluation may result in either less stringent or more stringent requirements. However, the reevaluation is unlikely to seriously constrain off-site disposal options.

The biodegradation process after neutralization effectively removes the thiodiglycol so that back-reaction to produce new mustard is not possible. Further, the Army has shown that after biological treatment in the SBRs, the toxicity of the hydrolysate has been substantially reduced; the remaining low toxicity (to aquatic organisms) is primarily associated with the inorganic salt content (sodium chloride and sodium sulfate) of the SBR effluents. Table 7-1 shows the results from tests of the acute aquatic toxicity of SBR feed and effluent from laboratory and bench-scale operations. EC50 represents the solution concentration at which a negative response was obtained in 50 percent of the test population. Test populations included *Photobacterium phosphoreum* (a bioluminescent marine bacterium used in the Microtox assay), brine shrimp, Daphnia magna (water fleas), sheepshead minnows, and fathead minnows. These

[10] An order-of-magnitude estimate of the partial pressure of sulfur mustard above an aqueous solution can be made based on pure-component vapor pressure and aqueous solubility (Mackay and Shui, 1981). The Henry's law coefficient is estimated as the pure-component vapor pressure (0.0872 mm Hg at 22°C) divided by the aqueous solubility (920 mg/l at 22°C). The vapor pressure of interest is then calculated as the product of the Henry's law coefficient and the solution concentration. This analysis for a 200 ppb solution of sulfur mustard indicates a vapor pressure of 1.896×10^{-5} mm Hg, which is the equivalent to 0.176 mg/m^3. (The actual vapor pressure may be somewhat lower because of partial dissociation of HD in water.) This estimated concentration can be compared with the 3X standard of 0.003 mg/m^3 after 1 hour and the airborne exposure limit of 0.003 mg/m^3 (8-hr time-weighted average). Thus, according to this preliminary analysis, air in equilibrium with the 200-ppb aqueous standard would exceed both the 3X standard and the airborne exposure limit. A more detailed analysis is warranted to determine the actual equilibrium vapor pressure above a 200-ppb solution of HD.

TABLE 7-1 Aquatic Toxicity of Bioreactor Feed and Effluent from Laboratory and Bench-Scale SBR Testing[a]

Test Population	Test Duration	SBR Feed (diluted hydrolysate)[b]	SBR effluent (laboratory-scale)[b]	SBR Effluent (bench-scale)[b]
Photobacterium phosphoreum	5 min.	3.9%	95.4%	44.3%
Brine shrimp	24 h.	91.0%	n.t.	n.t.
Daphnia magna	48 h.	5.4%	27.0%	23.5%
Sheepshead Minnows	48 h.	84.2%	82.1%	87.8%
	96 h.	70.0%	77.2%	82.5%
Fathead Minnows	48 h.	n.a.	n.a.	57.3%
	96 h.	n.a.	n.a.	40.0%

n.t. = nontoxic when not diluted. n.a = not analyzed.
[a]All results presented as EC50 values (the concentration that induces a response in 50 percent of the test organisms) in volume/volume percentages.
[b]Concentration of inorganic salts in all three test materials was 2 wt%.

Source: Haley, 1996

results indicate that SBR effluent is a good candidate for discharge to the FOTW at Aberdeen. The TPC anticipates additional tests of toxicity in support of regulatory permitting, if the process is selected for pilot testing.

Process Flow Diagrams and Overall Mass and Energy Balances

Process flow diagrams with corresponding overall process mass and energy balances for configurations 1 through 4 are presented in Appendix G. Individual unit operations and inputs for each configuration are summarized in Table 7-2. The number of separate unit operations required to complete processing for each configuration can be used as an indication of overall process complexity. Configuration 4 requires only three unit operations to complete processing to a point suitable for disposal at a commercial TSDF, whereas the other configurations require six or more unit operations.

Table 7-3 summarizes the waste streams and quantities for each configuration. The panel considers it advantageous to minimize the total amount of waste that requires subsequent off-site treatment and disposal. Each configuration produces the following waste quantities for disposal, in addition to the decontaminated (3X) ton containers. Configuration 1 produces 9 kg of solid waste per kg of agent destroyed. Configurations 2 and 3 produce 1 kg of solid waste and 88 kg of wastewater for discharge to a FOTW per kg of agent destroyed. The wastewater is anticipated to contain approximately 0.03 kg of residual organic contaminants and 1.6 kg of salts per kilogram of agent destroyed. In addition, configuration 3 produces 2.4 kg of aqueous waste containing VOCs for disposal at a TSDF. Configuration 4 produces less than 0.01 kg of solid waste and 29 kg of hydrolysate (aqueous) per kg of agent destroyed for disposal at a TSDF. Thus, configuration 4 results in the least quantity of solid wastes that must be shipped off site for disposal. Of the three configurations with the option for off-site disposal of liquid residuals, configuration 4 produces the least aqueous waste to be treated and discharged. Configuration 1 results in the lowest overall mass of waste (solid and liquid) that must be shipped off site but requires the most complex on-site processing.

All configurations require air for ventilation and drying during the ton container cleanout process. In addition, on-site biodegradation requires air to supply oxygen for biodegradation of the hydrolysate, biomass digestion, and sand filtration backwashing (configurations 1, 2, and 3). Biodegradation processes result in the release of carbon dioxide from the microbial oxidation of organic constituents in the hydrolysate (primarily thiodiglycol). Release of organic and inorganic

TABLE 7-2 Summary of Unit Operations and Inputs Required for Each Process Configuration

Unit Operation	Inputs	Configuration 1	2	3	4	Notes
Ton container (TC) drain and cleanout	TCs, agent, water, steam, air	✓	✓	✓	✓	Required for all alternative technologies
Neutralization reaction	agent, water, sodium hydroxide	✓	✓	✓	✓	
Vent gas scrubber	TC cleanout and neutralization vent gas, water, sodium hydroxide	✓	✓	✓	✓	Vent gas cleanup from TC cleanout will be required for all alternative technologies
VOC separation	hydrolysate (with VOCs)	✓	✓	✓		VOCs either treated by photochemical oxidation (configs. 1 and 2) or shipped to TSDF (config. 3)
Biodegradation (SBRs)	hydrolysate, water, air, sodium carbonate, nutrients	✓	✓	✓		Dilution required to biodegrade thiodiglycol in hydrolysate
Biomass digester and filter press	biomass from SBRs, air, conditioning chemicals	✓	✓	✓		
Water recycle evaporator	effluent water from biodegradation	✓				
Photochemical oxidation	sodium hydroxide, hydrogen peroxide	✓	✓			Requires dilution water for config. 2
Residual salt solidification/ stabilization	concentrated salts from water recycle	✓				
Number of unit operations		9	7	6	3	

✓ = required unit operation.

contaminants through atmospheric emissions is anticipated to be negligible because of scrubbing and multistage activated-carbon filtration prior to release.

On-site energy requirements for each process configuration are presented in Table 7-4. These values do not include the energy required for office and laboratory operations or for normal building ventilation. Configuration 1 requires much more energy (356 MJ per kg of agent destroyed) than the other configurations because of the evaporator required to recycle water. Configuration 4 requires significantly less energy (49 MJ per kg of agent destroyed) than the other configurations. The on-site energy needs were used by the panel as a measure of process complexity rather than a metric for discrimination because the energy requirements can easily be met by existing power sources. The energy requirements associated with the many off-site disposal options were not considered by the panel and are judged to be of little or no significance in discriminating among the configurations.

An evaluation of process complexity in conjunction with the above mass and energy balances indicates that

TABLE 7-3 Summary of Waste Streams and Quantities for Each Process Configuration

		Waste Quantity (kg/1000 kg agent destroyed) Configuration			
Waste Stream	Disposal Method	1	2	3	4
Air Emissions					
Ton container clean-out and neutralization	scrub and filter through carbon before atmospheric discharge	1,651	1,651	1,651	1,651
Biodegradation and VOC separation	filter through carbon before atmospheric discharge	71,339	71,339	71,339	n.a.
Liquid Wastes					
Hydrolysate	to TSDF (biodegradation)	n.a.	n.a.	n.a.	29,468
VOCs (aqueous solution)	to TSDF (incineration)	n.a.	n.a.	2,352	n.a.
Biodegradation aqueous effluent	to FOTW	n.a.	87,947	87,947	n.a.
Solid Wastes					
Ton containers, valves, plugs, and metal cuttings	to Rock Island Arsenal for smelting				
Solidified salts	to TSDF (landfill)	8,754	n.a.	n.a.	n.a.
Biomass filter cake	to TSDF (landfill)	972	972	972	n.a.
Activated carbon	to TSDF (landfill)	6	6	6	2

n.a. = not applicable.

configuration 4 (neutralization followed by off-site biodegradation) is the most advantageous. If biodegradation must be carried out on site, configuration 3 (neutralization and on-site biological treatment, with discharge of effluent to a FOTW and shipment of VOCs to a TSDF) is the most advantageous configuration.

Operational Modes

The punch-and-drain operation and the subsequent decontamination and packaging of ton containers for off-site shipping is essentially a batch process. This process is planned for 24-hour-per-day operation. Although it could be conducted on an 8-hour-per-day-basis, the current restriction on storing no more than 1.9 m^3 (500 gallon) of agent inside a facility will probably require close coupling of the punch-and-drain operation with the subsequent neutralization reaction.

The primary neutralization process is a semibatch process that requires approximately four hours to complete a batch, including the testing time to verify complete agent destruction. To meet the overall schedule requirements, it is best to operate this portion of the plant 24 hours per day. However, it is technically feasible to process batches only 8 hours per day. The process could also be redesigned as a continuous, rather than semibatch, process. But this change would require more complicated process controls and would have a higher probability of failures and upsets. A continuous process

TABLE 7-4 Summary of Energy Requirements for Each Process Configuration

Energy Input	Energy Required, by Configuration (MJ/kg agent destroyed) 1	2	3	4	Notes
Steam (for heating)	300	37	37	28	Config. 1 requires substantial steam heat for the evaporator to recycle water
Electricity					
Motors, fans, etc.	42	40	33	12	Substantial power input is required for fans to aerate bioreactors (configs. 1, 2, and 3)
For cooling	14	14	14	9	Required primarily for condensers
Total Energy Input	356	91	84	49	

would have the added disadvantage of making it more difficult to confirm complete agent destruction prior to release of material from the toxic control area.

The bioreactors operate semicontinuously; that is, there are cyclic additions of feed from an intermediate surge tank. Although the bioreactors require little nighttime monitoring, they must operate 24 hours per day. The water recycling facility can be easily started and stopped, assuming surge capacity is provided, and could be operated either 8 or 24 hours per day. The optional solidification plant for stabilizing the effluent from biodegradation is operated intermittently during daylight hours.

Emergency Startup and Shutdown

Because the total process consists of a sequence of batch or semibatch operations with holding capacity between them, there are separate startup and shutdown procedures for each operation.

Ton Container Handling. In cold weather, the ton containers must be preheated to melt the mustard. Melting HD with ambient room temperature would probably take several days and would therefore require a substantial inventory in storage. Thus, for thawing ton containers in cold weather, procedures must be developed that are consistent with the limitations on the total agent inventory in the processing facility (currently established as the 1.9-m^3 limit for the agent storage tank).

Punching and draining full containers; removing sludge, and decontaminating, cutting, testing, and shipping empty containers are sequential steps in a mechanical handling system with no significant surge capability between them. Thus, the whole sequence must be started up and maintained as a mechanical production line. Shutdown requires stopping the feed and allowing the in-process containers to continue to completion, followed by decontamination of the work area as necessary. These front-end procedures, which are common to all systems for treating ton containers regardless of the HD destruction technology, reflect extensive Army experience at JACADS (NRC, 1994a).

Neutralization. The primary neutralization process can be stopped in an emergency by stopping the feed or chemicals. Under these conditions, the reactions continue until the reactants are depleted. Only a very small quantity of agent is present in the reactor at any time because of the slow feed rate and the rapid reaction in the reactor.

VOC Stripper and Oxidation System. These systems can be started or stopped by initiating or stopping feed from intermediate storage tanks.

Bioreactors. The SBRs can be filled and started up using either the hydrolysate or a surrogate material to establish an equilibrium composition and distribution of the microorganisms, and to stabilize functioning support systems such as the various nutrient feed streams. Tests have shown that no further acclimatization of the microbial population is required prior to introducing the hydrolysate for biodegradation.

Although the bioreactors must continue to operate to maintain their active microbial populations, several SBRs will be operating in different phases, providing flexibility in operational modes. Further, the feed to each SBR can be shut off for a short time without significantly harming the microbial population. Alternative feeds can be provided to maintain the population during more extended shutdowns.

Sludge Dewatering. This independent system has feed and waste storage containers so that it can be independently started and stopped. The feed flocculation tank would ordinarily be drained prior to shutdown. If an extended shutdown is expected, the filter press would be cleaned.

Water Recycling. Startup requires turning on the system and internally recycling water until the desired water quality is achieved. The system can then be operated in standby recycling mode until effluent water to be recycled is added or until recycled water is needed. The system can be simply shut down. The VOC photochemical oxidation system can be shut down with or without continued water circulation by turning off the power to the ultraviolet lamps and stopping the hydrogen peroxide feed.

Waste Solidification. This batch operation is similar to concrete mixing and pouring, with the usual need for cleaning out the system periodically to avoid the buildup of solids that would impede the flow of the slurry.

Reagents and Feed Streams

Decontamination and storage container processing will require caustic solutions prepared by dilution on site of commercial, 50 wt pct NaOH solutions. Large volumes of process water are required for these operations. Neutralization also requires dilute NaOH solutions and large volumes of water. Much of the water may come from the ton container processing area and from water recycled from the bioreactor effluents (in configuration 1).

The biodegradation operations require sodium bicarbonate to control pH, as well as aqueous ammonia and mineral nutrients for the growth of the microorganisms. For startup of the SBRs, significant quantities of biomass are needed. This can be obtained from a local wastewater treatment facility. Various solids-conditioning chemicals are needed to facilitate operation of the filter press.

The waste solidification operation, if included, requires Portland cement, lime, some additives, and water. Water recycling, if incorporated in the overall process, may require ferric chloride as a flocculating agent and polymeric water conditioners. If the photochemical oxidation process is used to destroy VOCs, hydrogen peroxide must be supplied.

PROCESS STABILITY, RELIABILITY, AND ROBUSTNESS

Neutralization

The system will use standard industrial components that have been used extensively in conventional applications. Although feed rates are important, most of the process phenomena occur relatively slowly, so response time should not be critical.

The hydrolysis and acid neutralization reactions are mildly exothermic (420 kJ/kg and 700 kJ/kg agent, respectively). The heat of reaction is removed by cooling coils and by evaporation of water with subsequent condensation from the offgas in a reflux condenser cooled by chilled water. Failure of the cooling system would cause a temperature excursion estimated to heat a full batch of agent and water from 90°C (1 atm gauge) to 108°C (1.4 atm gauge). The design pressure for the neutralization system is 6.8 atm gauge. There should be no catastrophic thermal excursions.

The stored thermal energy in each neutralization reactor is 420,000 kJ at 90°C. The maximum agent present in each reactor is approximately 275 kg of HD, diluted to 4 wt pct in water. The actual quantity of agent present in the reactor will be much less because of the slow agent feed rate and rapid reaction when the agent is added to the large volume of hot water present in the reactor at the beginning of each process cycle.

The system is operated nominally at 1 atm gauge pressure. This headspace pressure is used to minimize loss of VOCs when the reactor is being filled and headspace gases are displaced. The principal headspace gas is nitrogen.

There is a possibility of excessive heat being generated if agent were to be inadvertently introduced into concentrated caustic solution. Although this situation is unlikely, if it occurred, it would be detected quickly by the amount of heat released. Reliability (e.g., on-line availability) of the integrated process for the treatment of agent can not be directly assessed without pilot-scale performance data. However, the reliability of equipment components (e.g., pumps, valves, and reactors) and subsystems can be assessed on the basis of their performance at JACADS and in other industrial chemical processing environments. The design assumes continuous operation (24 hours per day) while the facility is on line, with 6 hours per day allowed for slack time (when agent feed or other throughput operations are halted for maintenance, testing, etc.), but the remainder of facility is still on line. The design schedule also allows for 30 percent down time (facility off line) on an annual basis (U.S. Army, 1996b).

This level of availability has been designed into the process through the incorporation of multiple neutralization processing lines and the installation of redundant components (pumps, valves, etc.) in critical flow paths within the area where toxic materials are handled or processed. The design basis for the on-site biodegradation process is treatment of the hydrolysate from 3,175 kg of HD per day. This requirement can be met with a down time of one month for two of the three SBRs during the planned 12-month operating interval. The on-line availability of biodegradation increases for off-site processing because an off-site TSDF would have to be continuously available to handle other feedstreams.

Biotreatment

The bioreactors have proven to be stable when properly operated. Possible improper operations include an upset in a feed (air, nutrients, or hydrolysate) or improper mixing. Because several SBRs are used in parallel, a failed reactor could be readily restarted with biomass from another operating SBR.

Waste Solidification

This operation is very similar to mixing concrete. It is only mildly exothermic, and no stability problems are likely. Standard industrial components with proven reliability are used in this operation. A typical problem might be the accumulation of hardened cement, which could inhibit the flow of solids. Handling solids is usually the operation most subject to problems in any processing plant.

Water Recycling

There are no inherent instability problems in this operation. The sand filter and evaporator are common operational equipment, usually very reliable and robust. The photochemical oxidation system uses lamps and peroxide injection systems that are proven and reliable. Polymerization of residual organic contaminants during oxidation, resulting in opaque deposits on illumination surfaces, may require periodic cleaning of these surfaces. The lamps also will require periodic monitoring and replacement.

MATERIALS OF CONSTRUCTION

When agent is first hydrolyzed, the solution in the reaction vessel becomes acidic. This solution is moving at a relatively high velocity because it is being vigorously mixed. Depending on how much oxygen is available, corrosion of the reactor and components at the vapor-liquid interface may be accelerated. The extent of these combined effects is uncertain because sulfur and chloride in the solution may also influence the corrosion rate. The possibility of accelerated corrosion resulting from the combination of these conditions should be considered. In addition, the contents of the neutralization reactor go from a low pH near the end of hot-water hydrolysis to an alkaline pH from the addition of sodium hydroxide just prior to discharge to the bioreactor system. This pH change dictates that the reactors be made from a versatile metal or one lined with glass or a plastic like Kynar, but these are not unusual materials of construction. Most other systems are standard or have already been developed for the baseline technology.

The TPC has initiated an extensive program for testing materials of construction, including metallic and nonmetallic materials to be used in vessels, piping, seals, gaskets, and other components. Specific testing for metallic materials includes weight loss (immersion) testing, U-bend stress corrosion cracking tests, and electrochemical testing using solutions representative of the process environments. Nonmetallic materials are being tested with standardized immersion testing procedures. Additional information for selecting materials comes from the experience gained during agent destruction operations at JACADS.

OPERATIONS AND MAINTENANCE

Operational Experience

All of the unit operations included in the overall process have been used extensively in commercial operations with related, proven equipment. In addition, the TPC has acquired a substantial body of experience from bench-scale testing at Aberdeen. There has also been significant prior experience with a similar method of agent neutralization, albeit with agent GB.

There has also been significant prior industry experience with the design, construction, and operation of SBRs for industrial waste disposal. However, there has been no prior commercial experience with this specific grouping of unit operations or with the use of the unit operations for agent destruction. The TPC and the panel anticipate no unusual problems with process integration.

Maintenance

Maintenance requirements for all systems but the bioreactors are similar to requirements for the baseline technology systems. Maintenance requirements for SBRs have been established through industrial experience. The panel foresees no unusual problems. Maintenance manuals and documented procedures are not yet available for this combination of unit operations.

The lifetimes of major equipment items should all exceed the duration of plant operations, although some small items, such as small pumps or instrumentation, may require replacement. The downtime for replacing conventional components will probably be governed by preventive safety precautions rather than by actual failure of the equipment. Critical items (e.g., pumps) have redundant systems installed as part of the process design.

Worker safety practices to prevent exposure to chemical agent will be similar to practices used at JACADS. Additional safety precautions will be required for handling hydrogen peroxide, if photochemical oxidation is part of the process. The handling of caustic solutions should be similar to the handling of decontamination solutions at JACADS.

UTILITY REQUIREMENTS

Heating, Ventilation, and Air Conditioning System. Normal building heating will be required for cold weather operations. In addition, the ton containers require heating to melt frozen HD in the winter. Although ton-container heaters will be provided, the cold containers may require extra heating for the container storage area. Steam heating will be required for various small-scale operations. Steam will be provided by four boilers, each rated at 32,000 kJ/hour at 10 atm gauge and 185°C.

Electrical Systems. The systems will be standard for the chemical industry with a 3,400 kVA load.

Plumbing and Piping System. These systems will be standard for the chemical industry with the exception of floor drains in the ton-container and neutralization-process areas to drain spent decontamination fluids.

Fire Protection Design Requirements. The requirements will be normal for the chemical industry. Special hazards include hydrogen peroxide.

Other Systems. Chilled water is needed for the reactor cooling coils, the condensers, and other cooling systems. (Power for the refrigeration systems is included in the electrical systems load.) Compressed air is supplied by approximately three 186-kW (250 hp) compressors (power required is included in the electrical systems load).

SCALE-UP REQUIREMENTS

Bench Scale to Pilot Plant

The following key unit operations are required for configuration 1: (1) ton container processing, (2) reagent preparation and storage, (3) agent neutralization, (4) biodegradation of neutralization hydrolysate, (5) biosolids digestion and filtration, (6) evaporation, (7) photochemical oxidation, and (8) solidification (optional). The critical new unit operations are agent neutralization, biodegradation including the SBRs, and biosolids digestion and filtration. Scale-up is considerably simplified if one of the simpler configurations is selected—especially configuration 4, which requires only the first three unit operations.

Agent neutralization tests have been conducted in 110-liter (30 gallon) batch reactors. This test size will be extrapolated to 8.7 m^3 (2,300 gallon) reactors for pilot testing. The principal variables to be extrapolated are likely to be mixing and heat exchange. No catalysis is involved. Heat exchange is not likely to be a problem if adequate mixing takes place.

Mixing is of concern since the neutralization process involves intermediate reactions and requires an adequate ratio of water to agent to avoid the production of sulfonium salts. Mass transfer studies are currently under way by the TPC, including designing impeller baffles and eliminating dead spots.

Bioprocessing of the hydrolysate has also been limited to bench-scale reactors, typically 80-liter volumes. However, similar scale-up is usual in the development of biotreatments for many waste materials and is not expected to be a problem.

Pilot Plant to Full-Scale Facility

The TPC plans to develop the full-scale facility using the pilot plant neutralization reactors as one module. The full-scale plant will essentially require the addition of two more modules. Thus scale-up should be straightforward, apart from matters such as scheduling.

The SBRs for biodegradation of the neutralization hydrolysate will also be built by adding modules like the ones used for pilot tests. Other systems may or may not be modules, but they are relatively standard unit operations that should not be difficult to scale.

As with the other technologies, process-safety risk factors for the neutralization process can be divided into factors inherent in handling agent prior to its introduction to the neutralization process and factors related to the neutralization technology itself. The risk factors inherent in handling agent prior to neutralization are the same as the risks for the other technologies; they include storage risk, transportation risk, and risks associated with punching, draining, and cleaning ton containers. Storage risk can vary among the configurations because they may require different processing schedules. A simpler set of operations, such as configuration 4, can increase process reliability (e.g., less downtime) and increase the intrinsic safety of the process (less training needed, fewer things to go wrong).

The process-safety risk factors inherent in the neutralization process seem to be minor. The process operates in a batch mode at low pressure and low temperature. The purity of the neutralization products and scenarios for loss of cooling have been explored. It is difficult to envision safety being threatened except by an external factor of some sort.

Based on the panel's preliminary and qualitative evaluation, the most significant off-site risk appears to be associated with handling agent prior to neutralization. In particular, the principal risk factors appear to be mishaps in the punch-and-drain operation or damage to agent holding tanks from an airplane crash or other external event.

PROCESS SAFETY

Two scale-up considerations are involved, namely, scale-up from the existing bench-scale operation to the pilot level and scale-up from the pilot level to the full-scale operation. The TPC currently plans to conduct pilot testing in a single reactor line of a multireactor production facility. This approach should significantly reduce the difficulties and time involved in the second scale-up.

Worker Safety Issues

There are some worker safety issues associated with handling ton containers and handling associated chemicals used in the process. Specific concerns include manual handling of agent-filled ton containers (both injury from manual manipulation of large, heavy objects and the potential for agent exposure in case of a leak), concentrated sodium hydroxide, and hydrogen peroxide. These risk factors need to be addressed in the final operational design.

Specific Characteristics that Reduce Inherent Risk of Design

The system operates in a batch mode at near-atmospheric pressure and low temperature. The process proceeds extremely slowly. Temperature transients, if they occur, appear to be very mild. Safety risks from the destruction process appear to be minimal.

SCHEDULE

The proposed schedule is based on constructing a full-scale, multitrain facility at the Aberdeen site (U.S.

Army, 1996b). There would be three two-reactor trains, one of which would be used initially as a pilot facility to demonstrate the effectiveness of the neutralization and biodegradation processes. This approach would facilitate scale-up to full scale. It should also eliminate delays associated with previous plans to pilot test the technology at the Chemical Agent Munitions Disposal System facility in Utah before construction of a production-scale facility at Aberdeen. Consequently, the TPC forecasts that destruction of the HD stored in bulk at Aberdeen can be completed by August 1, 2003. The TPC projects that plant closure would occur late in 2004.

The pilot plant design reached the design level required for a RCRA permit in April 1996 and should be at the 90 percent stage by the end of November 1996. Modifications to the design that simplify the process are not likely to cause delays. If a decision to pilot this technology at Aberdeen is made in October 1996, permit applications would be submitted in January 1997. The Army's plan allows one year for permit acquisition, which seems reasonable considering the generally favorable reception of neutralization technology by the Maryland Citizens Advisory Commission (see Chapter 9).

During the latter part of the permit acquisition period, a contract for construction of the facility would be let and orders would be placed for equipment with long lead times for delivery. Construction would begin in June 1998 and be completed about October 30, 2000. Initial systemization of a single production line would take about nine months. Pilot operations in this reactor train would be carried out until February 1, 2002, at which time systemization of the other reactor trains would start and a low-rate production operation would begin. Full-scale operations beginning in August 2002 would continue for about nine months. The production schedule assumes treatment of six ton-containers per day, with the facility operating on a two-shift basis. Plant decommissioning and decontamination is estimated to require one year.

8

Neutralization Technology for Nerve Agent VX

The neutralization of VX nerve agent, like the neutralization of HD described in Chapter 7, can be carried out under mild conditions to give products with greatly reduced toxicity. If alkaline reagents like aqueous NaOH are used, the hydrolysis conditions for HD and VX are similar. The reactions can be carried out in commercially available chemical reactors at temperatures below 100°C and near atmospheric pressure. However, the conditions for hydrolyzing VX with neutral water differ from the conditions for HD. Hydrolysis of VX also results in a different set of reaction products than does hydrolysis of HD, and therefore different subsequent treatment is required prior to disposal (Yang et al., 1995).

The Army has explored several approaches to the hydrolysis of VX, including neutralization in a reactor vessel and an innovative in situ reaction in which the nerve agent is treated with a small amount of water while still in the original storage container (Brubaker et al., 1995). The research has opened some potentially attractive options for the detoxification and disposal of VX (U.S. Army, 1995b). Although the processes for VX neutralization are not as well developed as they are for HD, they have good potential for the safe, timely, and cost-effective disposal of the nerve agent.

In April 1996, the Army selected a process based on alkaline hydrolysis of VX as its preferred candidate for development (U.S. Army, 1996f). This neutralization process is closely analogous to the process for HD except that the reaction conditions are alkaline rather than neutral to acidic. The reasons for choosing this process rather than the superficially attractive in situ process are discussed in the Technology Status section below.

If this technology is selected for pilot demonstration, the TPC states that the neutralization products will be shipped to a commercial TSDF that uses biological oxidation for further treatment prior to final disposal. If further biological treatment at a TSDF is not available, other forms of treatment at commercial TSDFs may be evaluated.

As is true for the neutralization technology for HD in Chapter 7, the Army Alternative Technology Program has been the proponent for the VX neutralization technology for the purposes of the AltTech Panel's study. Therefore, the Alternative Technology Program will be referred to throughout the remainder of this chapter as the TPC.

PROCESS DESCRIPTION

Although the neutralization of VX with aqueous NaOH resembles the first part of the HD treatment described in Chapter 7, there are significant differences in detail. The current version of this technology is sketched in Figure 8-1. Removal of the nerve agent from ton containers will use the same punch-and-drain process described for mustard agent in Chapter 7, although no solid heels have been found during nonintrusive testing of the VX containers stored at the Newport site.

The drained agent is transferred to a holding tank. From the holding tank, the VX is fed slowly through a recirculation loop with an in-line static mixer to a vigorously stirred reactor containing hot (about 90°C) aqueous sodium hydroxide solution (20.6 wt pct). The total amount of VX added to the reactor is equal to 21 wt pct of the hydrolysate prior to addition of sodium hypochlorite. The mixture is heated for approximately six hours to destroy the VX and a similarly toxic by-product present in trace amounts, labeled EA-2192. After cooling, an equal volume of dilute (5 wt pct) sodium hypochlorite (bleach) solution is added to oxidize a malodorous reaction product and make the hydrolysate more amenable to subsequent biological treatment. After sodium hypochlorite is added, the amount of VX processed is equal to 10 wt pct of the final hydrolysate. The hydrolysate will be analyzed to ensure that the concentrations of both

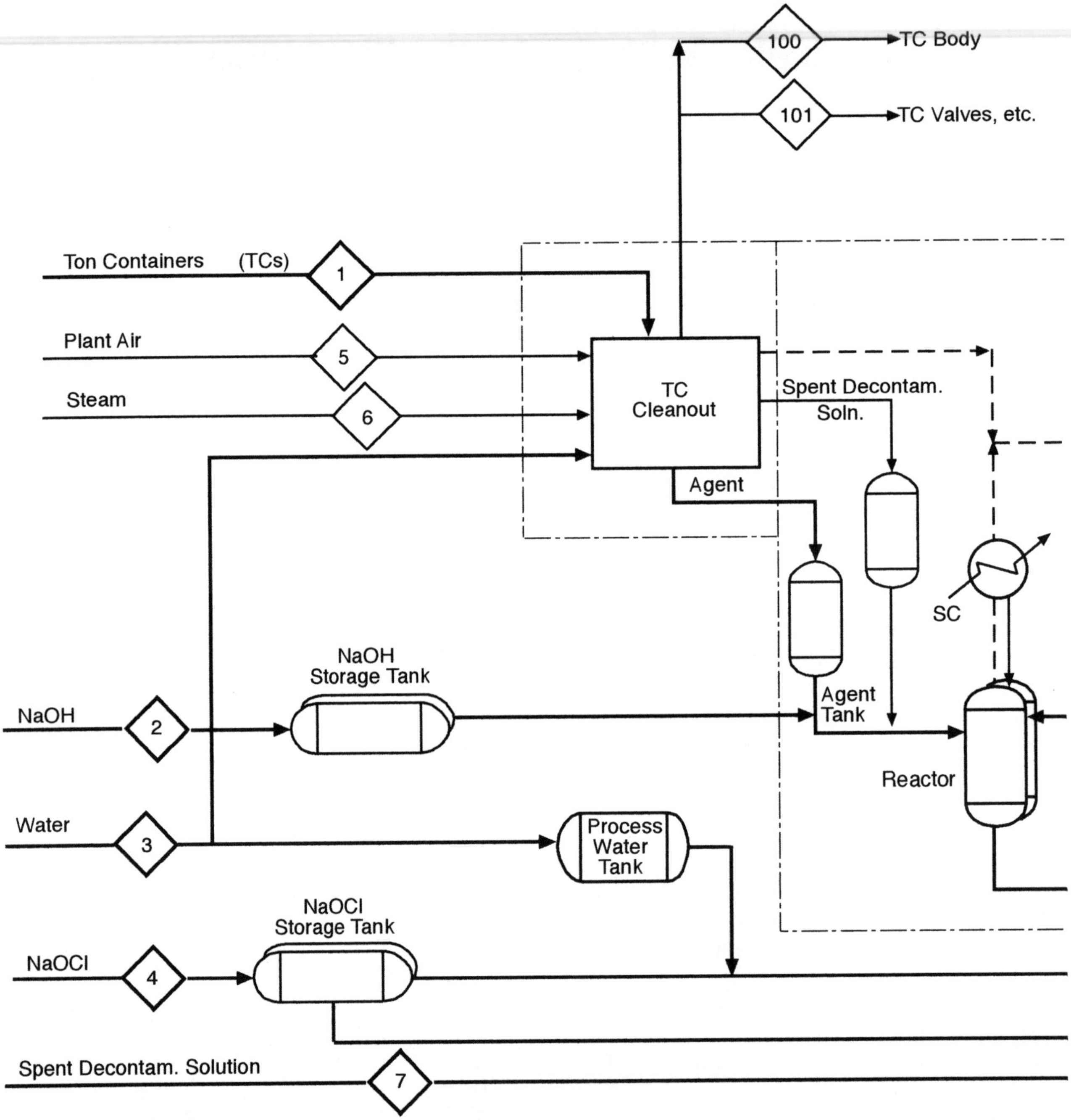

FIGURE 8-1 Block flow diagram of VX neutralization with sodium hydroxide and sodium hypochlorite.

agent and EA-2192 are below 20 ppb[1] before release from the toxics control area.

The hydrolysate is to be shipped for further treatment and final disposal to a commercial waste treatment facility that uses biological treatment to biodegrade organic contaminants. The products could also be treated by incineration (federally owned or commercial), by other existing treatment technologies at

[1]Destruction of VX to less than 20 ppb in the hydrolysate solution results in a DRE of greater than 99.99998 percent. Determination of the DRE is limited by the tenfold dilution during treatment and the analytical detection limit of 20 ppb in hydrolysate. The calculated DREs allow for the tenfold dilution.

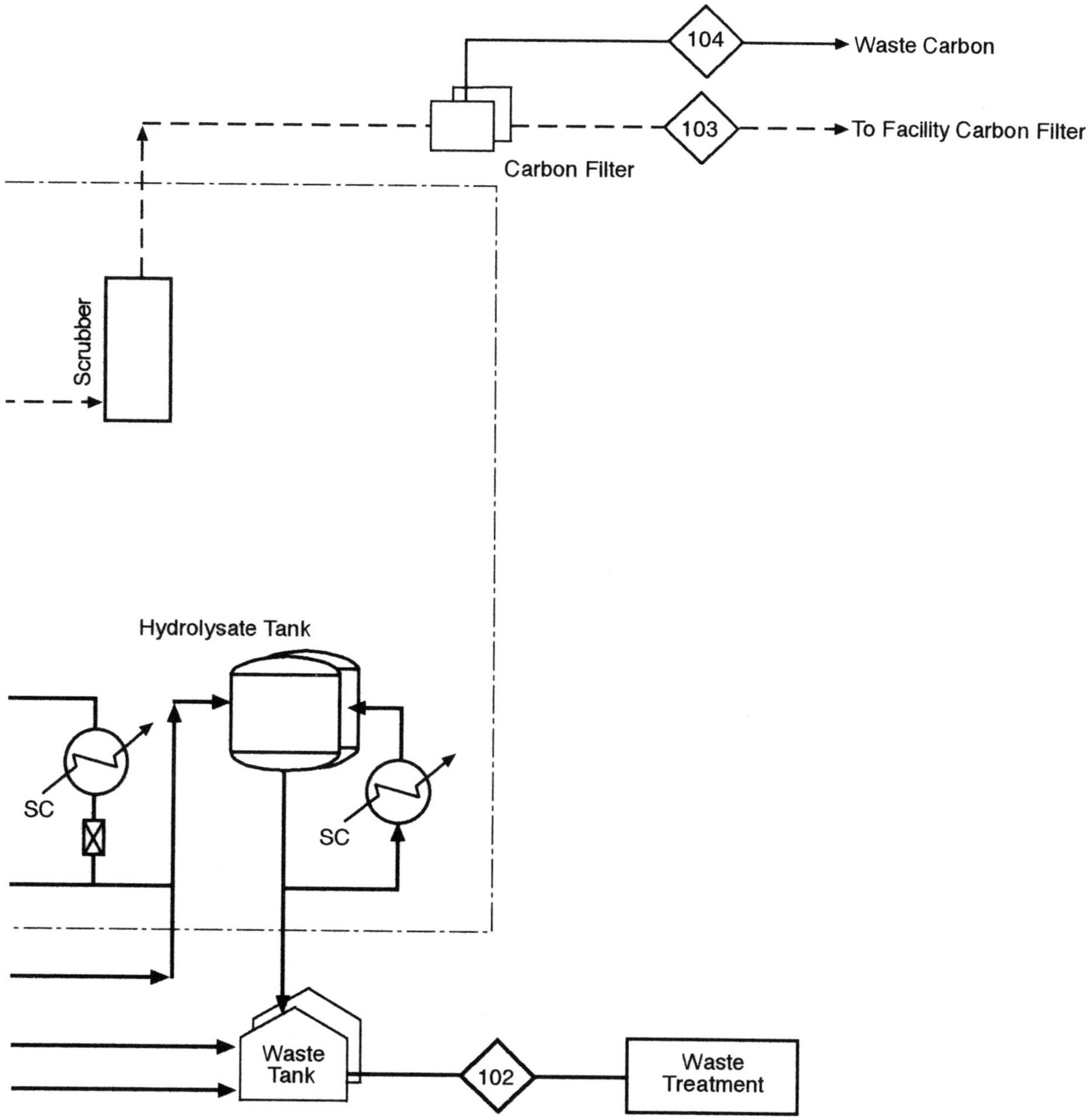

commercial TSDFs, by an on-site treatment facility using one of a variety of technologies, or potentially by any of the technologies discussed in Chapters 4, 5, and 6.

The process for cleaning ton containers resembles the process for HD containers, although no solid residues are anticipated with VX. Drained ton containers are rinsed with hot water to dissolve residual agent. The resulting solution is drained and the rinsing process repeated. The container is then cut open, steam cleaned, and tested for the presence of agent vapor. If no agent vapor is detected by a standard ACAMS monitor, the container is packaged and shipped to Rock Island Arsenal, Illinois, for metal reclamation, as proposed for HD containers (U.S. Army, 1996f).

The process for cleaning VX ton containers was demonstrated using a container from which agent had

The TPC's concept design package (U.S. Army, 1996f) presented the VX neutralization process as described in this chapter. The design included treatment of the hydrolysate with sodium hypochlorite to (1) reduce the odor associated with Thiol, (2) enhance the biodegradability of the hydrolysate, and (3) destroy any residual EA-2192. Because the EA-2192 hydrolyzes, although more slowly than VX, the third reason was no longer important after the duration of the hydrolysis procedure was extended to six hours.

Recently, the TPC has reconsidered this post-hydrolysis treatment option because VX has been detected at concentrations of several parts per million in some upper-layer samples of hypochlorite-treated hydrolysate (Lovrich, 1996). The VX appears to reform during the hypochlorite treatment, since it is not detected (at a detection limit of ≤60 ppb) in untreated hydrolysate prior to the addition of hypochlorite. The TPC now intends to eliminate the hypochlorite treatment option and instead ship the hydrolysate to an off-site TSDF, either without further treatment or with the addition of isopropanol. Adding isopropanol converts the hydrolysate to a single phase for ease of shipment and enhances its biodegradability by providing additional carbon.

The panel believes that problems associated with the disposal of VX hydrolysate can be resolved in a timely manner. However, the recommendations in Chapter 11 provide alternative disposal options in the event that shipment of hydrolysate to an off-site TSDF is not a viable option. The panel encourages the TPC to make additional efforts to control the hydrolysate odor because it could cause significant concern to the public in the event of a spill during handling or transportation. The TPC can draw on the experience of TSDFs that routinely handle malodorous organosulfur compounds.

been drained. A 3X condition was achieved after 2 hours of spraying with high-pressure hot water and steam (U.S. Army, 1996f).

SCIENTIFIC PRINCIPLES

The neutralization processes evaluated for disposal of VX involve hydrolysis of the P-S bond, which is essential to the toxicity of this nerve agent. Figure 8-2 presents the reaction scheme for hydrolysis mediated by sodium hydroxide. The reaction with sodium hydroxide produces the relatively nontoxic ethyl methylphosphonic acid (EMPA), which is present as its sodium salt, and an aminothiol compound (Yang, 1995). The aminothiol, which has a very unpleasant odor but low toxicity, is often referred to as "Thiol," as it will be here (not to be confused with methyl mercaptan, which is also called "thiol"). Much the same reaction occurs during hydrolysis with neutral water, but the resulting EMPA is present as the corresponding acid rather than the sodium salt.

A major advantage of the alkaline hydrolysis process became evident during neutralization studies on impure munitions-grade VX that contained small amounts of a compound containing two P-S bonds. This material (known as "VX-bis"[2]) reacts with water to form EA-2192 (which is present as the sodium salt under alkaline conditions). EA-2192 is almost as toxic as VX itself and is resistant to further hydrolysis by water alone. The concentration of EA-2192 is low, but it contributes significantly to the toxicity of the hydrolysate. During hydrolysis mediated by sodium hydroxide, EA-2192 is also hydrolyzed by an analogous reaction to form thiol and the sodium salt of methylphosphonic acid (MPA), which has low toxicity. At low temperatures (20° to 25°C), small quantities of EA-2192 may be present for a prolonged period because of slow reaction rates. At higher temperatures (75°C to 90°C), both VX and EA-2192 hydrolyze at acceptable rates to form relatively nontoxic products. The alkaline hydrolysis of VX is exothermic, releasing 32.3 kcal/mole.

The products of VX hydrolysis mediated by sodium hydroxide form two liquid phases. The large, dense, aqueous layer holds nearly all (98 mole-percent) of the phosphorus-containing products (predominantly sodium salts of EMPA and MPA) and about 80 percent of the sulfur-containing products, largely the sodium salt of Thiol. The small upper phase contains the rest of the Thiol and its secondary products, along with a mixture of compounds derived from the stabilizer (diisopropylcarbodiimide) added to the agent during manufacture. When the crude hydrolysate is treated with sodium hypochlorite to destroy the malodorous Thiol, the product continues to have two phases, but the nature and distribution of the sulfur-containing products changes significantly. The Thiol largely disappears from the bottom layer, and its concentration in the upper layer sharply decreases. With a small excess of bleach, the Thiol is largely converted to the disulfide, which becomes the major component of the upper layer. As the amount of bleach is increased, more of the Thiol is converted to the corresponding sulfonic acid, which appears in the lower layer as its sodium salt, along with the sodium salts of MPA and EMPA.

[2]VX-bis is S,S-bis-(2-diisopropylaminoethyl) methylphosphonodithioate.

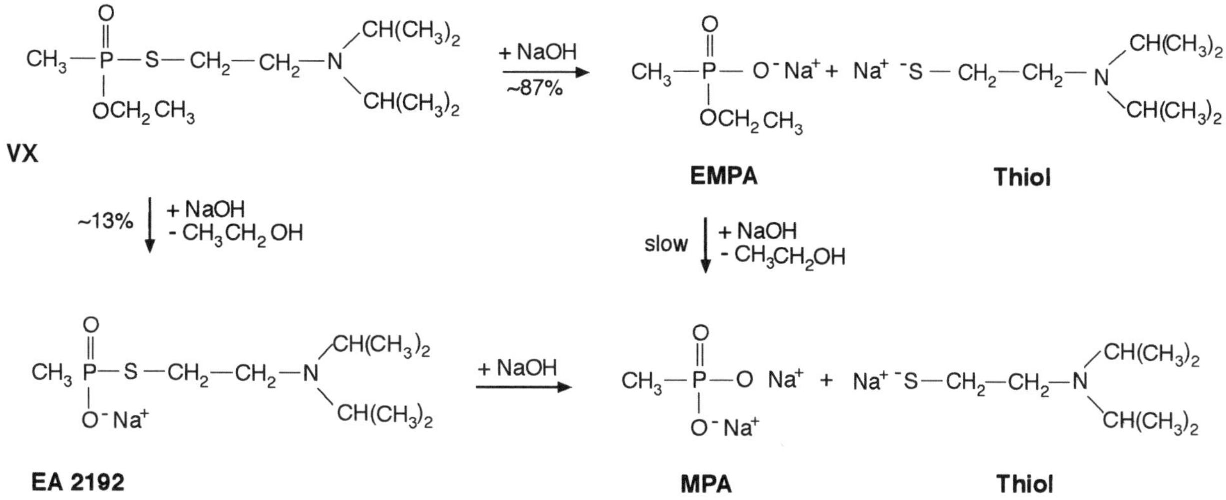

FIGURE 8-2 Reaction scheme for neutralization of VX with sodium hydroxide.

TECHNOLOGY STATUS

Alkaline Hydrolysis

The neutralization technology chosen for destroying VX agent stored at Newport was developed on the basis of previous experience and ongoing research. Although not much has been reported about the alkaline hydrolysis of VX, caustic (sodium hydroxide) has been used to destroy GB (Sarin) nerve agent on a substantial scale. Between 1973 and 1976, the U.S. Army destroyed 4,188 tons of GB at the Rocky Mountain Arsenal by treating it with aqueous sodium hydroxide (NRC, 1993; Flamm et al., 1987). United Nations teams used similar processes to destroy about 70 tons of GB-based agents in Iraq in 1992–1993 (NRC, 1993). The lessons learned from those operations facilitated the development of a VX hydrolysis process through research at Aberdeen by the TPC since 1993.

The VX research was carried out in three stages: (1) laboratory-scale scouting to establish reaction conditions for complete destruction of the agent; (2) process optimization studies in 2- and 12-liter Mettler reactors designed to acquire precise thermodynamic and kinetic data; and (3) bench-scale testing in a 114-liter (30-gallon) stirred tank reactor previously used for HD hydrolysis (described in Chapter 7). Parallel research was conducted on VX hydrolysis under neutral conditions (described below) and on the alkaline hydrolysis method that was ultimately chosen for development. The alkaline hydrolysis was initially tested on a small scale in the laboratory but was extended to testing in 1-liter glassware in runs that destroyed up to 265 g of agent at a time. These tests established satisfactory processing conditions and provided hydrolysate for developing new analytical procedures, as well as for toxicity testing. A new analytical procedure based on sequential liquid chromatography and mass spectrometry permits detection of both VX and EA-2192 at levels of 10 ppb in aqueous solution (U.S. Army, 1996f). As described below in Agent Detoxification, a 40,000-fold reduction in toxicity is accomplished by alkaline hydrolysis.

The bench-scale studies in Mettler reactors yielded reliable data on heats of reaction and reaction rates. The tests monitored the disappearance of VX and secondary products, such as EA-2192, as well as optimizing the test conditions for the 114-liter reactor. Five tests using the 114-liter reactor destroyed 25 to 30 kg of VX in a typical run, but neutralization of as much as 39.4 kg was demonstrated (Lovrich, 1996). More important, the effects of reaction times and mixing (e.g., stirring rate and the effect of adding a static mixer) were evaluated on a large enough scale to extrapolate to pilot- or production-scale reactors. The basic data generated in these tests will facilitate the design of the reactors to be used at the Newport site. More than 351 kg of VX was destroyed in these bench-scale studies.

The bench-scale studies demonstrated the effectiveness of alkaline hydrolysis and provided valuable operating experience under conditions similar to those proposed for full-scale operations. In addition, the tests yielded large volumes of hydrolysate for biodegradation studies and for testing at a TSDF that was being considered for the off-site biotreatment of hydrolysate. About 7,000 pounds of hydrolysate from the experiments on alkaline and neutral VX hydrolysis, which were performed at Aberdeen, was delisted under Maryland regulations and shipped off site for disposal at the TSDF. The treatment at this facility seems to have been satisfactory, but additional studies of the treatability of the alkaline hydrolysate are being carried out at several potential disposal facilities (U.S. Army, 1996c).

In Situ Neutralization

The concept of detoxifying VX in its storage containers (hence referred to as in situ neutralization) is the subject of ongoing research and development by the Army (U.S. Army, 1996f). For in situ processing of VX, a small amount of water (7 to 10 percent) is injected into the storage container. The water reacts over a period of days or weeks and hydrolyzes the VX to a product mixture with substantially lower toxicity. In principle, the viscous liquid product can then be prepared for disposal at a commercial TSDF. On first view, this approach appears attractive for the following reasons:

- Handling agent may not be necessary because the neutralization occurs in the agent storage containers.
- Substantial reduction of the toxicity of the stockpile could be achieved rapidly because all the ton containers could be treated in rapid succession with little lead time.
- The process is conceptually simple, requiring minimal processing equipment and capital costs.

Upon closer evaluation, the potential advantages of in situ neutralization have not been realized because of three obstacles. First, additional handling of agent is in fact required because the ton containers do not have sufficient excess capacity to hold the necessary amount of water for reaction without first removing a significant volume of agent (a few gallons of agent out of about 180 gallons in an average container). Removal of several gallons of agent is probably a simple operation because the valves on the ton containers stored at the Newport site appear to be operable. Even so, the need to transfer agent entails a small (but controllable) risk to plant personnel. Second, although the in situ process does substantially reduce the agent toxicity, further treatment is required to destroy residual EA-2192 completely and reduce toxicity to levels suitable for disposal at a commercial TSDF. Third, the in situ reaction is difficult to control because of poor mixing and inadequate temperature control within the ton containers.

At the time of the panel's evaluation, in situ neutralization of VX had been tested on three ton containers. Several difficulties were encountered during this testing. The low solubility of water in VX resulted in poor reproducibility during preliminary tests. The reaction proceeds slowly for some time after adding the water. As the reaction proceeds, the solubility of the water in VX increases because the initial hydrolysis product is a mutual solvent. In the first full-scale test of the reaction, which was conducted without heat or agitation, the VX and water initially formed two layers in the cylinder. After about three days, the water dissolved, and a rapid reaction ensued. The temperature inside the cylinder rose to a maximum of 98°C, and the VX concentration decreased to 5.4 percent at the end of the first week. Subsequently, the temperature gradually fell; less than 400 ppb of VX remained after four weeks.

Another potential difficulty was found in the second test of in situ neutralization. A change in the procedure for adding water led to an excessively rapid reaction and the formation of solid hydrolysis products that might be difficult to remove from the storage cylinder.

Two approaches to dealing with the immiscibility of VX and water are possible. The simplest is to mix the liquids by rolling the cylinder. The second approach is to add some hydrolysate from a previous VX hydrolysis to the ton container, along with the water needed to react with the VX. The advantage of the second approach is that it produces a smooth, rapid reaction. The major disadvantage is that it requires additional handling of agent and hydrolysate because more agent must be removed from the container to accommodate the volume of the added hydrolysate. The additional handling of the hydrolysate is disadvantageous because the liquid is still toxic, although a thousandfold less toxic than VX, based on intravenous testing in mice. Most of the residual toxicity appears to be due to the presence of EA-2192.

As a consequence of these complications in the in situ process, the TPC elected to concentrate further development work on the alkaline hydrolysis carried out in a conventional reactor. Testing of the in situ process in ton containers continues at the Chemical Agent Munitions Disposal System facility in Utah. Based on the difficulties described above, the AltTech panel does not recommend additional tests on the in situ process beyond the tests currently in progress.

OPERATIONAL REQUIREMENTS AND CONSIDERATIONS

Process Operations

Mass and Energy Balances

The TPC provided flow sheets indicating the major equipment, piping, and controls, as well as material and energy flow rates (U.S. Army, 1996f). These flow sheets were derived from operating experience gained in bench-scale testing.

A simplified block flow diagram and corresponding overall mass balance for the VX neutralization process using sodium hydroxide (NaOH) and sodium hypochlorite (NaOCl) are presented in Appendix H. Overall, approximately 8 kg of water, 0.4 kg NaOH, and 0.7 kg of NaOCl will be required per kilogram of VX neutralized. Energy requirements for this process are estimated to be 28,600 kJ/kg of VX neutralized, including 14,000 kJ/kg for steam heating, 9,500 kJ/kg for electricity, and 5,100 kJ/kg for cooling.

Draining, Cleaning, Packaging, and Shipping Ton Containers

Draining, cleaning, and decontaminating the ton containers can be done with the same system described in Chapter 7 for HD containers. The punch-and-drain system is essentially identical to the proven JACAD system (NRC, 1994a). Empty ton containers are cleaned, cut open, decontaminated to a 3X level, and packaged for shipment to Rock Island Arsenal, Illinois, a government metal recycling plant. The principal difference in this operation between the HD and VX facilities is the offgas scrubber. For VX, the caustic scrubber used for HD is replaced with a two-stage scrubber that provides both acid and alkaline scrubbing.

No major difficulties are foreseen with this operation, although some contamination of the equipment and surrounding areas can be expected from the high-pressure spray system. Although no heel is expected in VX ton containers, high-pressure spray decontamination is used to ensure that no agent remains in microscopic crevices in the container surface. It is difficult to predict the rates of use of water and decontamination solution, the requirements for their subsequent interim storage, and their dilution effect on the neutralization system. The method for treating water and decontamination solution, either separately or with agent batches, has not been defined. The sodium content of the spent caustic solution is not expected to be of concern to either the neutralization or subsequent treatment processes because it replaces some of the sodium hydroxide required by the agent neutralization process. Current monitoring methods developed for the baseline system should be adequate for these operations.

Agent Storage System

The drained agent is pumped to an interim holding and surge tank system, analogous to the proven JACAD system. The capacity could easily be adjusted for local external hazard conditions as needed. Gases vented from these tanks pass directly through a carbon filter bed located in the storage area. Consideration should be given to processing these vent gases through the scrubber prior to carbon filtering and release, along with the vent gas from the neutralization reactor.

Neutralization Reaction

The stored agent is fed in batches to one of two independent neutralization trains. Each train consists of a 2.5-m^3 (650 gallon) stirred neutralization reactor with both internal and external mixers, overhead offgas condensers, a reactor cooling jacket, and an external heat-exchanger cooling system; a 5-m^3 (1,320 gallon) hydrolysate storage tank with mixer; a 152-m^3 (40,000 gallon) waste storage tank; and an offgas treatment system.

The neutralization reactor is first partially filled with 11 wt pct caustic and brought up to the operating temperature of 90°C (194°F). The VX is then slowly added in the external recirculation loop just ahead of the static mixer. The mixture is heated for about six hours, as required to destroy the toxic EA-2192 by-product. The temperature is controlled by removing the exothermic heat of reaction through the reactor cooling jacket, the heat exchanger in the external cooling system, and an offgas reflux condenser.

As liquid agent is added to the reactor vessel, the overhead gases are vented through the reflux condenser to condense water vapor and volatile organic compounds generated during the reaction. The condensate is recycled to the reaction tank. The noncondensable gases pass through a dual scrubbing system. The first scrubber contains acid to absorb organic amines produced by the reaction. The second contains caustic to neutralize the acids formed in the first. A heat exchanger removes the heat of neutralization. The gas then passes through a chiller to reduce the water vapor content, a gas heater to elevate the gas temperature above the dew point, and an activated-carbon filter system.

The hot hydrolysate is analyzed for residual agent and, if acceptable, is transferred to the hydrolysate storage tank, where it is combined with water and hypochlorite to oxidize some of the organic products. The primary reasons for adding hypochlorite are to reduce the foul odor of the hydrolysate and to make the hydrolysate more amenable to subsequent biological treatment. Upon completion of the hypochlorite oxidation, the waste is analyzed to ensure complete agent destruction and pumped to an external storage tank, where additional water and hypochlorite are added to prepare the effluent for off-site disposal.

Biodegradation

Laboratory testing of the biodegradation of VX hydrolysate has been of limited success to date. The products of hydrolysis do not readily serve as the primary substrate for biological oxidation. Substantial quantities of co-substrate (i.e., other waste with a high carbon content but low in phosphorus) are required to force the microbial utilization of phosphorus from the methyl phosphonic acid present in the hydrolysate. Because of this need for high-carbon cofeed and because only limited success has been achieved in biodegrading Thiol, the hydrolysate is not a good candidate for treatment by on-site biodegradation prior to final disposal.

The very limited data available as of May 1996 suggest that off-site biodegradation is likely to succeed if the treatment facility receives sufficient quantities of high-carbon waste from other sources to force microbial degradation of the VX hydrolysate products as a source of nutrient phosphorus. Laboratory testing of biodegradation with SBRs has demonstrated significant biodegradation of organophosphonate and organosulfate constituents in VX hydrolysate (U.S. Army, 1996k). When the VX hydrolysate was the only phosphorus source available to the microbial population and isopropanol was provided as an additional carbon source (at about 2,900 mg TOC/liter), greater than 90 percent of the organophosphonate constituents and up to 51 percent of the organosulfate constituents were biodegraded. These results were obtained in laboratory-scale reactors operating in semibatch mode (periodic partial decanting of clear supernatant and removal of settled sludge, followed by refilling) over extended intervals. Hydrolysate biodegradation with a carbon cofeed has not yet been tested at bench scale.

Because of limitations in the available information, the panel is concerned that off-site treatment to date may have involved primarily dilution of the hydrolysate to an acceptable level rather than complete destruction by biodegradation of the products of concern. However, the preliminary toxicity testing described in the next section suggests that oxidized hydrolysate (VX hydrolysate after being treated with sodium hypochlorite or a similar oxidizing agent) may have sufficiently low toxicity that further degradation of organic constituents is not needed.

If further toxicity testing demonstrates that the hydrolysate poses no threat to human health or the environment, total biodegradation of the organic components during disposal at a TSDF may not have to be demonstrated. Otherwise, the alternative of off-site treatment by biodegradation at a TSDF will require appropriate treatability studies to substantiate that complete biodegradation of the hydrolysate constituents does in fact occur. Such treatability studies would need to be conducted at the TSDFs that are candidates to receive the hydrolysate. The presence of Thiol and methylphosphonic acid derivatives, which are scheduled precursors under the CWC (Chemical Weapons Convention), would subject a TSDF receiving the hydrolysate to destruction verification requirements (including inspection) under the terms of the CWC.

Agent Detoxification

Table 8-1 provides a summary of toxicity testing carried out on hydrolysates from VX neutralization with water and VX neutralization with sodium hydroxide. Intravenous exposure testing was performed on mice. An LD50 is the dose required to kill 50 percent of the test population within 24 hours. The LD50 for VX is included for comparison. Neutralization of VX with sodium hydroxide results in greater than a 40,000-fold reduction in toxicity, compared with a 970-fold reduction achieved by neutralization with water only.

TABLE 8-1 Toxicity of VX and VX Hydrolysates as Measured by 24-Hour Intravenous LD50 in Mice

	VX	VX hydrolysate (water only)	VX hydrolysate ($NaOH_{(aq)}$)
LD50 at 24 hours (mg substance per kg body weight)	0.014	13.6	587

As discussed earlier, laboratory and bench-scale tests have shown that the primary neutralization process can destroy the chemical agent to less than 20 ppb in the hydrolysate. The Army has considered a residual of less than 20 ppb VX in hydrolysate to be safe enough for release of the hydrolysate from the on-site toxics control area and transport off site for final treatment. The basis for defining this level as acceptable and its consistency with the 3X release standard for solid materials needs to be demonstrated. The Army also needs to define the standards to be used for transporting and disposing of the hydrolysate, as well as any related restrictions that would limit the pathways for human contact with the hydrolysate. The panel believes that defining these standards will not seriously constrain the off-site disposal options or the disposal schedule because significant quantities of hydrolysate have already been approved for off-site treatment and successfully shipped to an off-site TSDF. These standards and restrictions are therefore not anticipated to impede the successful operation of a VX neutralization facility.

Operational Modes

The operations of punching and draining ton containers and then decontaminating and packaging them for off-site shipping constitute a batch process. Although it might be conducted on an eight-hour-per-day basis, the current size limit on the agent storage tank is likely to require a close coupling of the container draining operation with subsequent agent processing. Thus, operations are anticipated to be conducted 24 hours per day, 7 days per week, except for scheduled maintenance periods.

The primary neutralization process is a semibatch process that requires about 8 hours per batch. Thus, although it would probably be more economical to operate this part of the plant around the clock, it could be operated on an 8-hour-per-day basis. Currently, the TPC anticipates operating the neutralization process 24 hours per day, 7 days per week, except for scheduled maintenance periods.

The semibatch mode of operation requires simpler process controls than a continuous operation and reduces the probability of control failure and resultant process upset. In addition, the semibatch mode permits the operators to hold the hydrolysate for confirmation of complete agent destruction.

Emergency Startup and Shutdown

Handling Ton Containers

Draining, washing, cutting, decontaminating, testing, and shipping are sequential steps in a mechanical handling system with no significant surge capability between steps. The steps are started in sequence, and the sequence must be maintained as a mechanical production line. Shutdown simply involves stopping the first step and allowing the in-process containers to continue to completion; normally this is followed by decontamination of the work area. These procedures are common to the front-end systems of all VX destruction technologies and reflect extensive experience at JACADS.

Neutralization

The primary neutralization process can be stopped in an emergency by halting the addition of agent or chemicals. Under these conditions, the reactions will continue until the reactants have been depleted. The maximum temperature rise that could occur from a loss of cooling water and cessation of agent feed is less than 10°C.

Feed Streams

Commercial 50 percent sodium hydroxide solution is stored in a 28-m3 (7,400 gallon) tank and then diluted to 33 percent for feed to the neutralization reactor and to 18 percent for feed to the offgas scrubber and for decontamination fluid. The TPC provided no estimate of the decontamination and gas scrubbing requirements. Based on processing agent at 3,200 kg/day, 3,300 kg/day of 40 wt pct caustic is required for neutralization. Treatment of the resulting hydrolysate before shipping requires 15,900 kg/day of 15 percent sodium hypochlorite solution, which is diluted for use to 5 percent concentration.

Process water is preheated to 90°C for use in the ton container cleanout systems, for which the estimated requirement is 19 liters per minute. Provision is made for using an abrasive in the water jets, but no estimate of quantity was provided to the panel.

Residual Streams

Processing Ton Containers

The ton container cleaning process produces a variety of liquid or slurry waste streams: (1) washdown solutions of contaminated hot water; (2) decontamination solutions, including spent solution and floor washings; and (3) water cutting slurry (contaminated water plus abrasive), which is mostly recirculated. These waste streams will be destroyed by adding them to the neutralization reactors. The residual caustic in them is taken into account as partial replacement for the caustic otherwise added for neutralization. Some of the washings may contain too many solids, which may be separately bagged for off-site shipment along with the ton containers or the hydrolysate. Because the quantities are not known, the exact methods of feeding the waste streams to the reactor can only be approximated and have not been developed.

Small solids, such as valve fittings and removal tools, are also bagged for off-site shipment with the ton containers. Vent gases will be ducted to the scrubber and carbon filter system. Building ventilation air will be treated by an HVAC system with carbon filters.

Neutralization

All liquids are shipped off site. These liquids are mixed residuals that may not be characterized except for testing to ensure that the VX and EA-2192 concentrations are below acceptable limits. The quantity (including decontamination fluids) is estimated to be 58 metric tons per day at the average agent-destruction rate.

All vent gases are treated in the scrubber and carbon filter system. All liquids and slurries proceed to off-site biotreatment.

Secondary wastes include personal protection equipment, rags, small metal parts, etc. These wastes are cleaned with decontamination solution, tested to ensure that they meet the 3X standard, and disposed of as hazardous waste.

PROCESS STABILITY, RELIABILITY, AND ROBUSTNESS

Stability

The neutralization reactions are mildly exothermic with removal of the heat of the reaction by a reactor cooling jacket, an external cooling system, and condensation of the offgases in a reflux condenser cooled by chilled water. Failure of the cooling system would cause a temperature excursion that is estimated to heat a full batch of agent and water from 90°C (1 atm gauge) to 98°C (1.1 atm gauge). The design pressure for the neutralization system is 6.8 atm gauge, but the system runs at 1 atm gauge. Thus, there should be no catastrophic thermal excursions.

The maximum agent present is a full load of 180 kg of VX. An upset of feed-stream flow rates could cause minor changes in the reactions and less efficient agent destruction. This condition can be countered by holding up a batch until it has been checked and extending the reaction duration if necessary.

Reliability and Robustness

The system will use standard industrial components for which there is extensive good industrial experience. Materials of construction have not been proposed, but their selection probably does not present serious problems. Although feed rates are important, most of the process phenomena occur relatively slowly so that response time should not be critical.

OPERATIONS AND MAINTENANCE

Operational Experience

All the unit operations included in the overall process have been extensively used in commercial operations with related proven equipment. There has also been significant prior experience with a similar method of agent neutralization, albeit the agent neutralized was GB

(NRC, 1993). An exception may be the analytical techniques, which may be newly developed, depending upon the disposition of process residuals and the related standards. As noted in the Technology Status section, the TPC has gained considerable operational experience with this neutralization process through repeated bench-scale tests using 114-liter reactors. If a contractor is hired to run the pilot plant and full-scale facilities, the contractor should have an established record of experience in operations on a similar scale with hazardous materials.

Maintenance

Maintenance requirements for all systems are similar to the requirements for the baseline technology systems. The TPC and the panel foresee no unusual problems. Maintenance manuals and documented procedures are not yet available for this process.

The equipment lifetimes should all exceed the duration of plant operations, with the exception of the replacement of small pumps and instrumentation. The downtime to replace conventional components is more likely to be governed by agent-related safety precautions than by failure of the equipment.

Experience at pilot scale for this process does not exist. Thus it is not possible to provide an estimate of operational time versus downtime.

SCALE-UP REQUIREMENTS

Two considerations are involved in scaling the process to a full-size facility: scale-up from the existing bench-scale work to the pilot plant and scale-up from the pilot plant to the full-scale facility.

Bench Scale to Pilot Plant

The following key unit operations are involved in the complete plant: (1) ton container processing, (2) reagent preparation and storage, (3) agent neutralization, and (4) optional hydrolysate oxidation with hypochlorite. The critical new unit operations are agent neutralization and hydrolysate oxidation. Although off-site treatment and disposal of hydrolysate is also a critical operation, it is considered external to on-site processing. Demonstrating its feasibility will require detailed treatability studies carried out by one or more off-site TSDFs that would be candidates to receive the hydrolysate. Off-site TSDFs appear to be available and are willing to receive the hydrolysate, so availability of a treatment facility is not a constraint.

Agent Neutralization

Agent neutralization tests have been conducted in 114-liter reactors. This size will be extrapolated to the 2.5-m^3 reactors for the pilot plant. The principal variables to consider in this extrapolation are likely to be mixing and heat exchange. No catalysis is involved. Heat exchange is not likely to be a problem if the mixing is adequate. Mixing is a concern because the reaction does involve intermediate reactions and requires the proper ratios of agent, caustic, and water to avoid producing undesirable residuals.

Hydrolysate Oxidation

The oxidation of VX neutralization products by sodium hypochlorite is similar to current methods used by the Army to dispose of VX wastes. Although the method has not been piloted for this specific purpose and design details are yet to be worked out, neither the TPC nor the panel anticipates a major risk in scaling up this method for treatment of process wastes.

Pilot Plant to Full-Scale Facility

The TPC plans to develop the full-scale facility using the pilot-plant neutralization reactors as one of two modules. The full-scale plant will require the addition of another line of equipment. Thus, scale-up of this technology is straightforward, except for scheduling and other matters. Although unit operations other than neutralization may or may not be modular, they are relatively standard unit operations that should not be difficult to scale.

PROCESS SAFETY

The process-safety risk factors for neutralization of VX are the same as for HD. The discussion in Chapter 7 applies equally well to this technology with respect to risk factors inherent in handling agent prior to neutralization, risk factors inherent in neutralization operations, worker safety issues, and specific characteristics that reduce the inherent risk of the design.

SCHEDULE

Although a regulatory question concerning an Indiana statute remains to be resolved (see Chapter 9), the TPC has proposed a plausible schedule for destroying the VX stored at Newport using the neutralization process followed by off-site biotreatment. The TPC proposes that pilot testing be done in one reactor line of a production-scale facility to be built at the Newport site. If the state of Indiana changes its legal requirements to permit this pilot test, developing and implementing the process for neutralization and off-site biotreatment of VX could be accomplished on the following schedule. The remaining issues about off-site treatment of the hydrolysate by biodegradation should have no effect on the schedule because off-site TSDFs using other treatment methods, such as incineration, could handle the hydrolysate without difficulty.

A pilot plant design is expected to reach the design level required for a RCRA permit in December 1996. If a decision to pilot-test this technology at Newport is made in October 1996, permit applications would be submitted in April 1997. The TPC's plan allows two years for permit acquisition, which seems conservative considering the generally favorable reception of the neutralization technology by the Indiana regulators in preliminary discussions with the panel (see Chapter 9).

During the latter part of the permit acquisition period, a contract for construction of the facility would be let and orders would be placed for equipment items with long lead times for delivery. Construction is scheduled to begin on October 10, 1999, and to be completed about February 27, 2002. Initial systemization of a single production line is scheduled to take nine months. Pilot operations in this reactor train would be carried out until May 28, 2003, during which time systemization of the other reactor trains would start and a low-rate production operation would begin. Full-scale operations beginning in November 2003 would continue for about nine months. The production schedule assumes treatment of five ton containers per day, with the facility operating on a two-shift basis. Plant decommissioning and decontamination is scheduled to begin in late 2004 and is estimated to require one year.

9

Community and Environmental Regulator Views Concerning the Alternative Technologies

This chapter discusses the processes and the results of the AltTech Panel's interactions with the public in communities near the Aberdeen and Newport sites, the CACs (citizens advisory commissions) for those sites, environmental regulators for the states of Indiana and Maryland, and managers of the CSEPP (Chemical Stockpile Emergency Preparedness Program) in both states. Also covered are past efforts by the TPCs and the Army to work with the communities.

BACKGROUND AND APPROACH

The 1994 report of the NRC Stockpile Committee, *Recommendations for the Disposal of Chemical Agents and Munitions*, urged the Army to increase public involvement across a wide spectrum of activities (NRC, 1994b, Recommendation 6).

> The Army should develop a program of increased scope aimed at improving communications with the public at the storage sites. In addition, the Army should productively seek out greater community involvement in decisions regarding *the technology selection process,* oversight of operations, and plans for decommissioning facilities. Finally, the Army should work closely with the Chemical Demilitarization Citizens Advisory Commissions, which have been (or will be) established in affected states. There must be a firmer and more visible commitment to engaging the public and addressing its concerns in the program.

In response to this recommendation, the U.S. Army's *Alternative Demilitarization Technology Report for Congress* documented increased efforts by the Army to obtain CAC comments on the NRC *Recommendations* report (U.S. Army, 1994, pp. 5-1 to 5-2 and Appendix G). Of the CACs that responded during the extended comment period, a majority favored a neutralization-based alternative over the baseline incineration technology (U.S. Army, 1994, p. 5-2). The views of the CACs were consistent with the NRC's recommendations concerning a neutralization-based alternative. The Army also decided to pursue a neutralization R&D program to determine the feasibility of neutralization as a technology for destruction of the stockpiles at sites with bulk storage of chemical agent, namely, the Aberdeen and Newport sites.

In the NRC *Criteria Report Evaluation*, the Stockpile Committee again emphasized the importance of public involvement in the selection of alternative technologies. A key aspect of this emphasis was that public acceptance was not viewed as one criterion among others but rather the end result of a meaningful process of public involvement in the critical decisions concerning the program (NRC, 1995, pp. 36–37).

Battelle Pacific Northwest Laboratories conducted a study for the Army in which focus groups in the affected communities were used to identify and characterize sources of support and opposition to the baseline (incineration) system. Battelle recommended that the Army "define the role of the public in decisions about technology choices and implementation" (Bradbury et al., 1994, p.68). The study further concluded, "In today's political and social context, program managers must take the initiative in engaging their stakeholders in a mutual, cooperative problem-solving approach" (Bradbury et al., 1994, p.69).

Finally, in *Review of Systemization of the Tooele Chemical Agent Disposal Facility*, the NRC Stockpile Committee recommended, "A substantial effort should be made by the Army to enhance interactive communications with the host community and the Utah Citizens Advisory Commission on issues of mutual concern . . ." (NRC, 1996, p. 6).

In short, the AltTech Panel had before it a long and important series of recommendations and findings from reports of the NRC, the Army, and Army contractors, all of which emphasize the importance of seeking public input to the CSDP, as well as gathering information about what the public considers to be the important criteria in evaluating the alternative technologies. Consistent with the above recommendations, the panel sought to obtain public input on the criteria to be used in the evaluation, as well as other factors that stakeholders in the affected communities identified as important. As a starting point in developing the panel's own criteria, the panel adopted criteria that are related to the public perspective and had been accepted by the Army in its *Criteria Report* and by the Stockpile Committee in the NRC *Criteria Report Evaluation* (see Chapter 2).

The panel also followed the lead of the Stockpile Committee by adopting a variety of approaches to find out how the affected communities viewed the alternatives and what criteria they thought were most important. First, the panel scheduled a series of information-gathering public forums in Indiana near the Newport site and in Maryland near Aberdeen. Second, the panel decided to precede these open forums with CAC meetings in both states. (Unfortunately, the expected meeting with the CAC for Newport had to be canceled because of scheduling difficulties. Subsequent conversations with members of that CAC have been used to augment the CAC's written views.) Third, the panel scheduled meetings in both states with state and local regulators and permitting authorities to learn about the regulatory and permitting hurdles for each of the technologies and to receive answers to written questions the panel had sent them.

Fourth, the panel visited the Newport storage facility to learn more about the storage situation there from its administrators. A similar visit to the Aberdeen site was considered unnecessary because of time constraints and because many panel members had already visited that site. Fifth, the panel met with representatives from CSEPP and other emergency managers to determine if any of the alternative technologies under consideration might affect existing response plans and preparations and if so, how. Finally, the panel requested information from the TPCs on their past efforts at community involvement. (Panel members were already familiar with the Army's community-involvement efforts related to the neutralization options, so a similar request to the Army was not needed.) The remainder of this chapter describes what the panel learned.

PUBLIC FORUMS

The members of the public who offered verbal or written testimony at the public forums held by the panel cannot be assumed to be *representative* of the affected communities in a statistical sense. The panel does not have public opinion survey data that would provide a statistical cross-section of community views. The panel has been informed that opinion surveys have been conducted by an Army contractor at other chemical stockpile sites—Tooele, Anniston, and Pine Bluff—and will be conducted at the Aberdeen and Newport sites (Gibbs, 1996; Morales, 1996). The public forums were obviously attended by the most concerned residents and by representatives of organizations that are actively interested in and affected by the decisions concerning alternative technologies. In fact, representatives of groups that had opposed the baseline system were invited to attend the public forums and meet with members of the panel to ensure that the panel fully understood the criteria these organizations believed were most important in differentiating among alternative technologies.

As will become clear below, the panel heard an array of views concerning both the alternative technologies and the criteria by which they should be evaluated, as well as comments supporting or opposing the baseline system. In reporting on views expressed during the public forums or in correspondence, the panel does not claim that these views represent a consensus or even a majority view within the communities affected by chemical demilitarization activities at the Newport and Aberdeen sites. The panel does assert, however, that these views are important for understanding the intensity of feelings of an active and vocal segment of the affected communities and are therefore worthy of Army and panel consideration.

Context

The context for the open forums is important for interpreting the comments received. In late January 1996, the Army's Office of the Product Manager for Alternative Technologies and Approaches (OPMAT&A), with representatives of the TPCs for the three technologies other than neutralization, had held a series of public meetings cosponsored by the Indiana and Maryland CACs. The meetings were intended to "provide information to the public on the alternative technologies being considered for APG [Aberdeen] and NECA [Newport], to solicit public input, and to establish a dialogue between Army, public and CACs" (U.S. Army, 1996j). One meeting was held at South Vermillion High School, northwest of Clinton, Indiana, on January 27. Two meetings were held in Maryland, one at the Kent County Courthouse in Chestertown on January 25 and the second at Edgewood High School on January 26. According to a newspaper report, about 200 people attended at the Indiana meeting (Clinton Herald, 1996), which received fairly extensive coverage by the local media.

The panel initially scheduled its public forums for February. However, to ensure that residents had enough time to prepare for the meetings and to absorb the information provided by the Army and the TPCs, the meetings were rescheduled for mid-March. This allowed the Army time to provide the communities with more detailed TPC and Army information on the technologies. The information was placed in repositories that had been established at libraries and public offices in the affected communities.

The OPMAT&A further cooperated by providing the panel with copies of the public sign-up sheets from the January meetings, as well as summaries of the question-and-answer periods from those meetings. The sign-up sheets, along with lists provided by the two CACs, were used to augment the panel's mailing list of individuals and organizations who were notified of the public forums planned for March 1996. Major environmental groups were also notified, and their attendance or written input was solicited. The OPMAT&A also placed announcements of the panel's schedule of public forums in local newspapers serving the communities around the Newport and Aberdeen sites. The NRC study director for the panel spoke with the director of the Chemical Weapons Working Group (CWWG) to ensure that the CWWG and its member groups were aware of the planned forums. The CWWG was invited to provide written and oral testimony on the alternative technologies. Representatives from the TPCs were present at all of the public forums, but the panel asked them not to speak during the formal portion of the forum. The TPCs did have displays outside the meeting rooms and were encouraged to talk with interested attendees before or after the formal part of the program.

The panel's public forum in Indiana was held at North Vermillion High School in Cayuga on the evening of March 12, 1996. Approximately 75 people were present, and 15 signed up to offer verbal comments. The panel received 20 letters from area residents commenting on the alternative technologies, the criteria for their evaluation, or the importance of public involvement in the final decision between using the baseline system or an alternative technology for the destruction of the Newport stockpile. Two of the letters were from members of the organization Citizens Against Incineration at Newport (CAIN) and contain about twenty cosigners. The introductory remarks of the panel chair indicated the purpose of the forum, invited everyone in attendance to comment on the technologies and the criteria being considered, and reminded them that the forum was not a debate on the baseline system.

The first of two Maryland public forums was held on the evening of March 15 at the Kent County Courthouse in Chestertown. About 50 residents attended, filling the meeting room to capacity. The second forum was held on the morning of March 16 at the Edgewood High School in Edgewood and also attracted about 50 residents. Wayne Gilchrest, the congressman from the district that includes the Kent County area, attended the Chestertown forum to listen and to offer his views on the alternatives.

Issues Common to Communities at Both Sites

Table 9-1 shows that many of the concerns and views expressed at the public forums were stated by residents of both the Newport and Aberdeen communities. These common issues are discussed first; subsequent sections focus on issues expressed by just one community.

Public Health and Safety and the Environment

The first two issues listed in Table 9-1 are not surprising and need little explanation. The public requires that any alternative technology protect their health and safety and not endanger the environment. These are necessary but not sufficient conditions for public acceptance of whichever technology is used to destroy the chemical agent stockpile. At Newport, the concerns focused on accidental airborne releases. Residents near the Aberdeen site also expressed particular concern about damage to Chesapeake Bay from airborne emissions or aqueous discharges.

The panel considers both issues—public health and safety and protection of the environment—to be of paramount importance in evaluating alternative technologies. The evaluation criteria associated with these issues are discussed in the section on Safety, Health, and the Environment in Chapter 10.

TABLE 9-1 Summary of Community Issues Raised in Public Meetings with the AltTech Panel

Issue	Description
Issues Raised at Both Sites	
Public health and safety	Any alternative technology must ensure public health and safety.
Environment	The environment must be protected, including protection of the atmosphere from accidental releases and protection of sensitive ecosystems such as the Chesapeake Bay from discharges.
Opposition to the baseline system	There was considerable opposition to incineration and the baseline technology. The Army's credibility was questioned on the basis of a perception that the Army was not committed to finding and evaluating an alternative.
Closed-loop or batch process	The alternative technology should be a closed-loop or batch system that can be shut down quickly; these processes are perceived as intrinsically safer than others.
Low temperature and pressure	The alternative technology should operate at low temperature and low pressure; these conditions are perceived as intrinsically safer than others.
Use of the facility after stockpile destruction	No matter which technology is chosen, members of the affected communities want a guarantee that, once the stockpile at their site has been destroyed, the facility will not be used to destroy hazardous materials from elsewhere.
Schedule driven by safety, not external commitments	The schedule for destruction of the chemical agent stockpile should not be driven by external pressures such as treaty obligations or legislated deadlines, especially if risks to health, safety, or the environment are increased as a result. Public health and safety are the most important considerations.
Public involvement in decisions and oversight	The decision-making process regarding the alternative technologies should be open to public participation and scrutiny to offset the belief that the Army is biased and remains committed to the baseline system.
Appropriate role for cost control	Cost should not be the decisive factor in selecting an alternative, but it should be a consideration.
Issues Specific to the Newport Site	
All-in-one process with minimal process residuals	The alternative technology should be capable of destroying the stockpile in a "complete process" that does not produce large amounts of toxic or hazardous waste.
Issues Specific to the Aberdeen Site	
Consider shipping off-site for final treatment	The feasibility of processing the chemical agent to a less toxic condition and then shipping it off-site for final destruction at already existing toxic-waste facilities should be thoroughly investigated.
Toxicological assessment of alternatives needed	The evaluation of alternative technologies should include an assessment of their potential health (toxicological) effects.

Opposition to the Baseline System

At the beginning of each public forum, the chair indicated that the purpose of the meeting was not to discuss the Army baseline system. Despite this admonition, residents near both sites expressed opposition to the baseline system. The presence of opponents to that technology was not surprising at a forum considering alternatives to the baseline system. Although the panel heard criticism of the baseline system at the forums for both sites, the criticism was not universal.

The public was reminded during the meetings that the panel had not been asked to compare the alternatives to the baseline incineration system and was present only to receive feedback on alternative technology preferences. Despite the panel's statement, the citizens continues to voice their concerns about incineration. The panel concluded that public objections to incineration would nevertheless be useful to the panel to help determine the objective characteristics of an agent destruction technology that would be opposed or supported by the communities near the sites.

There were a number of negative comments on the Army's credibility, and these comments reflected two distinct themes. One theme was distrust that the Army was truly committed to a full assessment of an alternative to the baseline system and doubt that, if one or more alternatives were recommended by the AltTech Panel, the Army would diligently pursue them. A second theme was that, even if an alternative were pursued, the Army did not have the management capacity or commitment to implement it adequately. Newport residents expressed concern that the Army was not committed to a fair evaluation of the alternative technologies offered by the TPCs because it was continuing to promote the baseline incineration system it had developed.

These feelings of distrust are outside the panel's task of providing a technical review of specific alternative technologies. Nevertheless, public mistrust may affect the Army's ability to carry out any alternative technology program, which would affect the implementation schedule and ultimately increase the storage risk.

Closed-Loop or Batch Process

Because the panel members were uncertain what was meant, in engineering terms, by public testimony favoring a *closed-loop* system, they pursued the point in subsequent discussions with some participants who mentioned it.[1] To some of the participants it appears to mean "controlled emissions," i.e., a system in which, once the agent enters, there is no route by which any emissions can escape to the environment until they have been held and tested. To others, closed loop seems only to mean a process with fewer emissions, or perhaps fewer unknowns in the emissions, than they perceive as resulting from an incineration process. Several participants said TPC representatives had emphasized a closed-loop feature as an advantage of their technology. Those who favored a closed-loop system at the Newport forums believed it would be safer and, more protective of the environment and could avoid accidental releases of agent or dangerous process residuals.

The public apparently feels that an advantage of the alternative technologies over incineration is that all process-residual streams can be captured, held, and analyzed prior to release to their ultimate destination. If testing shows that some material of concern has gotten through, that batch can be recycled through earlier stages for retreating.[2] From a technical standpoint, therefore, the preference for a closed-loop process is closely akin to the preference, stated by other forum participants, for an alternative technology that uses batch processing and can be quickly and safely shut down if testing shows a batch has not been fully treated.

In formulating criteria for evaluating alternative technologies (see Chapter 10), the panel represented these

[1] In standard engineering parlance, a system is a "closed loop" with respect to a particular material if that material is completely recycled *internal to the system*. To the extent that some of the material is degraded and lost or otherwise escapes, the system is not perfectly closed. For example, a common automobile battery is a closed loop with respect to the lead and sulfate in it, even through several years of charging and discharging cycles. Modern automotive batteries are coming close to being a closed loop with respect to water, but they are not perfectly closed if they require an occasional topping-up of the cells. In this technical sense, none of the alternative technologies (or the baseline system) is a closed loop with respect to the materials in the chemical agent to be destroyed.

[2] To meet the "hold and test" condition in a continuous process, the process stream must be uniformly and continuously sampled, the analytical results from the sampling must be available and action on them taken while the sampled material is still within the system, and the stream of positive-test material must be diverted and somehow returned to an earlier process stage for retreatment. These are tough requirements to meet without having a batch step in the otherwise continuous process.

community concerns for a closed-loop or batch process in the criterion of test prior to release.

Low Temperature and Pressure

Two other process characteristics that the public strongly associates with safety and health issues are the temperature and pressure at which agent-destruction processes operate. Many community members near both sites, commented that processes that work at low temperature and low pressure are intrinsically safer than processes that require high temperature or high pressure or both. The panel members in attendance were not always certain that the participants who offered these comments correctly identified which of the alternative technologies had the desirable characteristics and which did not, but it was clear that the speakers viewed these characteristics as important.

From a technical perspective, a high pressure process may have a large inventory of releasable energy. This energy itself could be a hazard, or it could increase the risk associated with other hazards in the process. For example, a rupture or leak in a high-temperature, pressurized reaction vessel could disperse a larger amount of hazardous material over a wider area or into the atmosphere than a containment failure would in a low-temperature, low-pressure system. Safety engineers assess the entire inventory of hazards associated with a process, including the thermal energy (heat), pressure, and material hazards. Thus, the public concern about high temperature and pressure is represented in the panel's evaluation criterion of the hazard inventory (Chapter 10). This criterion, moreover, includes hazards other than high temperature and pressure. With respect to process performance and engineering, a system that operates at high pressure may pose more of a challenge to stability than one at lower pressure, so the panel's criterion of stability, reliability, and robustness is also relevant to this public concern.

Facility after Stockpile Destruction

Another issue raised by many is whether the facility built to destroy the chemical agent stockpile at a site will continue to be used after the stockpile has been destroyed. Two currents of thought permeated the testimony and letters. In both states, the largest number of letters and statements strongly opposed using the facility to destroy hazardous waste after the stockpile had been eliminated. A smaller number of public comments suggested that the facility could be used for the destruction of other on-site wastes. These seemingly disparate views can be reconciled by understanding that they represent the views of people who do not trust the Army or Congress to carry through on promises that the facility will not be used for other purposes after stockpile destruction and will not be used to destroy wastes brought from other locations.

The policy issues related to the final disposition of the stockpile destruction facilities are outside the purview of the AltTech Panel. The NRC Stockpile Committee, which has addressed this issue, has supported congressional actions that currently prohibit the use of these facilities after the destruction of the stockpile has been completed (NRC, 1994b, 1995, 1996). To provide information relevant to the potential public acceptance of a technology, the AltTech Panel has indicated in Chapter 10 whether or not an alternative technology would, from a purely technical standpoint, be readily adaptable to treating other wastes.

Schedule Driven by Safety, Not External Commitments

Several people expressed the desirability of slowing down the evaluation process for alternative technologies. These comments seemed to reflect a belief that the congressionally mandated date of 2004 for the complete destruction of the unitary chemical weapons stockpile was no longer realistic and that more time was needed to ensure that alternatives had been thoroughly evaluated. Many residents felt that the entire examination of alternatives was being driven by the overall demilitarization schedule. They were concerned that the panel and the Army did not have sufficient time to analyze all of the relevant information.

The AltTech Panel is not in a position to assess the flexibility of the 2004 date, but members did respond to comments at the forum by saying that the panel did have sufficient time and information to evaluate the alternatives under consideration.

Public Involvement in Decisions and Oversight

A considerable concern to the public and to this panel is maintaining or increasing the involvement of citizens and communities in the process for selecting an alternative technology for each of the two sites. Many residents criticized what they perceive as a decision process that will be largely closed to their participation once the panel issues its report. Several members of the public stated their opinion that an evaluation of alternatives to the baseline system would not have been undertaken at all if some of the communities and various members of the public had not organized against the baseline system. The panel cannot assess the accuracy of these views, but it is aware that public involvement has been and continues to be a source of contention between the Army and the public.

Comments from the general public, as well as written statements from CAIN and CWWG, urged that members of the affected communities be included in the decision-making process that will continue after the panel makes its recommendations. The testimony heard by the panel in Indiana favored increased participation by including either the Indiana CAC or a representative of the public from the Newport area. In either case, the intention was to increase participation in *making the decisions*, not merely to increase the release of information from the Army about decisions that had been made without public participation. This frequently and strongly expressed desire for community involvement in the process of deciding about alternative technologies is consistent with several of the Stockpile Committee recommendations cited at the beginning of this chapter.

Public acceptance grows out of public involvement in which the affected communities are active partners in the evaluation and decision process (see the NRC *Criteria Report Evaluation*). In turn, public acceptance of the process can avoid costly scheduling delays and ultimately provide the Army with a strong base of support for the effective implementation of the disposal program. In the communities visited by the panel, opposition to the baseline system is obvious. Yet, it was equally clear that the destruction of the stockpile is a shared goal. The panel believes the Army can increase the probability of public acceptance of its evaluation of alternative technologies—and acceptance of the entire stockpile demilitarization program—by ensuring adequate opportunities for participation in the decision making by the residents of the affected areas. The panel chair told the forum participants that panel members would return to explain as fully as possible the panel's report and recommendations, but at the time the panel could not provide answers on the Army plans for continued community involvement and participation in the selection process.

Appropriate Role for Cost Control

Based on verbal comments and written communications from the public, the panel recognizes there is concern about the rising projected cost of the baseline system. A related concern is that, if an alternative is recommended for pilot-testing, it should not be eliminated because of cost projections. At the forum in Indiana, some residents commented that they are also taxpayers and that they wanted the most agent destruction (effectiveness) for their dollar (efficiency) without compromising safety. The apparent conflict between not wanting to eliminate an alternative because of cost but wanting an economically efficient destruction technology indicates that the desire for an alternative to incineration is primary and that cost control is secondary.

Although the panel received many comments about cost, assessment of the relative costs of the alternative technologies is outside the panel's charge, and it will not evaluate cost data. Therefore the evaluation criteria developed by the panel do not address this issue.

Specific Concerns of the Newport Community

The community near the Newport site raised one issue that was not raised by the Aberdeen community. The Newport community wanted the alternative technology selected for Newport to be capable of destroying chemical agent in a "one-step or complete process" and not produce large amounts of toxic or hazardous waste. The panel interprets the reference to a "one-step or complete process" to mean that the alternative should be capable of complete destruction of the agent and not require shipping by-products or wastes for additional treatment or additional steps to complete the processing of residuals before they are released for reuse or into the environment. The comments again indicated that various TPC representatives, in their meetings with citizens, had emphasized these features as advantages of their technologies over neutralization.

The panel's evaluation criterion for environmental burden (Chapter 10) directly addresses public concerns about the amount of waste, hazardous or otherwise, from the treatment process. In response to the concern that a process be "one-step or complete"—which is difficult to assess directly because all of the alternatives involve consist of multiple unit process operations—the panel has included summary information in Chapter 10 on the condition and amounts of *all* process residuals associated with each alternative technology.

Specific Concerns of the Aberdeen Community

The public forums in Maryland revealed that the public was well informed and that participants had acquainted themselves with the alternatives under consideration. A large majority of the comments favored neutralization as the technology to use at Aberdeen. A preference for neutralization is also the stated position of the Maryland CAC (Nunn, 1996a, 1996b). The Aberdeen community expressed two concerns that were not raised by the Newport community. One was that the feasibility should be investigated of processing the HD at Aberdeen to a less hazardous state and then shipping the process residuals off site for further treatment. The second concern was that the AltTech Panel (or perhaps another NRC committee) and the Army should include a toxicological evaluation of the alternative technologies before making a decision about which, if any, technology to pursue. Although this concern was not raised by the Indiana citizens, it was mentioned at the Indiana forum by a representative of the Kentucky-based CWWG.

Consider Shipping Off Site for Final Treatment

Several Maryland residents commented on what they perceived to be a logical solution for destroying the stockpile at Aberdeen: neutralizing the agent and shipping the waste products elsewhere for further treatment and disposal at a permitted TSDF. As noted earlier in the report, partial treatment onsite followed by shipping off site for final treatment and disposal were not originally options in evaluating alternatives to the baseline system, but they are now being considered by the Army. The "treat and ship" option for neutralization hydrolysate is addressed by the panel in this report.

Toxicological Evaluation of Alternatives Needed

Several participants at the Maryland forums voiced their desire to have a toxicologist on the AltTech Panel so that the health effects of each of the alternatives could be evaluated. The panel's view on this issue, which the panel chair expressed at the forums, is that, because of the early stage of development of some of the alternatives, it would be premature to attempt a thorough and effective toxicological evaluation. However, the panel fully expects that any alternative(s) that might be pursued would undergo a thorough health risk assessment.

PANEL MEETINGS WITH THE CACs

The meeting with the Indiana CAC had to be canceled because of scheduling difficulties. However, panel members did subsequently meet with the chair of the Indiana CAC and with state officials. The panel met with the Maryland CAC prior to the first public forum in Chestertown.

Meeting with the Chair of the Indiana CAC

On April 11, 1996, several panel members met in Washington, D.C., with the cochair of the Indiana CAC, Melvin Carraway. He was accompanied by a representative of the Indiana governor's office and several state environmental regulators. Speaking for the Indiana CAC, Carraway stated that the views expressed at the panel's public forum in Indiana in March strongly reflected the views of the Indiana CAC. In addition, both Carraway and the governor's representative clearly stated that they consider the best alternative technology option to be in-situ neutralization and are adamantly opposed to incineration at Newport.

Meeting with and Comments from the Maryland CAC

Several Maryland CAC representatives met with panel members on March 15, 1996, prior to the public forum in Chestertown. In addition, several Maryland CAC members spoke at the public forum. The cochair of the Maryland CAC gave the panel a copy of the CAC's 1994 comments, which put the CAC on record as recommending both neutralization of the HD at Aberdeen and reduction of the storage risk at that stockpile (Maryland CAC, 1994). These comments contained several CAC recommendations that were discussed at the March 15 meeting with the panel.

According to the 1994 CAC comments, of all the chemical stockpile sites, Aberdeen is located in the most densely populated area and in the state with the second highest cancer rate in the nation (Maryland CAC, 1994). The CAC further stated that incineration was an unacceptable method of destruction.

Consistent with the NRC *Recommendations* report, the CAC favored (and continues to favor) a low temperature, low pressure system that it believes is inherently safer than incineration or other systems that process agent at high temperature or high pressure (Maryland CAC, 1994, pp. 6–7).

Having recommended in 1994 that the Army pursue a neutralization program, the CAC is pleased that neutralization is one of the alternatives being considered (Maryland CAC meeting with panel, March 15, 1996). The 1994 comments recommended neutralization that could transform the mustard agent into a nonhazardous waste; if this could not be accomplished, the recommended action was that the Army use neutralization followed by shipment of the hydrolysate to an existing hazardous-waste management facility (Maryland CAC, 1994, p.8). At the March 15 meeting, the CAC representatives said that they view neutralization as a closed-loop or batch process, which the CAC favors. In a letter to the AltTech Panel, the Maryland CAC reiterated support for neutralization followed by biodegradation, if that technology is found to be effective, but stated it would accept any other safe alternative technology (Nunn, 1996a).

The Maryland CAC members who spoke with the panel raised several issues related to the process of evaluating alternative technologies. First, they were concerned that because of the dual role required of the head of the OPMAT&A (U.S. Army Office of the Product Manager for Alternative Technologies and Approaches), the neutralization technology was not being thoroughly explained to Maryland citizens and the state regulatory community. The head of OPMAT&A is in charge of both the evaluation of all alternative technologies and the Army's neutralization program. The CAC members feared that, in an effort to be completely neutral, neutralization would not be as strongly promoted by the Army as other alternatives were being promoted by the TPCs.

Second, CAC members were concerned that the information on the alternatives being provided by the TPCs was not entirely accurate and that some TPCs had been actively lobbying state environmental regulators.

Third, one CAC member expressed concern that the Army was not committed to a fair evaluation of the alternative technologies. This member believed that the criteria proposed by the Army for evaluation (see Army *Criteria Report*) could be scored any way the Army wanted.

Fourth, CAC members were concerned that citizen involvement in the Army's evaluation, once the AltTech panel report was complete, might be limited. It was clear that they want to play a role in subsequent government decisions regarding the selection of an alternative technology at Aberdeen.

Although the panel is neither knowledgeable about nor charged with exploring some of these concerns, there is clearly a substantial credibility gap between the Maryland CAC, which is a well-informed group, and parts of the Army. With respect to the panel's role in recommending an alternative technology for the Aberdeen site, the panel has taken note of the preference, stated by CAC members and others in the community, for a technology that has the characteristics associated with a closed-loop or batch process and that processes chemical agent at low temperature and low pressure.

ENVIRONMENTAL REGULATORS

A critical element in the implementation schedule for disposing of a chemical agent stockpile is the environmental permitting process. Each state establishes its own permitting process with the aim of ensuring public health and safety and protecting the environment. To determine if there are unique permitting issues in Maryland or Indiana for any of the alternative technologies, the panel pursued two information-gathering approaches. The panel was particularly interested in whether there are any potential show-stoppers, i.e., obstacles that could halt implementation of a technology indefinitely.

First, the panel developed a series of 25 questions to solicit information on regulatory and permitting matters that could affect the implementation schedule of an alternative technology (see Appendix J). The questions were sent to state regulators, and the answers were compiled as a source of information for the panel.

The second source of information was a series of direct meetings with groups of Indiana and Maryland environmental regulators. The meeting with the Maryland regulators was held on March 15, 1996; meetings with Indiana regulators took place on March 13 and April 11. Issues that arose during those meetings were clarified in materials sent to the panel by the regulators.[3]

Regulators from both states indicated to the panel that, based on their current knowledge of the proposed alternative technologies, *any of the technologies* could be permitted in their states. That is, there did not appear to be any show-stoppers for any of the technologies being evaluated. However, there are five particular regulatory and permitting issues that bear directly on the panel's evaluation of the alternative technologies: (1) requirements for permitting under the Resource Conservation and Recovery Act (RCRA), (2) time to obtain permits, (3) off-site shipping of process residuals, (4) treatment of synthesis gas combustion, and (5) pilot demonstration of an alternative technology.

Permitting Requirements under RCRA

Two of the TPCs have approached regulators in both states to explore the possibility of obtaining recycling designations for their technologies.[4] A recycling designation would obviate a RCRA hazardous-waste permit, which would otherwise be necessary. The panel believes that a recycling designation may prove difficult to obtain in either state (Hosseinzadeh and Sachdeva, 1996; Ray, 1996). The TPC for CEP (catalytic extraction process) has indicated, according to regulators in both states, that the company is reluctant to accept a requirement to obtain a RCRA hazardous-waste permit because of potential negative consequences for applying the technology elsewhere.

Time to Obtain Permits

Regulators from both states gave the panel estimates of the time required to obtain the necessary permits in their respective states, once formal applications for permits are received. In Indiana, the state may spend a maximum of 365 days conducting a RCRA permit review. This review allows for two "clock-stops," when the applicant may be required to submit additional information at the state's request. (Until the applicant submits the requested information, the time after the request does not count against the 365-day limit or "clock.") Hence, the actual time depends on two factors: the quality and completeness of the initial application and the complexity of the technology being reviewed. The more complex a technology is, the more likely regulators are to require additional information from the applicant, leading to a longer clock-stop.

In Maryland, the state has assigned two full-time engineers to assess technology packages on an ongoing basis, even before the Army makes its recommendation to the Defense Acquisition Board. Maryland regulators told the panel that the actual time to obtain permits will depend on the relative complexity of the technology and the familiarity of the regulators with it, but the range would probably be one to two years (Bowles, 1996).

Off-Site Shipping of Process Residuals

For reasons discussed in Chapter 3, shipping process residuals off site for further treatment and disposal is now being considered by the Army. This option was discussed with regulators in each state, and no insurmountable problems were identified. However, Maryland regulators noted that there currently are no in-state landfills with RCRA permits for disposing of residuals from a hazardous-waste treatment process. Therefore, solid process-residuals from the Aberdeen site would have to be shipped to another state. At present, the liquid hydrolysate from the Army program to develop the neutralization technology is shipped out of state. Indiana regulators said that, because VX residuals would be hazardous wastes, the existing TSDF permits of facilities accepting or receiving process residuals from VX neutralization would have to be amended (Indiana, 1992a). However, the regulators did not think amending the permits would be a problem.

Treatment of Synthesis Gas Combustion

Two of the alternative technologies, CEP and GPCR (gas-phase chemical reduction), produce a synthesis gas

[3]The panel is indebted to these individuals for the thoroughness and timeliness of their efforts and for their willingness to share their experience and expertise regarding potential regulatory issues that may apply to these evolving alternative technologies, even prior to receiving final design information from the TPCs.

[4]The issue of a recycling designation for these technologies was raised by state regulators in meetings with the AltTech Panel on March 12, 1995, in Indianapolis, Indiana, and on March 15, 1996, in Baltimore, Maryland.

as part of the processing of VX or HD. The syngas is burned within the on-site facility to produce energy. The environmental regulators said that they have not yet determined whether the products from synthesis gas combustion require regulation under existing hazardous-waste rules.

Pilot Demonstration of an Alternative Technology

A potentially significant issue exists for piloting an alternative technology for chemical agent destruction at the Newport site. An Indiana law prohibits the permitting of any facility to destroy chemical munitions unless the technology has been in operation at another facility at a scale comparable to the proposed facility and for a time sufficient to demonstrate that the operating facility destroys or treats at least 99.9999 percent (six 9's) of the chemical agent (Indiana, 1992b). In effect, this statute prohibits the permitting of any pilot plant for an alternative technology in Indiana. Until another facility using an alternative technology is operated elsewhere, only the baseline incineration system would qualify for permitting. However, Indiana regulators were reasonably sure that this law could be amended to accommodate pilot testing of an alternative technology, *if the technology had strong community support*.

EMERGENCY MANAGEMENT

Although state laws in both Maryland and Indiana have differing requirements for chemical-weapon destruction facilities, they share one feature. Both states require that, prior to issuing permits to such a facility, the state must plan and be prepared for emergency incidents involving the release of a chemical agent and a threat to state residents. The permitting of any technology can be delayed until adequate preparations have been made (Maryland 1987a; Indiana, 1992b). Given the requirements for emergency preparedness and the difficulties related to emergency management that the baseline system has encountered at the Tooele, Utah, facility (NRC, 1996), the panel sought information on what, if any, impediments might be encountered by the alternative technologies.

To determine if the technologies under consideration might affect the status of emergency planning in either Indiana or Maryland, the panel scheduled meetings with state and local personnel responsible for the CSEPP for the Newport and Aberdeen sites. At a meeting attended by state CSEPP directors from Indiana and Illinois (at least one county in Illinois is part of the Immediate Response Zone for the Newport site) and by emergency management directors and CSEPP personnel from several Indiana counties, two salient points emerged.

First, as long as an alternative technology did not change the geographic area encompassed by the response and planning zones that were already established, emergency management personnel believed that their preparations would not be adversely affected by selecting an alternative to the baseline system (panel meeting with CSEPP personnel at the Parke County Emergency Operations Center, Indiana, March 12, 1996).

Second, there was some concern that an alternative technology might change the assumptions about the relation between the probability of an emergency incident and its severity. Specifically, it was noted that whichever alternative technology was used, it should not increase the probability of emergency incidents, even if it decreased the consequences of those events (Keane, 1996).

The panel met with the Maryland director of emergency management on March 15, 1996. The planning and response zones for the Aberdeen site have recently been reduced in area by national CSEPP personnel because risk-mitigating actions have been implemented that reduce the risk associated with continued storage at the Aberdeen site. The director's concern was that an alternative technology might alter the new planning or response zones, with negative consequences for the densely populated area around the Aberdeen site.

TPC EXPERIENCE WITH PUBLIC INVOLVEMENT AND ENVIRONMENTAL REGULATORS

The panel asked each TPC to provide information about recent experience with and approach to (1) involving local communities in siting decisions and (2) working with environmental regulators. The panel already had information about the Army's program and plans in these two areas. Two of the three TPCs and the Army have experience that bears directly on these issues.

The TPC for the CEP technology described an aggressive and thorough public information program that targets "key leaders," special interest groups, and a variety of stakeholder groups to identify the issues they believe are important (M4 Environmental L.P., 1996f). The company has had active outreach programs in both the Aberdeen and Newport communities since January 1996 and has met with environmental regulators in both states.

In 1995, the developer of the CEP technology was awarded an EPA Merit Award for community activities associated with the Molten Metals facility at Fall River, Massachusetts. The TPC states that it is committed to a "totally open public involvement policy," which the company credits for several successful sitings and public support (M4 Environmental L.P., 1996f, pp. 3–4).

The TPC for GPCR appears from the information submitted to have less experience with public involvement programs. However, the TPC described itself as successful in both winning public acceptance and working with Canadian environmental regulators. The TPC described a public outreach program that is less detailed and has fewer outreach activities than the program described by the first TPC, but the approach has been used with successful sitings (ECO LOGIC, 1996b). The company's efforts at its Toronto project, which will provide additional experience, include a number of information meetings with the public. The TPCs for CEP and GPCR participated in all the public education meetings on alternative technologies hosted by local CACs and the Army at Aberdeen and Newport.

Based on the information provided, the TPC for Silver II appears to have less experience than the first two TPCs with public involvement, particularly in the United States (Gill, 1996). Most of the activities cited by a representative of the company have been in providing testimony at hearings for its projects. Representatives of this TPC were not present at the panel's public forums. Given the relative lack of prior experience, this TPC may require more lead time to obtain the necessary information about community needs and environmental regulations pertaining to the Aberdeen and Newport sites.

The Army has been active in the Newport and Aberdeen communities for a long time, although the character of its interactions with the communities concerning stockpile issues has evolved considerably since the mid-1980s. The Army recently expanded the public affairs program of the CSDP (Chemical Stockpile Disposal Program) and has begun to plan outreach activities for each of the stockpile sites (NRC, 1996). At the Tooele site, the Army has worked extensively with environmental regulators and the community and has succeeded in getting siting permits for a major baseline-system facility, although that facility is still undergoing systemization tests for agent disposal. The continuing presence of the Army at Aberdeen and Newport is positive and negative for future activities. The Army appears to understand the needs and concerns of the communities far better now than in the past, but it has also inherited some citizen distrust (Bradbury et al., 1994). It is not clear if the Army is willing to adapt and make a long-term commitment to more public involvement in decision-making, as it appears to be doing in the alternative technology program. However, the Army certainly possesses the necessary staff and planning capacity to implement a successful public participation program.

10

Technology Comparisons

HOW THE COMPARISON CRITERIA WERE DERIVED

This chapter provides a succinct account of how the five technologies discussed in Chapters 4 through 8 compare with one another. As noted in Chapter 2, the panel began the process of deriving these criteria by adopting three of the four critical factors identified and applied by the NRC Stockpile Committee in its *Criteria Report Evaluation*. The panel adapted those factors and the associated subfactors for use in the questionnaire sent to the technology TPCs and the Army (Chapter 3). The panel also used them to set the agenda for meetings with community groups and regulators (Chapter 9). From the questionnaires and meetings, the panel learned which aspects of the original factors were most important for characterizing and differentiating among the technologies selected for review and which were most important for expressing community concerns or regulatory issues. The panel has abstracted the most relevant aspects of the original factors and reorganized them to emphasize issues and relative differences that the panel believes are most important for supporting decisions on pilot demonstration of one or more technologies. These decisions may lead to operational implementation of an alternative agent-destruction technology at one or both of the bulk-storage sites.

Some of the evaluation subfactors presented in Chapter 2 are important but are satisfied almost equally by all the technologies selected for the panel's review. An important example is the capacity to destroy agent. All TPCs supplied test results to the panel indicating that they had successfully destroyed both HD and VX. Because of time constraints, the panel was not able to do an in depth review or analysis of the data from these important tests. The panel emphasizes that these tests were conducted under conditions that varied from conditions in a pilot-scale or fully operational facility. In addition, the tests for different technologies were not conducted under comparable conditions. Thus, it is inappropriate to infer that the particular DREs (destruction removal efficiencies) attained in these tests would be attained in an operating facility or to compare technologies on the basis of the number of 9's in the DREs calculated from these tests. Consequently, the panel has used the DRE results only to ascertain, in yes-or-no fashion, whether the technology can destroy agent. Because all the technologies have successfully demonstrated that they can destroy agent, this extremely important criterion is not included in the comparison criteria below. For a given technology, the total time to destroy agent at each site is covered under Implementation Schedule.

The next section describes the criteria for comparison as they emerged from the panel's deliberative process. Then, each of the five technologies is assessed with respect to the criteria.

THE COMPARISON CRITERIA

The panel has continued to use three headings to organize comparison criteria into groupings that are similar but not identical to groupings used in the NRC *Criteria Report Evaluation*. The headings used here are Process Performance and Engineering; Safety, Health, and the Environment; and Implementation Schedule. Some subfactors that had been located under the old heading of Process Efficacy appear among the criteria for Safety, Health, and the Environment to emphasize their relevance to those issues rather than to a narrower, process-engineering evaluation of a technology. Other subfactors are included under Implementation Schedule to emphasize community concerns and acceptance. The following brief descriptions of the criteria are intended to orient them with respect to the discussions in Chapters 2 and 9 and to indicate why the panel considers each criterion relevant for comparing the alternative technologies.

PROCESS PERFORMANCE AND ENGINEERING

This heading includes two comparison criteria taken from the Process Efficacy section of Chapter 2:

(1) technology status and (2) stability, reliability, and robustness. Table 10-1 summarizes basic engineering data for each of the evaluated alternatives, including general process description, operating conditions, and the fate of the elements from destroyed agent (that is, the form of the process residuals containing elements from the agent).

Technology Status

Except for neutralization, none of the alternative technologies has been used on a significant scale to destroy chemical agent. Only incineration and neutralization technologies have been used on agent at practical scales. However, for other wastes, the status of the technology varies from laboratory-scale to full-scale commercial operation. Furthermore, pilot designs must be sufficiently documented in TPC submissions to enable an assessment of hazard inventory and intrinsic safety. Incomplete designs required the panel either to apply its best "engineering judgment" based on the information provided or to state that significant uncertainties remain with respect to the technology's ability to meet cross-cutting requirements or to achieve the claimed capabilities.

Stability, Reliability, and Robustness

Processes that function effectively and reliably are desired. Such processes minimize unit operations, use proven components, and can be constructed from materials that are compatible with residual streams and with process conditions—including startup, shutdown, and emergency response. Frequently these processes have slow reaction rates, are operated at low temperature and low pressure, and are simple to operate and control. Although slow reaction rates are perhaps more reliable and less prone to process upsets, they may also imply greater costs because they require longer agent destruction campaigns. Slow reaction rates may also increase storage risk because the stockpile remains for a longer time.

SAFETY, HEALTH, AND THE ENVIRONMENT

The panel identified five criteria under this heading as important for differentiating among the alternative technologies or for addressing issues of major importance to decision-making. The criteria are safety interlocking, hazard inventory, test prior to release, environmental burden, and worker safety.

Safety Interlocking

The safety interlocks should be simple and proven. Process monitoring that can tolerate long time constants for appropriate response is safer and contributes to steadier plant operations, with fewer unnecessary stoppages for false alarms than monitoring that requires immediate response. For example, monitoring that would stop operations as a result of a momentary anomaly such as a temperature spike is less desirable than monitoring that responds only after the elevated temperature has been detected for a longer duration. Under the latter condition, a true process upset is more likely to exist. Also, a plant becomes inherently safer when its safety performance depends less on add-on devices and more on safety interlocks that are integral to the plant design.

Hazard Inventory

The potential for a process upset or failure seriously affecting human health or the environment increases as the inventory of hazards increases. Relevant hazards include the quantity of agent, the quantities of other reactive or toxic materials (feed materials, intermediates, and process residuals), the presence of acids or combustible gases, thermal energy, and pressure. The potential for material failures can be assessed on the basis of characteristics of the feed and residual streams integral to the processes; examples include the localized corrosion of even highly alloyed materials, stress corrosion cracking, hydrogen embrittlement, and fatigue from stress cycling. Proper selection of materials of construction will be more difficult for some technologies, those that require high-temperature corrosive environments, for example, than for others.

Test prior to Release

Members of the public have indicated a strong preference for batch processes or end-of-process operations that allow for the sampling and analytical testing of all process residuals prior to discharge (see the discussion of "closed loop" processes in Chapter 9). In general, a

TABLE 10-1 Process Engineering Data for Alternative Technologies

Engineering Paramater	Catalytic Extraction Processing	Mediated Electrochemical Oxidation	Gas-Phase Chemical Reduction	Neutralization of HD (Configurations 1, 2, 3)	Neutralization of HD Configuration 4)	Neutralization of VX
Process Description						
Medium of Treatment	molten iron or nickel	8 M nitric acid	gas phase H_2 and steam	hot water	hot water	33% aqueous NaOH
Batch or Continuous	continuous	semibatch	continuous	semibatch	semibatch	semibatch
Operating Conditions						
Process Temperature (^oC)	1600	90 (max)	850	90	90	90
Process Pressure	<10 atm. at injection; 12 atm above bath	near atmospheric	near atmospheric (slight positive pressure)	near atmospheric	near atmospheric	near atmospheric
No. of Unit Operations	>12	10	>10	7 (config. 1) to 5 (config. 3) on site	2 on site	3 on site
Electrical Power (kW•h/1000 kg)	7,400 net (HD), 25,000 excluding cogeneration	72,600 (HD) 134,900 (VX)	1,020	99,000 (conf. 1) to 23,000 (config. 3)	13,600	8,000
Fate of Agent: Ultimate Form (kg/1,000kg agent)[a]						
Carbon in HD	705 kg CO from HD; 166 kg CO from CH_4 cofeed	1,100 kg CO_2	30 kg C soot; remainder CO_2*b*	163 kg biosolids 4,563 kg CO_2 67–77 kg organics	624 kg thiodiglycol 125 kg other hydrolysis products	
Sulfur in HD	201 kg elemental S	900 kg Na_2SO_4	201 kg elemental S	799 kg Na_2SO_4 67–77 kg organics 163 kg biosolids	624 kg thiodiglycol	
Chlorine in HD	460 kg HCl	750 kg NaCl	460 kg HCl	747 kg NaCl	715 kg NaCl; remainder in other hydrolysis products	
Carbon in VX	1,152 kg CO from HD; 136 kg CO from CH_4 cofeed	1,900 kg CO_2	52 kg C soot, remainder CO_2			463 kg EMPA-Na 49 kg MPA-2Na 931 kg S/P/N organics
Phosphorus in VX	116 kg P in iron alloy	600 kg Na_3PO_4	326 kg H_3PO_3*c*			554 kg P-containing organics
Sulfur in VX	120 kg S in alloy	550 kg Na_2SO_4	127 kg elemental S			864 kg S-containing organics
Nitrogen in VX	52 kg N_2 in offgas	300 kg $NaNO_3$	56 kg N as N_2/NH_3			828 kg N-containing organics

[a]The elemental composition of 1,000 kg of HD is 302 kg of carbon, 50.3 kg of hydrogen, 201 kg of sulfur, and 446.5 kg of chlorine. The elemental composition of 1,000 kg of VX is 494 kg of carbon, 97 kg of hydrogen, 116 kg of phosphorus, 120 kg of oxygen, 120 kg of sulfur, and 52 kg of nitrogen.

[b] Total carbon for GPCR includes carbon from natural gas reformed to CO and H_2, as well as carbon from agent.

[c] Appearance of phosphorus as H_3PO_3 or its salts is hypothesized by the panel as the most likely product from the reactor, based on thermodynamics. The actual P-containing process residuals from GPCR have not been demonstrated.

hold-and-test operation prior to release is more readily implemented for liquid and solid residuals than for gaseous residuals.

Environmental Burden

Processes can vary greatly in the composition and quantity of process residuals produced during agent destruction operations. These residuals, whether in a gaseous, liquid, or solid waste stream, will ultimately be discharged into the environment. The focus should be on minimizing the overall environmental burden (composition and quantity) resulting from an agent-destruction technology.

Worker Safety

Plant safety and health risks are of particular concern to workers directly involved in agent destruction operations. The risk of worker exposure to agent or other hazards is a function of technology maintenance requirements, the degree of process automation, the duration of destruction campaigns, the quality of in-plant monitoring, and the intrinsic safety of the technology.

IMPLEMENTATION SCHEDULE

The panel identified four criteria under this heading by which the alternative technologies could be assessed for their potential impact on the implementation schedule for stockpile destruction at Aberdeen and Newport. The four criteria are technical development, processing schedule, permitting requirements, and public acceptance. These four factors can interact in various ways to shorten or lengthen the overall schedule.

Technical Development

The alternative technologies under evaluation are at various stages of design, development, and demonstration. The time for each of them to reach pilot-plant status will vary, thus affecting the overall schedule. The technology status, as discussed above, has a direct impact on the time to reach pilot-plant status. This criterion considers the likely implications of technology status on the implementation schedule.

Processing Schedule

The size of the plant, the agent-processing rate, and consequently the duration of the agent destruction campaign at a given site will vary from one technology to another.

Permitting Requirements

A major component of implementation for any technology is obtaining the necessary regulatory approvals and permits, particularly RCRA permits. Lack of complete information can considerably increase the time required. Regulators familiarity with and ability to comprehend the details of a technology can affect the RCRA permitting process. Community and governmental receptivity to a proposed technology can also influence the speed of the process.

Public Acceptance

Public acceptance of a technology can speed up regulatory decision-making. Public opposition, even by a small but determined minority, can impede implementation at many stages through litigation, extending regulatory timelines, seeking legislative redress, and other delaying actions. Public acceptance results from a program that involves affected communities meaningfully in the decision process and from decisions that reflect, at the very least, the factors the public believes are most important.

For example, the selection of a technology that is capable of treating a wide range of industrial and military wastes is a concern of the communities around the agent stockpile sites. Although current law precludes other uses of an agent destruction facility, community members fear that the facility may be used to treat wastes imported from elsewhere. Hence, technologies that are designed to treat specific agents or are otherwise not available for other uses are more acceptable to the public than technologies with wide applicability.

SUMMARY OF KEY COMPARATIVE DIFFERENCES

Table 10-2 summarizes the discussions below of how the AltTech panel evaluated each alternative technology with respect to the 11 comparison criteria. Unless otherwise noted, table entries apply to both HD at Aberdeen and VX at Newport. The table provides a quick overview of the panels evaluations, with emphasis on the differences among them. However, table entries must be interpreted not only in the context of the summary evaluations in the remainder of this chapter but also in light of the detailed analyses of the technologies in Chapters 4 through 8 and the discussion of public concerns and permitting requirements in Chapter 9. Readers are urged to study these more detailed presentations so they can better understand the table entries.

CATALYTIC EXTRACTION PROCESSING

Process Performance and Engineering

Technology Status

The following points support the panels evaluation of this technology as being ready to begin commercial operation.

- The CEP technology has had more than 15,000 hours of testing to date.
- More than 12 bench-scale units have been operated, and two commercial-scale demonstration units are in operation. A third commercial-scale demonstration unit is scheduled to start operation in the summer of 1996.
- Commercial units are ramping up to full-scale operation at two sites in Oak Ridge, Tennessee, for volume reduction of low-level radioactive wastes.

Stability, Reliability, and Robustness

CEP is an example of a complex process that has been engineered to provide a high level of stability and reliability despite its inherent complexity. It uses proven components that are tightly integrated into a continuous process with numerous unit operations. For HD destruction, the unit operations include modules for the recovery of hydrogen chloride and sulfur. The panel believes the materials of construction are compatible with the process streams that will be involved in HD or VX destruction. For example, the design and materials selection for refractories are based on an intensive development program. However, decision-makers and other concerned parties should note that the reactors operate at high temperature (1425 to 1650°C) and that the agent is injected into the reactor at moderately high pressure (less than 10 atmospheres).

Safety, Health, and the Environment

Safety Interlocking

Because CEP consists of numerous unit operations that are tightly integrated in a continuous process, a high degree of integrated process control and safety interlocking is required. Commercial-scale demonstration units have proven control systems, including safety interlocks. Two commercial-scale units designed for treating low-level radioactive waste have been started and are ramping up to full-scale operation. Process control loops and control logic, including process monitoring, have been proven. The panel believes the response times demonstrated for the monitoring and control system are adequate for safe operation with HD or VX.

Hazard Inventory

The primary hazard inherent in the CEP system is the energy stored in the high-temperature molten-metal baths. The integrity of the refractory confinement and the proximity of the molten bath to water cooling become important safety considerations. Because these issues have been addressed by the TPC, it appears the hazard inventory does not present an insurmountable impediment to the safety of the process.

The combustible offgas and the agent are also part of the hazard inventory. Prior to the introduction of agent to the reactor, the quantity of agent is identical for all the alternative technologies. The TPC plans to operate the baths in agent-destruction facilities with operators at a remote location. For HD destruction, the process includes proven, commercially available modules for recovering hydrogen chloride and sulfur (acid-management operations).

Test prior to Release

Although CEP is a tightly integrated continuous process, the TPC has provided for gaseous residuals

TABLE 10-2 Summary of Comparison Criteria for VX at Newport and HD at Aberdeen

Comparison Criterion	Catalytic Extraction	Electrochemical Oxidation	Gas Phase Reduction	Neutralization
Process Performance and Engineering				
Technology status	VX and HD destruction demonstrated at bench scale.	VX and HD destruction demonstrated at bench scale.	VX and HD destruction demonstrated at laboratory scale.	VX and HD destruction demonstrated at bench scale.
	Entering commercial operation for low-level radioactive waste.	No commercial or pilot-scale operation for other wastes.	Applied commercially (full-scale) to chlorine-containing organics.	For VX, low toxicity/burden of hydrolysate or treatability needs validation.
			For VX, phosphorus-containing products and subsequent scrubbing yet to be determined.	
Stability, reliability, robustness	Proven components tightly integrated into a well-controlled process.	Easily controllable oxidation at very low agent concentrations	Several (≥ 10) proven unit operations that require tight integration. No strongly exothermic reactions.	Low temperature, low pressure semibatch process. Standard equipment.
Safety, Health, and the Environment				
Safety interlocking	High degree of integrated process control and safety interlocks are required and have been developed.	Minimum interlocking required; reactions can be stopped easily by shutting off power.	High degree of integrated process control and safety interlocks are required; high-temperature hydrogen; temperature and pressure control are critical.	Because of interstage storage, minimal interlocking required.
Hazard inventory	Agent under pressure in delivery system. A high temperature, moderately high pressure process. High thermal mass. Combustible and reactive offgases.	Large volume of reactive reagents (HNO_3, H_2O_2, NaOH).	High temperature agent and combustible gas. Difficulty of preventing buildup of hydrogen and containing agent within a building.	Concentrated sodium hydroxide.
	For HD, large volume of toxic by-product H_2S.		For HD, large volume of toxic by-product H_2S.	
Test prior to release	Provision made for testing gases prior to combustion. Solids and liquids can be tested before shipment. Combustion gases released without analysis through stacks.	All aqueous and solid residual streams can be tested prior to release. Gases treated extensively prior to release.	All product streams can be stored and analyzed before release. Combustion gases released without analysis through stacks.	Main (aqueous) waste stream is tested.
Environmental burden	Relatively low because of high degree of recycling, especially if synthesis gas-to-energy is considered recycling.	Released residuals are common gases or salts in their most stable forms.	Low. Sulfur recycled. HCl and/or NaCl in stable, diposable, or recyclable form. State and disposition of all secondary wastes must still be defined.	For HD, aqueous discharge contains salts (NaCl, Na_2SO_2). Biomass.
				For VX, same as HD except aqueous discharge also has Na_3PO_4.

Worker safety	Process is complex but well developed. Preliminary FMEA indicates process meets safety standards.	Low temperature and low pressure. Requires handling reactive chemicals.	Hazard analysis for chlorine wastes developed. Analysis of more complex recovery/scrubbing systems required for agent.	Low temperature and pressure. Mild caustic. HD configs. 1 and 2 have H_2O_2.
Implementation Schedule				
Technical development	For HD, advanced. For VX, advanced but not as far as for HD.	Appears to be straightforward, but technology least developed of those evaluated and much engineering development remains to be done.	Advanced. Well-developed process for destroying organic wastes. For VX, phosphorus-containing products need to be determined. Integration of phosphorus recovery into process not demonstrated.	HD: 60% design status. Ready for permit application. VX: 60% design status. 6 months to resolve toxicity/treatability of hydrolysate.
Processing schedule	Approximately one year to destroy agent at each site, after systemization. Operations at Newport will not begin until Aberdeen activities are completed.	Controlled by number of modules in facility.	For each site, approximately one year to construct and systematize, one year to destroy stockpile.	15 months to systemization. Less than one year to destroy stockpile, operating at full rate.
Permitting requirements	Current documentation adequate for timely review (unclear whether RCRA permit required).	Process novelty could lengthen permit review time.	Full-scale facilities permitted outside U.S. Permitting strategy submitted to panel.	Regulators indicate technology would take the least time to permit.
Public acceptance	Attractive if seen as recycling. High temperature and pressure not attractive to public. Stack emissions may be a concern. BDAT designation for incinerable wastes may be positive (proven technology) or negative (versatile for other wastes).	Meets key preferences of public: low temperature, atmospheric pressure, closed loop.	Perceived as a closed loop system with provision for testing before release. Low pressure but high temperature. Hydrogen may be perceived as a risk. Stack emissions may be a concern.	Concern about Army management of the technology. Low temperature and pressure. Closed loop batch process. Testing before release. CACs favor it. Agent-specific, not easily applied to other wastes.

(synthesis gas) to be held for analytical testing prior to combustion. The products of combustion, however, are not tested prior to release to the atmosphere. Metal, sulfur, and ceramic process residuals are solids and will be tested prior to shipment off site. Recovered HCl will be tested prior to shipment off site. Process cooling water will be tested.

Environmental Burden

The TPC proposes to minimize the environmental burden by producing metal, HCl, and elemental sulfur as by-products that will be offered on the commercial market for reuse. The technical and economic feasibility of marketing these by-products is yet to be established. The process design also includes burning the offgas, which is rich in H and CO, in a gas turbine to generate electricity for in-plant use. The ceramic slag will require disposal.

Worker Safety

Although CEP is complex, the panel judges it to be robust enough and sufficiently developed that worker safety and health risks are satisfactorily low. A preliminary FMEA (failure modes and effects analysis) based on the conceptual design for destroying chemical agents has not revealed any unacceptable or abnormal risks.

Implementation Schedule

The TPC provided the panel with a detailed schedule. The panel judges this schedule to be reasonable for the complete destruction of the stockpiles at Aberdeen or Newport by 2004, provided there are no unforeseen delays.

Technical Development

Development efforts by the technology developer and the TPC are sufficiently advanced that, as of May 1996, they were ramping up to commercial operation of a facility (see Technology Status). The panel views this and the other status factors as a strong indication that technical development will not delay the TPC's schedule.

Processing Schedule

The TPC's schedule mentioned above includes approximately one year for HD processing at the Aberdeen site and one year for VX processing at Newport, once facilities are ready for full-scale operation with agent. However, because the TPC intends to use the same equipment at both sites, operations will not begin at Newport until agent destruction at Aberdeen is complete.

Permitting Requirements

The TPC has extensive experience in dealing with regulators and the public; it has obtained permits for CEP facilities in Massachusetts and Tennessee. The EPA has granted the companys technology the status of a best demonstrated available technology (BDAT). This designation means it has been judged to be equivalent in performance to incineration (the other BDAT). The EPA has also determined that CEP is not incineration.

State regulators have not decided whether the technology requires a RCRA permit when used to destroy chemical agent. In other applications, the TPC has not been required to obtain a RCRA permit on the ground that the process was in those instances judged to be resource recycling rather than waste treatment.

Public Acceptance

The discussion above in the Permitting Requirements section is relevant to public acceptance of this technology. In addition, the TPC has mounted a public education program in the communities around Aberdeen and Newport to explain the beneficial aspects of its technology, with particular emphasis on its recycling characteristics. To date, comments at public meetings have been generally positive.

However, the panel is not sure about longer-term reactions as the communities gain a fuller understanding of all the alternative technologies reviewed here. CEP is a high-temperature, moderately high-pressure process. The combustion of the offgas does entail stack emissions. (The current design provides for testing of the offgas prior to combustion and the release of combustion products to the atmosphere.) The EPA designation as a BDAT alternative to incineration is likely to affect some in the community positively because it shows the technology has passed a significant standard of governmental review and acceptance. However, it also means the process has demonstrated versatility for treating a wide variety of materials, including other hazardous wastes. As explained in Chapter 9, various community members have voiced opposition to processes with high temperature and pressure, processes that involve the release of combustion gases, or processes that could be used to treat a variety of hazardous wastes. From an engineering perspective, the panel views the CEP technology as well engineered to protect the public and the environment. Whether the interested communities will concur is an open question.

ELECTROCHEMICAL OXIDATION

Process Performance and Engineering

Technology Status

The TPC has demonstrated destruction of HD and VX in laboratory tests with a 4-kW cell consisting of a single anode-cathode pair. A facility for tests at larger scale, processing approximately 250 g of agent per hour, has been built and is undergoing commissioning tests with an agent surrogate. This facility includes a 4-kW electrochemical cell, anolyte and catholyte feed circuits, an anolyte offgas condenser, an NO_x reformer system, and a modified version of the combined offgas treatment circuit. Tests with VX and HD at this facility are planned. A small-scale version of the silver recovery system will be tested on the anolyte and catholyte solutions from the 4-kW facility.

Stability, Reliability, and Robustness

The agent-destruction system operates at low temperatures and atmospheric pressure. The processes in the unit operations of the system are not sensitive to small excursions in composition or temperature. Rapid or runaway changes that might create emergency conditions are highly unlikely. Therefore, the response time required for control instrumentation is not very demanding. However, compositions of some constituents will change substantially during the course of a campaign, and a test program is needed to verify that the planned control systems are adequate to ensure stability over the full range of composition that will occur during operation. For processing HD, removal of precipitated silver chloride is essential for reliability during a 5-day campaign, and the equipment proposed to accomplish this will need to be tested at loadings and conditions like those in full-scale operation.

Although a runaway condition is unlikely, the system does produce large heat loads in relatively small volumes. Temperature control in each of the unit operations and in the system as a whole must be tested and validated.

Safety, Health, and the Environment

Safety Interlocking

The electrochemical oxidation process consists of several unit operations, but they do not have to be tightly integrated. Temperature, pressure, and chemical concentrations will be monitored closely. If the monitoring data signal a malfunction, the cell current can be rapidly shut down. Once the problem has been found and corrected, restarting the process is straightforward.

Hazard Inventory

The agent feed rate of about 0.01 m^3/h for each 180-kW cell implies that 12.7 kg/h of HD is added to each cell, or 10 kg/h of VX. Because the agent will be rapidly hydrolyzed by the concentrated nitric acid, the panel expects that the inventory of agent in the anolyte circuit at any given time would be far less than 12.5 kg (equivalent to 5,000 ppm in a 2.5-m^3 anolyte volume).

The process requires handling highly corrosive or reactive materials such as nitric acid, concentrated sodium hydroxide solutions, hydrogen peroxide, and 90 percent oxygen gas. Worker-safety training and chemical containment are therefore paramount concerns, but harmful releases to the surrounding community are unlikely.

Test prior to Release

All liquid and solid reaction products will be tested prior to release. Gaseous products will not be tested prior to release but will be treated extensively to ensure the removal of any agent and of volatile organic contaminants formed in the electrochemical cell. Moreover, reaction conditions such as temperature, pressure, and the basic reaction mechanism ensure very low concentrations of agent and other organics in the feed to the gas-cleaning system. In the panel's judgment, the offgas circuit could be modified to accommodate hold-and-test prior to release, if that step is required.

Environmental Burden

The major liquid process residual is an aqueous solution of common salts: sodium chloride, sodium sulfate, sodium phosphate, and sodium nitrate. The solution will contain silver at a concentration below applicable regulatory standards. (The U.S. standard is 50 ppb).

Gaseous effluents are anticipated to be primarily CO_2, O_2, and N_2.

Worker Safety

The worker safety concerns for this technology relate to handling agent and to possible exposure to some highly reactive chemicals. The agent handling procedures will be the same as for other technologies under review. The reactive chemicals of concern have been listed above: nitric acid, nitrogen oxide gases, hydrogen peroxide, and sodium hydroxide.

The chemical of most concern is nitric acid, which is a particularly hazardous and reactive material. It is, however, a common industrial chemical, and the TPC has had experience handling it in the nuclear fuel processing industry. In addition, most of the equipment operates at near-atmospheric pressure. A sound safety program will ensure a high level of worker safety.

Implementation Schedule

Technical Development

The basic oxidation reactions of the Silver II process have been demonstrated at laboratory scale on many materials; there is little doubt that a high level of agent destruction is possible. The entire process is complicated by the recovery of all the reaction products, as well as the silver reagent. The resultant overall process thus requires a large number of unit operations. In the case of VX, these operations appear to be straightforward and raise no critical or difficult problems of control or operation. However, additional engineering and demonstration will be required.

The oxidation of HD raises a technical issue because of the high chlorine content of HD. In the working solution, the chlorine precipitates with silver as solid silver chloride. Whether the process can be operated satisfactorily with a large amount of solid precipitate accumulating in the cells during a campaign remains to be demonstrated. Some initial operability of the process with HD will be observed in an experimental program that started recently. This technology is the least developed of the technologies evaluated by the panel.

Processing Schedule

The facility design is based on a modular standard unit: two 180-kW cells of a commercial design form a 360-kW unit. As an example, one unit is capable of destroying 2 tons of mustard in approximately 3.5 days. The inventory of HD at Aberdeen could be destroyed by a facility of three standard units in 4 years; destroying the inventory of VX at Newport over the same 4-year period would require five units. The schedule for complete destruction of the stockpile at either site could be accelerated by increasing the number of modular units in the facility.

Permitting Requirements

Electrolytic oxidation processes have been used in industry but not for destroying hazardous wastes. The Silver II process would probably be viewed as novel by regulators, who would have little, if any, experience to rely on.

In addition, the offgas treatment process has features that are not extensively used in waste-treatment applications and would require validating demonstrations. These features include oxidation by hydrogen peroxide to clean up the final traces of NO_x and organic residuals. Again, lack of regulator familiarity with the technology might delay the permitting process.

Public Acceptance

The communities near the Aberdeen and Newport sites have stated their preference for low temperature, low pressure, "closed loop" processes. The Silver II electrochemical process comes close to meeting all of these preferences. Most of the heteroatoms in the agents

(P, S, and Cl) will be oxidized to stable acids or salts in solution, which can be analyzed before release. The carbon will be oxidized to CO_2 and will be released to the atmosphere on a continuous basis after a cleanup that includes scrubbing with hydrogen peroxide to remove any trace organic compounds in the gas, followed by filtering through activated carbon. Thus, the process comes close to being "closed loop," as well as destroying agent at low temperature and low pressure.

GAS-PHASE CHEMICAL REDUCTION

Process Performance and Engineering

Technology Status

The GPCR (gas-phase chemical reduction) process has been demonstrated at commercial scale for treating several organic wastes including chlorocarbons such as PCBs and hydrocarbons such as toluene. This commercial experience provides a substantial basis for assessing operational requirements and related considerations, mass balances (although the panel received little quantitative information on the actual PCB operation), gas recycling, secondary waste stream management for HCl, and operation of the catalytic reformer.

The reactor is clearly capable of destroying chemical agent. However, the presence of heteroatoms other than chlorine (sulfur, phosphorus, and nitrogen) in the agents increases the complexity of the total system because additional operations are needed to remove products containing these atoms from the process-gas stream. The sulfur in HD and VX will appear as H_2S in the process gas and will be recovered as elemental sulfur. The scrubbing and sulfur conversion require a number of additional, albeit commercially available, unit operations.

Because the fate and handling of phosphorus-containing materials are still uncertain on both a fundamental and practical level, this technology is not as mature for VX destruction as it is for HD. Two main uncertainties exist (which the TPC has acknowledged): (1) the principal phosphorus-containing products exiting the reactor have yet to be identified, and (2) a method must be demonstrated for scrubbing the phosphorus-containing products from the process gas and treating them to yield residuals suitable for disposal. Thermodynamic considerations suggest that oxyphosphorus acids and elemental phosphorus will be the predominant reaction products. The TPC reported little experience with these issues even at bench scale. Although the TPC has developed a plan for addressing these issues, the time needed to resolve them is unclear.

Another open issue is whether operation on agent will require monitoring stack gases from the combustion of fuel and process gas.

Stability, Reliability, and Robustness

The GPCR reactor has been used commercially and has proven to be reliable. The process operates at high temperature and near-ambient pressure. None of the reactions are strongly exothermic, and the methane-reforming reaction is strongly endothermic.

However, the entire process consists of a number of sequential unit operations that must be tightly integrated and controlled. The recovery of solid process residuals (those containing phosphorus, sulfur, chlorine, nitrogen, and solid carbon) and the manufacture of hydrogen via steam reforming are carried out continuously with the gas-phase reduction in a recirculating gas loop. For simple chlorocarbons, the information provided by the TPC indicates that the overall system has been stable and reliable in operation. The reliability of the more complex system required for processing HD or VX will need to be demonstrated; tighter controls will certainly have to be implemented. In such a tightly integrated system, failure in one unit operation could significantly affect others. For example, any carryover of sulfur or phosphorus from the scrubber train to the steam reformer can poison the reforming catalyst.

The current materials of construction, which have apparently worked reliably in the presence of chlorine, are to be used for agent destruction. In this different chemical environment, problems could develop that will have to be addressed by the TPC.

Safety, Health, and the Environment

Safety Interlocking

From a safety standpoint, the most important parameter for GPCR is maintaining a slightly positive pressure throughout the recirculating gas circuit to avoid potential explosions from oxygen leaking into the circuit. The TPC has demonstrated provisions for pressure controls and interlocks. Upgrading monitoring and control-room equipment to include new technology would further enhance the safety envelope.

Hazard Inventory

The inventory of high-temperature hydrogen in GPCR presents a number of potential safety issues in the context of an agent destruction facility. The entire system must be maintained at slightly positive pressure. In addition to standard safety protocols for working with hydrogen, additional procedures for managing leaks must be developed. Hydrogen leaks that occur in the open are generally manageable and present no inordinate hazard, but leaks in a secondary-containment building could cause an explosion if the gas accumulates. All of the TPC's existing facilities operate in the open without secondary containment and without area monitoring for hydrogen or feed material. For this technology to be used in an agent-destruction facility with secondary containment, future designs must address the difficulty of preventing H_2 buildup in the building while maintaining the integrity of this containment as backup protection against the accidental release of agent. Also, area monitoring for both H_2 and agent will be required for safe operation within the secondary containment building.

Strong acids and bases are used or created in the scrubbing systems. H_2S, which must be scrubbed from the process gas and converted to elemental sulfur, is extremely toxic.

Test prior to Release

The process gas stream that goes to the steam boiler for combustion is held in tanks and tested prior to combustion, although the products of combustion are not tested prior to release through the stack. Solid and liquid residuals are to be tested prior to release.

Environmental Burden

Aside from the uncertainties about phosphorus, all the inorganics derived from the heteroatoms present in the agent are ultimately converted to common salts, salt solutions, or elemental sulfur. The TPC's submission did not detail the final disposition of all these materials.

The process has two stacks for releasing combustion gases to the atmosphere: one for the propane burner that heats the SBV (sequencing batch vaporizer) and the other for the steam boiler, which burns a mixture of process gas and propane. Based on the design and the TPC's experience, these gas streams should be "clean." Nonetheless, if GPCR is selected for pilot-testing, the TPC and the Army will need to address the issues typically raised about trace products of combustion and the release of combustion products to the environment.

Worker Safety

The intrinsic safety of the technology was discussed above. In-plant monitoring must be upgraded for use with agent and hydrogen in a facility with secondary containment. Standard hydrogen safety procedures, which are well documented, must be employed.

Implementation Schedule

Technical Development

Work in progress should identify the fate and necessary treatment of the phosphorus products. Except for provisions for increased monitoring, a secondary containment, and the engineering necessary for managing the sulfur and phosphorus wastes, the technology is developed to the point that a system like the TPC's existing commercial systems could serve as a pilot operation. Still to be resolved are the schedule implications of accommodating secondary containment and providing related reengineering. The TPC's submission assumes that the Army will provide the secondary containment building and ancillary nonprocess facilities.

Processing Schedule

The TPC's schedule calls for processing 5 metric tons of agent per day. This rate of operation, with about a 20 percent downtime, would require about one year to destroy the stockpile at the Aberdeen site. The panel estimates that the processing time at each site would be closer to two years.

Permitting Requirements

The TPC has received environmental permits for commercial operations in both Canada and Australia. However, no commercial operations have yet been sited in the United States, so the TPC has not been through the permitting process here. Several demonstrations at pilot scale have been carried out in the United States. The TPC's personnel do have considerable experience with permitting issues in general (several come from a regulatory background), so the panel expects that the TPC can handle the necessary permitting and regulatory issues. For agent destruction, the need for secondary containment along with the potential for hydrogen leaks inside this containment could affect permitting requirements. This issue remains to be addressed in coordination with the Army.

Public Acceptance

The TPC states that the GPCR process has been well received and supported by the public. The process was tested by the EPA under the Superfund Innovative Technology Evaluation program. According to the TPC, several state departments of environmental quality and health have stated that the process is acceptable for treating sites contaminated by chemical wastes (e.g., the Colorado Department of Health for remediation at Rocky Mountain Arsenal).

However, the technology has characteristics that some members of the communities near the Aberdeen and Newport sites have stated to be objectionable or contrary to their preferences. The process operates at high temperature and slightly positive pressure, whereas a preference for low-temperature processes has been expressed in both communities. A portion of the process-gas stream is burned in a conventional boiler, and the products of combustion are released to the atmosphere through a stack. The existing design includes provisions for holding the process gas for analysis and confirmation of composition before combustion.

NEUTRALIZATION OF HD

Process Performance and Engineering

Technology Status

he TPC has demonstrated neutralization of HD with hot water at bench scale (114-liter reactors). The neutralization process is simple and uses conventional reactors common in the chemical industry. Additional complexity arises from treating the product of neutralization (hydrolysate) on site.

The biological oxidation of HD hydrolysate (primarily an aqueous solution of thiodiglycol) by mixed bacterial cultures in a SBR (sequencing batch bioreactor) has also been demonstrated at bench scale. SBRs are in commercial operation for other applications. The biodegradation of HD hydrolysate can also be carried out effectively off site at a commercial TSDF (treatment, storage, and disposal facility).

Stability, Reliability, and Robustness

The neutralization of HD is simple and easily controlled. Because the equipment is standard for the chemical industry, it should be reliable. The semibatch process operates at low temperature and atmospheric pressure, and the energy content of the reaction mixture is low. These characteristics preclude uncontrollable or runaway reactions. The biodegradation process should be similarly stable and reliable, except for possible upsets in microbial activity from loss of air or cooling.

Safety, Health, and the Environment

Safety Interlocking

The unit processes, such as neutralization and biodegradation, operate independently of each other with interstage storage of the aqueous process stream. Therefore, only minimal interlocks are required. The process is monitored by analyzing for residual agent before the effluent (hydrolysate) is released from the neutralization reactor.

Hazard Inventory

The inherent hazard potential, apart from the hazards associated with handling agent, is limited because the aqueous streams are nonflammable and at low temperature (90°C) and pressure (1 atm gauge).The process does require that workers handle sodium hydroxide at concentrations of 18 to 50 percent.

Test prior to Release

The hydrolysate from the neutralization reactor is tested for the presence of residual agent before release from the toxics control area. The hydrolysate will be released only if the agent concentration is below 200 ppb. (The analytical detection limit is 10 ppb.) The consistency of this standard with Army agent-treatment standards needs to be evaluated. Process vapors are monitored for agent. They are scrubbed through a sodium hydroxide solution and passed through multiple carbon filters before release.

Environmental Burden

The major liquid process residual after biotreatment (either on site or off site) is a large volume of a dilute aqueous solution of sodium chloride, sodium sulfate, and unbiodegraded organic compounds. Toxicity testing using bioassays has indicated that the remaining toxicity is low and primarily a consequence of total salt concentration. This effluent stream should be demonstrated to be of acceptably low toxicity before discharge. The major solid residual is biomass in the form of bacterial cell material that resembles municipal sewage sludge. The largest gaseous residual will be oxygen-depleted air from the bioreactors, which will be water-saturated and will contain carbon dioxide.

Worker Safety

The major potential for exposure of workers to agent is in handling the ton containers before and after the agent is pumped out. This operation is common to all of the technologies. Agent destruction and waste disposal are carried out at low temperature and pressure, conditions that limit the possibility of injury. Handling sodium hydroxide solutions requires care, but the requisite practices are standard in the chemical industry.

Implementation Schedule

Technical Development

The TPC has obtained considerable operating experience and some basic process data from bench-scale testing in reactors ranging up to 114 liters. The TPC plans to pilot-test the process in what would be a single module of a multimodule full-scale, full-rate facility. This approach reduces the risks, and should reduce the time, involved in scaling up from pilot test to full-scale operation. The design of the pilot/production facility appears to be completed to the point at which the technology is ready for permit applications.

Processing Schedule

The schedule proposed by the TPC calls for about 15 months of systemization and low-rate operations. Full-rate operation of the multimodule facility is projected to continue for nine months.

Permitting Requirements

There appear to be no statutory barriers to obtaining permits for the HD neutralization–biodegradation technology. The favorable reaction from the Aberdeen community to this technology should allow necessary permits to be issued in about one year. The Army constraints on shipping the hydrolysate need to be modified to allow either off-site biodegradation of the hydrolysate at a TSDF or on-site biodegradation followed by discharge of the liquid effluent to a FOTW (federally owned treatment works).

Public Acceptance

A neutralize-and-ship process, such as configuration 4, seems likely to gain public acceptance because it meets four important criteria supported by the Aberdeen community and the Maryland CAC: (1) a full-containment (closed loop) process with controllable emissions; (2) low-temperature, low-pressure processing; (3) simplicity; and (4) an agent-specific technology, in the sense that the facility would require extensive modification to process a wide range of other wastes.

NEUTRALIZATION OF VX

Process Performance and Engineering

Technology Status

The TPC has demonstrated neutralization of VX with aqueous sodium hydroxide at a bench scale (114-liter reactors) This neutralization process closely resembles the water hydrolysis or caustic hydrolysis of HD. More than 350 kg of VX were destroyed in the bench testing. The neutralization process is simple and uses conventional reactors common in the chemical industry.

The hydrolysates from the bench tests were oxidized with sodium hypochlorite (bleach) and then were treated and disposed of, within permit requirements, by a TSDF, which used biodegradation in its processing. The efficacy of off-site biodegradation has not been validated through detailed treatability studies. However, the panel's preliminary assessment suggests that the toxicity of the hydrolysate may be sufficiently low that complete biodegradation is not necessary during disposal at a TSDF. As an alternative, existing commercial processes other than biodegradation could be used either at an off-site TSDF or on site, if further treatment is necessary.

Stability, Reliability, and Robustness

The low-temperature, low-pressure, semibatch processing should be stable and reliable. The hydrolysis reaction is mildly exothermic (heat-releasing), but the relatively low energy content of the hydrolysis mixture precludes uncontrolled or runaway reactions. The simple unit processes and standard equipment closely resemble well-tested counterparts in the chemical industry.

Safety, Health, and the Environment

Safety Interlocking

The unit processes, such as ton-container processing and VX neutralization, operate independently with interstage storage of the aqueous process stream. Therefore, only minimal interlocks are required. The hydrolysate from neutralization is analyzed for residual agent before it is released from the toxics control area.

Hazard Inventory

The inherent hazard potential, except for the hazards associated with handling agent, is limited because the aqueous streams are nonflammable and at low temperature and pressure. The hydrolysate retains some non-agent toxicity. The process requires handling corrosive caustic and bleach solutions, but the procedures for doing this are standard in the chemical industry.

Test prior to Release

The hydrolysate from the VX neutralization reactor is tested for the presence of residual agent and for a toxic by-product (EA-2192) before release from the toxics control area. Process vapors, which are monitored for agent, are scrubbed through a sodium hydroxide solution and passed through carbon filters before release. Emptied storage containers are steam cleaned and tested for the presence of agent vapor before being shipped to the Rock Island Arsenal in Illinois for melting.

Environmental Burden

The major liquid process residual for off-site treatment and disposal is the hydrolysate, which appears to have low toxicity. As a consequence of dilution during the process, the volume of the hydrolysate is much greater than the volume of agent treated. The major solid residual is biomass. The nitrogen contained in the agent is incorporated into the biomass.

Worker Safety

The major potential for exposure of workers to agent is in handling the ton containers before and after agent is pumped out. This operation is common to all of the technologies. The hazards of neutralization are limited by the low temperature and pressure of the process. Handling sodium hydroxide and sodium hypochlorite solutions requires care, but the requisite practices are standard in the chemical industry.

Implementation Schedule

Technical Development

The TPC has considerable operating experience and some basic process data from bench-scale testing with agent in reactors ranging up to 114 liters. The TPC plans to pilot-test the process in what would be a single module of a multimodule facility. This approach reduces the potential for schedule delays—and should reduce the time—in scaling up from pilot test to full-scale operation. The design of the of the pilot/full-scale facility is advancing rapidly, and the technology appears ready for permit applications.

The panel estimates that a maximum of six months should suffice to resolve the issues related to toxicity of the hydrolysate and to perform detailed treatability studies of hydrolysate biodegradation, if further treatment is required to reduce toxicity. If these issues cannot be resolved quickly, another proven process for treating the hydrolysate prior to disposal can be selected.

Processing Schedule

The schedule proposed by the TPC calls for about 15 months of systemization and low-rate operations. Full-rate operation of the multimodule facility is projected to continue for nine months.

Permitting Requirements

Implementing the TPC's plan to pilot-test VX neutralization in one module of what would become the multimodule full-scale, full-rate facility at Newport will require modification of the Indiana statute that mandates prior success of the technology at a comparable facility elsewhere. There appear to be no other statutory barriers to acquiring permits for the neutralization pilot plant. It should be feasible to modify a TSDF permit to allow shipping and treating the hydrolysate. Based on discussions with regulators, the panel estimates that acquiring the permits for the neutralization facility may require one year.

Public Acceptance

The neutralization process seems likely to gain public acceptance because it meets four important criteria supported by the Newport community: (1) a full-containment (closed loop) process with controllable emissions; (2) low-temperature, low-pressure processing; (3) simplicity; and (4) an agent-specific technology, in the sense that the facility would require extensive modification to process a wide range of other wastes.

11

Findings and Recommendations

Destruction of the unitary chemical agent stockpile is a complex undertaking. However, the challenges at the Aberdeen and Newport sites are considerably lessened by the relative simplicity of the stockpiles at these sites: a single agent at each site, stored in bulk (ton) containers.

Because of concerns about emissions from incineration, the neighboring communities have insisted that alternative technologies be implemented at these sites. They are also concerned that the selected alternative be safe with respect to public health and the environment, cost-effective, and implementable within a reasonable time. Perhaps most important, they want to be meaningfully involved in the decision process.

The following findings and recommendations are based on the AltTech Panel's in-depth technical evaluation and assessment of five alternative technologies: catalytic extraction, electrochemical oxidation, gas-phase chemical reduction, stand-alone neutralization or neutralization followed by biodegradation for HD, and stand-alone neutralization or neutralization followed by biodegradation for VX. The panel evaluated these technologies for the particular application of destroying HD blister agent or VX nerve agent stored in bulk containers. The panel's findings and recommendations are specific to this application of the technologies and do not encompass other applications, including application to other agents or other storage sites. The panel believes that its efforts to obtain public views on the criteria used in these technical evaluations will result in public support and acceptance of its recommendations. Furthermore, the panel's findings and recommendations reflect information on environmental regulations relevant to the pilot-testing and eventual full-scale operation of an alternative technology.

GENERAL FINDINGS

General Finding 1. Since the 1993 NRC report, *Alternative Technologies for the Destruction of Chemical Agents and Munitions* (NRC, 1993), there has been sufficient development to warrant a re-evaluation of alternative technologies for chemical agent destruction. However, the developmental status of the technologies varies widely, the time required to complete pilot demonstrations will also vary.

General Finding 2. All the technologies selected for the panel to review have successfully demonstrated the ability to destroy agent at laboratory scale.

General Finding 3. Members of the communities near the Aberdeen and Newport sites want an alternative to incineration that has the following characteristics: operation at low temperature and low pressure; simplicity; the capability to test all process residuals prior to release; and minimal potential for detrimental effects, short term or long term, on public health and the environment. Although the communities do not want treaty or legislative schedules to drive decisions on technology options, they want the stockpiles at the two sites to be destroyed as quickly as possible.

General Finding 4. Based on the panel's discussions with state regulators, all the technologies appear to be permittable under the Resource Conservation and Recovery Act and associated state regulations within one to two years of application submission. The time will depend on the complexity of the technology and the regulators' familiarity with it.

General Finding 5. As complete processing systems for chemical agent, all the technologies reviewed are of moderate to high complexity. Although components of each process are standard and proven, no alternative is an off-the-shelf solution as an agent-destruction process. Any one of them will require extensive design review, hazard and operability studies, materials selection, and related work as it moves through the piloting stage to full-scale demonstration and operation. During this necessary preparation for implementing an agent-destruction system, everyone involved should bear in mind that most failures in complex, engineered systems occur not during steady-state, normal operations but during transient conditions such as startup, shutdown, or operator responses to deviations from design conditions.

FINDINGS AND RECOMMENDATIONS FOR THE ABERDEEN AND NEWPORT SITES

Specific Finding 1. The Army required each TPC (technology proponent company) to demonstrate the capacity of its process to destroy agent in a government-approved laboratory. Each TPC supplied test results to the panel indicating the process had successfully destroyed both blister (HD) and nerve (VX) agents. Because to time constraints, the panel was not able to review and analyze in depth the data from these important tests. However, two key issues stand out.

First, the tests were conducted under conditions that varied in different ways from conditions in a pilot-scale or fully operational facility. It is therefore inappropriate to expect that the particular DREs attained in the tests would be the same as DREs attained in an operating facility. It is also inappropriate to compare technologies on the basis of which attained more "9's" in the DRE results. Given the lack of comparability between the test conditions and the scaled-up facility for an individual technology and the differences in test conditions for different technologies, the panel has used the test results only to address, in yes-or-no fashion, whether a technology can destroy agent.

Second, the by-products of any agent-destruction process are of significant concern to the panel, the neighboring communities, and the regulators. A DRE gives no information on the composition and concentration of by-products that may be hazardous to human health or the environment. The panel had insufficient time to analyze the comparability of the tests with respect to methods of detection of by-products, completeness of coverage of potential products of concern (particularly those produced in trace quantities), limits of detectability under the test conditions, and other parameters essential to understanding the toxicologic and environmental hazards associated with residuals from the technology. An in-depth, independent analysis of these test data will be necessary to support future Department of Defense decisions about proceeding with pilot-testing. This analysis may show that further independent testing is needed.

Recommendation 1. For any technology that is to be pilot-tested, the Army should support an in-depth analysis of the agent-destruction test results by a competent, independent third party not associated with the Army or any of the TPCs. This analysis should address (1) the comparability of the test conditions to process conditions of anticipated pilot-scale and fully operational facilities, (2) the extent to which reported results for agent destruction and detection of by-products are comparable across the tests, and (3) weaknesses or omissions in the testing—whether for agent for the destruction or detection of by-products, including trace quantities of toxic by-products—that must be addressed in subsequent testing of the technology as an alternative for agent destruction at Aberdeen or Newport.

Specific Finding 2. Current Army prohibitions on off-site treatment and disposal of process residuals unduly restrict the options for stockpile destruction. No toxicologic or risk basis for the proposed Army release standards has been developed. In addition, there appears to be an inconsistency among limits for airborne exposure and for residual concentrations in liquid or solid materials that are to be released from agent handling facilities to off-site facilities for subsequent treatment and disposal.

Recommendation 2a. Standards for releasing wastes should be evaluated on a clearly defined regulatory and risk basis that takes existing practices into account. Standards should be revised or established as necessary.

Recommendation 2b. The Army should review and revise current restrictions on off-site treatment and disposal of process liquid and solid residual streams to allow treatment and disposal of the process effluents from agent destruction at permitted off-site treatment, storage, and disposal facilities and at permitted FOTW (federally owned treatment works) for wastewater.

Specific Finding 3. The panel determined that the development status of the technologies assessed and the lack of long-term experience with their use for the destruction of chemical agent necessitate a comprehensive design review of any selected technology prior to the construction of a pilot plant. Reliability of the facility, as affected by system design, control, operation, maintenance, monitoring, and material selection, must be thoroughly evaluated.

Recommendation 3. A detailed, comprehensive design review of any selected technology or technologies should be performed prior to starting pilot-plant construction. This review should examine reliability as affected by system design, controls, operation, maintenance, monitoring, and materials selection.

Specific Finding 4. The panel has found that, no matter which technology is selected for potential use at either site, the affected communities insist that they be included in a meaningful way in the process leading up to key decisions, including the decision to proceed to pilot demonstration. At a minimum, a meaningful community involvement includes:

- determining, with community input, the nature and extent of involvement the community wants and how it can be achieved
- ensuring that the infrastructure exists to support this involvement
- updating the TPC packages in the information repositories located in the affected communities to ensure that the public has access to the latest, most complete information
- seeking additional ways to sustain and deepen the dialogue between the Army and the communities and the exchange of views within the communities

Recommendation 4. The Army should take immediate steps, if it has not already done so, to involve the communities around the Aberdeen and Newport sites in a meaningful way in the process leading up to the Army recommendation to the Defense Acquisition Board on whether to pilot-test one or more alternative technologies.

Specific Finding 5. The risk assessment performed by MitreTek Systems, Inc., was not available to the AltTech Panel until very late in the preparation of this report. As was noted in Chapter 2, the panel assumes that more-rigorous, site-specific assessments will be done at an appropriate time before a full-scale facility for agent destruction is built and operations on agent begin. These required assessments include a quantitative risk assessment and a health and environmental risk assessment.

Recommendation 5. Before any technology is implemented at a stockpile site, an independent, site-specific quantitative risk assessment and a health and environmental risk assessment should be completed, evaluated, and used in the Army's risk management program.

Technology Selection

The panel's evaluation criteria presented in Chapter 10 favor technologies with the following characteristics:

- inherent process safety, which includes consideration of the required safeguards or engineering controls, the potential for process upsets, the requirements for process control interlocking, reliability, and the hazard inventory
- technical maturity, as shown by such factors as the scale of demonstrated ability to process agent and commercial industrial experience with the equipment, systems, and processes that would be required for an agent-destruction facility
- process simplicity, judged by such factors as the number of unit operations required and the ease of scale-up to a full-production facility

Based on these priorities, the panel reached consensus on the following findings and recommendations on alternative technologies to be pilot-tested for agent destruction at the Aberdeen and Newport sites.

HD at Aberdeen

Specific Finding 6. Aqueous neutralization of the chemical agent HD followed by biodegradation of the hydrolysate surpasses the priority criteria listed above. This technology has the following advantages:

- Among the alternatives reviewed, it has the largest-scale successful demonstration with agent.
- The equipment required has been proven through extensive use in industry for processes similar to those planned for use in agent destruction.
- The principal unit operations are independent batch processes that do not require elaborate safety interlocking.
- Because the process involves batch processing of liquids, hold-and-test analyses to determine batch composition can be readily performed at several points in the process.

- The process is performed at low temperature and near-atmospheric pressure; the hazard inventory in general is low.
- The selection of materials of construction appears to be straightforward.
- No step in the process involves combustion; therefore no combustion products are emitted.

Recommendation 6. The Army should demonstrate the neutralization of HD at Aberdeen on a pilot scale.

- The AltTech Panel recommends biodegradation of hydrolysate from HD at an off-site treatment, storage, and disposal facility (configuration 4 in Chapter 7) as the most attractive neutralization configuration. Of the four neutralization configurations described in Chapter 7, this one is the most reliable and robust; has little potential for process upsets; makes more use of existing facilities and trained staff, rather than requiring new facilities and newly trained staff; should be most rapidly permittable; should have the shortest implementation schedule; and should be the quickest and easiest to decommission.
- The second-best configuration is neutralization with biodegradation on site, followed by disposal of the aqueous effluent through a FOTW. If this option is selected, the panel recommends separating the VOCs (volatile organic compounds) prior to biodegradation, followed by off-site treatment and disposal of the VOCs.

VX at Newport

Specific Finding 7. Neutralization of chemical agent VX with sodium hydroxide solution destroys agent effectively and substantially lowers the toxicity of the process stream. With respect to the priority criteria listed under Technology Selection, this technology followed by off-site treatment and disposal of the hydrolysate has the same relative advantages as neutralization of HD. One difference, however, is the uncertainty about the appropriate disposal method for VX hydrolysate.

Although biodegradation of oxidized VX hydrolysate has been demonstrated in the laboratory, as of May 1996 limited treatability studies have not demonstrated biodegradation at a TSDF, even though a TSDF has disposed of VX hydrolysate from bench-scale testing within its permit requirements. It is possible, although not yet established by adequate testing, that the hydrolysate has sufficiently low toxicity associated with its organic products that complete biodegradation prior to discharge may not be necessary.

Furthermore, treatment of VX hydrolysate by existing commercial TSDF processes other than biodegradation is likely to be possible. Therefore, any treatment at a TSDF, whether by biodegradation or another proven and tested process, that results in appropriately low toxicity and low environmental burden in the discharge from the TSDF is a suitable disposal option for VX hydrolysate. As an on-site option for the disposal of hydrolysate, the panel believes that existing, commercially proven processes other than biodegradation could be used. The residual concentrations of agent allowable under chemical-weapon treaty negotiations are likely to be less stringent than the concentrations required by the environmental permits for the destruction and downstream disposal facilities.

Recommendation 7a. The Army should pilot-test VX neutralization followed by off-site treatment of the hydrolysate at a permitted TSDF (treatment, storage, and disposal facility) for potential use at the Newport site, but only if the effluent discharged from the TSDF has been shown to have acceptably low toxicity and results in minimal environmental burden.

Recommendation 7B. If on-site disposal of VX hydrolysate is preferred to shipping it off site for TSDF treatment, existing commercial processes other than biodegradation should be considered. The panel does not recommend on-site biodegradation because of the need for cofeeding a substantial amount of carbon substrate and because of limited success to date in testing on-site biodegradation.

Specific Finding 8. Electrochemical oxidation is the next best alternative for destroying VX at the Newport site. Although the developmental status of this technology is not as advanced as the status of other technologies considered, the panel is confident that the remaining development can lead to a successful pilot demonstration. Although power requirements for this technology are considerable, there is sufficient power available to operate a facility. All process residuals can be handled with commonly used procedures. With respect to the priority evaluation criteria listed under Technology Selection, electrochemical oxidation has the following advantages:

- The required equipment has been proven through extensive use in industry, although it has not been used for chemical agent destruction.
- The principal operations are performed independently and do not require elaborate safety interlocking.
- The semibatch operation can be halted quickly with little danger of a process upset or of stressing the equipment and materials.
- Because many of the process streams are aqueous solutions, hold-and-test analyses to determine stream composition can be readily performed.
- The process is performed at low temperature and pressure with aqueous reaction solutions.
- No step in the process involves combustion; therefore no combustion products are emitted.

Recommendation 8. If successful off-site treatment of VX hydrolysate at an existing TSDF is not confirmed by appropriate treatability studies, and successful on-site treatment of VX hydrolysate with existing commercial processes cannot be demonstrated, then the Army should pilot-test the electrochemical oxidation of VX for potential use at the Newport site.

References

Bowles, A. 1996. Personal communication from Alvin Bowles, Hazardous and Solid Waste Management Administration, Maryland Department of the Environment, to the AltTech Panel. April 8, 1996.

Bradbury, J., K. Branch, J. Heerwagen, and E. Liebow. 1994. Community Viewpoints of the Chemical Stockpile Disposal Program. Washington, D.C.: Battelle Pacific Northwest Laboratories.

Brenner, A., R. Chozick, and R.L. Irvine. 1992. Treatment of a high-strength mixed phenolic waste in an SBR. Water and Environmental Resources. 64: 128–133.

Bretherick, L. 1985. Handbook of Reactive Chemical Hazards. Third Edition. London: Butterworths. Pp. 1,100–1,124.

Brubaker, J., C. Maggio, G. Young, J. Henson, D. Grieder, L.L. Szafraniec, and W.T. Beaudry. 1995. Proceedings, ERDEC [Edgewood Research Development and Engineering Center] Scientific Conference on Chemical Defense Research, November 14, 1995, Aberdeen Proving Grounds, Maryland.

Clinton Herald. 1994. The Clinton Herald. Clinton, Indiana. January 26, 1994.

Douglass, J.J. 1996. Personal communication from J.J. Douglass, Dupont Environmental Treatment, Deepwater, New Jersey, to NRC Panel on Review and Evaluation of Alternative Chemical Technologies. August 1, 1996.

ECO LOGIC. 1996a. System Safety Program and System Hazard Analysis. The ECO LOGIC Gas-Phase Chemical Reduction Process for the Destruction of Bulk HD and VX Chemical Agent and Decontamination of Their Containers and Associated Materiel. March 27, 1996. Prepared for U.S. Army product manager for alternative technologies and approaches, Aberdeen, Maryland. Rockwood, Ontario: ECO LOGIC.

ECO LOGIC. 1996b. Report by ECO LOGIC, Inc., on Public Consultation Activities. April 4, 1996. Submitted to National Research Council, Panel on Review and Evaluation of Alternative Chemical Disposal Technologies. Rockwood, Ontario: ECO LOGIC.

Farmer, J.C., F.T. Wang, R.A. Hawley-Fedder, P.R. Lewis, L.J. Summers, and L. Foiles. 1992. Journal of the Electrochemical Society 139(3): 654–662.

Flamm, K.J., Q. Kwan, and W.B. McNulty. 1987. Chemical Agent and Munition Disposal: Summary of the U. S. Army's Experience. Report SAPEO-CDE-IS-87005. Aberdeen Proving Ground, Maryland: U.S. Army Program Manager for Chemical Demilitarization.

Gibbs, K. 1996. Personal correspondence to the panel from Kathy Gibbs, CSEPP Public Affairs Office. April 3, 1996.

Gill, D.M. 1996. Letter from D.M. Gill, business development manager, Defence-Land, AEA Technology, U.S. DOD Chemical Demilitarization Project-Alternative Technology Program. April 4, 1996.

Haley, M.V. 1996. Toxicity Testing by ERDEC [Edgewood Research, Development and Engineering Center]: HD Neutralization and Biodegradation Products. October. (Informal results presented to the AltTech panel, April 1996) Aberdeen Proving Ground, Maryland: Edgewood Arsenal.

Harvey, S.P. 1994. Report to the U.S. Army Program Manager for Chemical Demilitarization, Agent Neutralization. I. Hydrolysis of Sulfur Mustard. March 1994. Aberdeen Proving Ground, Maryland: Edgewood Research, Development and Engineering Center and Geo-Centers, Inc.

Harvey, S.P., T.A. Blades, L.L. Szafraniec, W.T. Beaudry, M.V. Haley, T. Rosso, G.P. Young, J.P. Earley, and R.L. Irvine. 1996. Kinetics and toxicological parameters of HD hydrolysis and biodegradation. Presented at NATO Advanced Research Workshop on Chemical Problems Associated with Old Arsenical and Mustard Munitions, March 17–19, 1996. Lodz, Poland.

Harvey, S.P., ERDEC Research Biologist, 1995. Oral communication to G.W. Parshall. May 5, 1995.

Hosseinzadeh, K. and N. Sachdeva. 1996. Personal communication from Kaveh Hosseinzadeh and Nand Sachdeva to Harold "Butch" Dye, and Alvin Bowles, all from Hazardous and Solid Waste Management Administration of the Maryland Department of the Environment. March 15, 1996.

Indiana, 1992a. Indiana Code, Section 13-7-8.5-3(b).

Indiana, 1992b. Indiana Code, Section 13-7-8.5-13(b).

Irvine, D.A., J.P. Earley, D.P. Cassidy, and S.P. Harvey. In press. Biodegradation of sulfur mustard hydrolysate in the sequencing batch reactor. To be published in Proceedings of the First International Water Quality Specialized Conference on Sequencing Batch Reactor Technology.

Irvine, R.L., and L.H. Ketchum. 1988. Sequencing batch reactors for biological wastewater treatment. Critical Reviews in Environmental Control. 18: 255–294.

Keane, P. 1996. Verbal communication from Patrick Keane, Illinois CSEPP, during an AltTech Panel meeting, March 13, 1996, Indianapolis, Indiana.

Koch, M., and Z. Wertejuk. 1995. Scientific advances in alternative demilitarization technologies. Presentation at the NATO Advanced Research Workshop, April 25, 1995, Warsaw, Poland.

Kosson, D.S., T.T. Kosson, and H. van der Sloot. 1993. Evaluation of Solidification/Stabilization Treatment Processes for Municipal Waste Combustion Residues. Springfield, Virgina: National Technical Information Service. (NTIS) Report PB93-229 870/AS.

Lehmani, A., P.Turq, and J.P Simonin. 1996. Oxidation kinetics of water and organic compounds by Silver (II) using a potentiometric method. Journal of the Electrochemical Society 143(6): 1860–1866.

Lovrich, J.W. 1996. Personal communication from J.W. Lovrich, coordinator for VX research, Office of the Program Manager for Chemical Demilitarization, to G. W. Parshall, NRC Panel on Review and Evaluation of Alternative Chemical Technologies. July 22, 1996.

Mackay, D., and W.Y. Shui. 1981. Critical review of Henry's Law constants for chemicals of environmental interest. Journal of Physical and Chemical Reference Data 10(4): 1175–1198.

Maryland. 1996. Annotated Code of the Public General Laws of Maryland, Environment, 7-239.3 (1987, as amended 1996).

Maryland Citizens Advisory Commission (CAC). 1994. Comments on the Recommendations for the Disposal of Chemical Agents and Munitions. Report issued February 21, 1994.

Maryland CAC. 1996. Meeting of the Maryland Citizens Advisory Commission with the NRC Panel on Review and Evaluation of Alternative Chemical Disposal Technologies, March 15, 1996, Chestertown, Maryland.

Massalski, T.B. 1986. Binary Alloy Phase Diagrams. Metals Park, Ohio: American Society for Metals.

Mentasti, E., C. Baiocchi, and J.S. Coe. 1984. Mechanical aspects of reactions involving AG (II) as an oxidant. Coordination Chemistry Reviews 54: 131–157.

Metcalf & Eddy, Inc. 1979. Wastewater Engineering: Treatment/Disposal/Reuse. 2nd ed. New York: McGraw-Hill.

Morales, M. 1996. Personal correspondence from Mickey Morales, public affairs specialist, Office of the Program Manager for Chemical Demilitarization, Alternative Technology. May 10, 1996.

M4 Environmental L.P. 1996a. News Release for Immediate Release, March 20, 1996. Oak Ridge, Tennessee.

M4 Environmental L.P. 1996b. Alternative Technologies for Chemical Demilitarization. Response to NRC Questions and Conceptual Design Baseline Updates. Submitted to National Research Council. Panel on Review and Evaluation of Alternative Chemical Disposal Technologies. Oak Ridge, Tennessee: M4 Environmental L.P. April 4, 1996.

M4 Environmental L.P. 1996c. Correspondence to Michael Clarke, study director, from T.J. Abraham, M4 Environmental L.P. May 10, 1966.

M4 Environmental L.P. 1996d. Alternative Technologies for Chemical Demilitarization: National Program Plan. Submitted to U.S. Army Chemical, Biological and Defense Command. Oak Ridge, Tennessee: M4 Environmental L.P. May 31, 1996.

M4 Environmental L.P. 1996e. Chemical Demilitarization Utilizing Catalytic Extraction Processing: Hazard Analysis. Prepared by H&R Technical Associates, Inc., Oak Ridge Tennessee. Submitted to U.S. Army Program Manager for Chemical Demilitarization. Oak Ridge, Tennessee: M4 Environmental L.P. March 1996.

M4 Environmental L.P. 1996f. Public Participation Document. Submitted to National Research Council Panel on Review and Evaluation of Alternative Chemical Disposal Technologies. Oak Ridge, Tennessee: M4 Environmental L.P. April 4, 1996.

Nagel, C.J., C.A. Chanenchuk, E.W. Wong, and R.D. Bach. 1996. Catalytic extraction processing: an elemental recycling technology. Environmental Science and Technology 30(7): 2155–2167.

Novad, J. 1996. Personal communication from J. Novad, coordinator for HD neutralization studies, U.S. Army Alternative Technology Program, Aberdeen

Proving Ground, Maryland, to G. W. Parshall, NRC Panel on Review and Evaluation of Alternative Chemical Technologies. July 23, 1996.

NRC (National Research Council). 1984. Disposal of chemical Munitions and Agents. National Research Council. Committee on Demilitarizing Chemical Munitions and Agents. Washington, D.C.: National Academy Press.

NRC. 1993. Alternative Technologies for the Destruction of Chemical Agents and Munitions. National Research Council. Committee on Alternative Chemical Demilitarization Technologies, Washington, D.C.: National Academy Press.

NRC. 1994a. Evaluation of the Johnston Atoll Chemical Agent Disposal System Operational Verification Testing, Part II. National Research Council. Committee on the Review and Evaluation of the Army Chemical Stockpile Disposal Program. Washington, D.C.: National Academy Press.

NRC. 1994b. Recommendations for the Disposal of Chemical Agents and Munitions. National Research Council. Committee on the Review and Evaluation of the Army Chemical Stockpile Disposal Program. Washington, D.C.: National Academy Press.

NRC. 1995. Evaluation of the Army's Draft Assessment Criteria to Aid in the Selection of Alternative Technologies for Chemical Demilitarization. National Research Council. Committee on Review and Evaluation of the Army Chemical Stockpile Disposal Program. Washington, D.C.: National Academy Press.

NRC. 1996. Review of Systemization of the Tooele Chemical Agent Disposal Facility. National Research Council. Committee on Review and Evaluation of the Army Chemical Stockpile Disposal Program, Washington, D.C.: National Academy Press.

Nunn, J. 1996a. Letter from John E. Nunn, III, cochair of the Maryland CAC, to the AltTech Panel. March 15, 1996.

Nunn, J. 1996b. Letter from John E. Nunn, III, cochair of the Maryland CAC, to the AltTech Panel. April 11, 1996.

O'Brien, G.J. and E.W. Teather. 1995. A dynamic model for predicting effluent concentrations fo organic priority pollutants from an industrial wastewater treatment plant. Water Environment Research 67(6): 935–942.

OTA (Office of Technology Assessment). U.S. Congress. 1992. Disposal of Chemical Weapons: An Analysis of Alternatives to Incineration. Washington, D.C.: U.S. Government Printing Office.

Po, H.N., J.H. Swinehart, and T.L. Allen. 1968. Kinetics and mechanism of the oxidation of water by Silver II in concentrted nitric acid solution. Inorganic Chemistry 7: 244–249.

Rao, Y. K. 1985. Stoichiometry and Thermodynamics of Metallurgical Processes. New York: Cambridge University Press.

Ray, A. 1996. Letter from A. Ray to W. McGowan, M4 Environmental L.P. April 1, 1996.

Satterfield, C.N. 1991. Heterogeneous Catalysis in Industrial Practice. 2nd ed. New York: McGraw-Hill.

SBR Technologies. 1996. Laboratory Feasibility Studies: Biodegradation of HD Hydrolysate in Sequencing Batch Reactors. Prepared for Edgewood Research, Development, and Engineering Center, Aberdeen Proving Grounds, Maryland. Contract DAAL03-91-0034. TCN No. 94353. South Bend, Indiana: SBR Technologies, Inc.

Solarchem Environmental Systems. 1996. Report on the Rayox® design test to mineralize organic by-products in HD-caustic hydrolysate. Markham, Ontario. March 13, 1996.

Steele, D.F. 1990. Electrochemical destruction of toxic organic industrial waste platinum. Metal Review 34: 10–14.

U.S. Army. 1988. Chemical Stockpile Disposal Program Final Programmatic Environmental Impact Statement (PEIS). Aberdeen Proving Ground, Maryland: U.S. Army Program Manager for Chemical Demilitarization.

U.S. Army. 1994. U.S. Army's Alternative Demilitarization Technology Report to Congress. 11 April 94. Department of the Army. Aberdeen Proving Ground, Maryland: U.S. Army Program Manager for Chemical Demilitarization.

U.S. Army. 1995a. Assessment Criteria to Aid in Selection of Alternative Technologies for Chemical Demilitarization. 26 April 95. Department of the Army, Alternative Technology Branch. Aberdeen Proving Ground, Maryland: U.S. Army Program Manager for Chemical Demilitarization.

U.S. Army. 1995b. Concept Design Package: Stand-Alone Neutralization. 21 November 1995. Aberdeen Proving Grounds, Maryland: U.S. Army Program Manager for Chemical Demilitarization.

U.S. Army. 1996a. Alternative Technology Program Summary Status Report, Period 1 October 1995 through 31 December 1995. Aberdeen Proving Ground, Maryland: U.S. Army Program Manager for Chemical Demilitarization.

U.S. Army. 1996b. Concept Design Package for HD Neutralization Followed by Biodegradation, 4 April 1996. Product Manager for Alternative Technologies and Approaches. Aberdeen Proving Ground, Maryland: U.S. Army Program Manager for Chemical Demilitarization.

U.S. Army. 1996c. HD and VX Effluent Management Interim Summary Report. 5 April 1996. (Draft version made available to the NRC.) Aberdeen Proving Grounds, Maryland: U.S. Army Program Manager for Chemical Demilitarization.

U.S. Army. 1996d. Memorandum from Robert B. Perry, chief, Risk Management and Quality Assurance Office, Chemical Demilitarization Program, for the chief of staff, ATTN: DACS-SF. March 18. Memorandum from James A. Gibson, senior safety manager, Army Safety Office, Office of the Chief of Staff, to the program manager for chemical demilitarization. March 29, 1996.

U.S. Army. 1996e. Mustard Agent, Neutralization Followed by Biodegradation: Findings Since 4 April 1996. May 31, 1996. Aberdeen Proving Ground, Maryland: U.S. Army Program Manager for Chemical Demilitarization.

U.S. Army. 1996f. Newport Chemical Activity Concept Design Package for VX Neutralization Followed by Off-Site Biodegradation. April 4, 1996. Product Manager for Alternative Technologies and Approaches. Aberdeen Proving Grounds, Maryland: U.S. Army Program Manager for Chemical Demilitarization.

U.S. Army. 1996g. Cumulative Sample Analysis Data Report, HD Ton Container Survey Results. Dated March 14, 1996. Aberdeen Proving Grounds, Maryland: U.S. Army Program Manager for Chemical Demilitarization.

U.S. Army. 1996h. U.S. Chemical Weapons Stockpile Information Declassified, News Release, Office of Assistant Secretary of Defense (Public Affairs). Washington, D.C. January 22, 1996.

U.S. Army. 1996i. Summary Report on Laboratory- and Bench-Scale Biodegradation Testing of Neutralized HD Mustard. Prelease draft dated March 13, 1996. Aberdeen Proving Ground, Maryland: U.S. Army Program Manager for Chemical Demilitarization.

U.S. Army. 1996j. Alternative Technologies Public Information Sessions. [undated white paper summarizing public sessions from January 25–27, 1966. Aberdeen Proving Ground, Maryland: Office of the Product Manager for Alternative Technologies and Approaches.

U.S. Army. 1996k. Nerve Agent VX: Neutralization Followed by Offsite Biodegradation. Concept Design Package Supplement: Findings since 4 April 1996. May 31, 1996. Product Manager for Alternative Technologies and Approaches. Aberdeen Proving Ground, Maryland: U.S. Army Program Manager for Chemical Demilitarization.

Valenti, M. 1996. Ironing out industrial wastes. Mechanical Engineering 118 (March): 106–110.

Yang, Y-C. 1995. Chemical reactions for neutralizing chemical warfare agents. Chemistry and Industry, n.9 (May 1): 334–337.

Yang, Y-C., J.A. Baker, and J.R. Ward. 1992. Decontamination of chemical warfare agents. Chemical Reviews 92: 1729–1743.

Yang, Y-C., L.L. Szafraniec, W.T. Beaudry, D.K. Rohrbaugh, L.R. Procell, and J.B. Samuel. 1995. Proceedings, ERDEC Scientific Conference on Chemical Defense Research, November 14, 1995. Aberdeen Proving Grounds, Maryland.

Zulty, J.J., J.J. DeFrank, and S.P. Harvey. 1994. Abstract 130, Scientific Conference on Chemical and Biological Defense Research, November 16, 1994. Aberdeen Proving Ground, Maryland.

Appendices

Appendix A

Commerce Business Daily Announcement
August 14, 1995

ALTERNATIVE TECHNOLOGIES FOR CHEMICAL DEMILITARIZATION

1. The U.S. Army, through the Office of the Program Manager for Chemical Demilitarization is responsible for the demilitarization and disposal of chemical agents and munitions. Eight demilitarization facilities are proposed for construction and operation in the continental United States.

2. The Army has demonstrated the operational effectiveness of incineration at its Johnston Atoll Chemical Agent Disposal System facility. The first demilitarization facility for the continental United States has been constructed at Tooele Army Depot and is scheduled to be operational in 1995. In the spring of 1981, the Army began testing at the Chemical Agent Munitions Disposal System (CAMDS) at Tooele, Utah. The mission of CAMDS is to test and evaluate equipment and processes proposed for chemical agent munitions demilitarization facilities.

3. The National Research Council's (NRC) Committee on Review and Evaluation of the Army Chemical Stockpile Disposal Program (Stockpile Committee) was formed in 1987 at the request of the Undersecretary of the Army to monitor the disposal program and to review and comment on relevant technical issues. The Stockpile Committee is a standing committee which remains in service with rotating membership until the demilitarization program is completed.

4. As a consequence of public concern over the use of incineration for chemical warfare agent disposal, the Army commissioned in November 1991, the National Research Council to conduct a study to evaluate alternatives to the reverse assembly (baseline) incineration process for use in destroying the U.S. chemical stockpile. In January 1992, the National Research Council established the Committee on Alternative Chemical Demilitarization Technologies (Alternatives Committee) to develop a comprehensive list of alternative technologies and to review their capabilities and potential as agent and munitions disposal technologies. The Defense Authorization Act for FY93 directed the Army to submit to Congress a report on potential alternative technologies.

5. The NRC report on recommendations for the disposal of chemical agents and munitions was published in 1994. The NRC recommended that the Army continue the current baseline incineration program, since, at that time, no other technologies were mature enough to meet the Army's requirements. However, the NRC did recommend that the Army investigate alternative technologies based on chemical neutralization for the bulk-only sites.

6. In August 1994, the Army initiated an aggressive RDT&E program to investigate, develop, and support testing of two technologies based on chemical neutralization for the destruction of mustard (agent HD) at Aberdeen Proving Ground, MD, and nerve agent VX at the Newport Chemical Activity at Newport, IN. The two alternative technologies are stand-alone chemical neutralization and neutralization followed by biodegradation. The purpose of the RDT&E program is to determine whether an alternative technology warrants pursuing a pilot scale facility based on one or both technologies. The decision to proceed to pilot testing will be made by the Defense Acquisition Board (DAB) in October 1996.

7. The NRC also was aware that there would be ongoing development of the various research programs involving

potential alternatives subsequent to the publication of the NRC report in 1993 on alternative technologies. Thus, the NRC recommended that the Army continue to monitor research developments.

8. The Army agrees with this NRC recommendation, and the Army has been exploring developments in technologies with potential application to chemical demilitarization as part of the RDT&E program.

9. The Army will be conducting a survey to determine if there are any technologies other than the two already being evaluated by the Army as part of the Alternative Technology Program which are capable, within the Chemical Stockpile Disposal Program (CSDP) schedule of meeting chemical demilitarization requirements for the HD (mustard) and VX (nerve) agents stored at the Aberdeen Proving Ground, MD, and Newport Chemical Activity, IN storage sites, respectively. This announcement requests information from industry on any alternative technology that a firm believes is mature enough to meet the needs of the Army program. The Army will conduct a preliminary 30 day screening to determine whether any of the technologies identified pursuant to this announcement warrant further review by the NRC. The Army will identify up to a maximum of three of the most promising technologies in addition to neutralization and neutralization followed by biodegradation. The evaluators will determine whether the technology meets the following screening criteria.

- Any proposed alternative technology should not resemble incineration (high-temperature oxidation) or produce effluents characteristic of incineration;
- The technology must utilize processes and equipment that are developed or capable of being developed in time to meet the requirements of the Chemical Weapons Convention;
- Laboratory-scale testing must have been completed with agent or chemicals with similar properties to agent. Data must be available to provide an initial indication of performance characteristics and destruction efficiency.

10. Interested firms are asked to provide information in the form of a conceptual design package within 60 days from the date of this announcement. The purpose of the conceptual design package is to demonstrate the feasibility of using an alternative set of process unit operations to conduct the total activities that are required to complete the program, and to provide a basis for its comparison with the baseline system. At a minimum, it should include the following:

- Process description. The information package should include a description of the total process, detailing how actual experience or test results have been used to project equipment performance, and how the various agent destruction, decontamination, and waste processing steps are conducted. The description should also provide an adequate basis for establishing that the process has a high probability of success, after pilot testing, to perform the necessary agent destruction and waste disposal functions.
- Process data. Chemical and physical properties of all process materials should be provided to the extent that data is needed to design each unit operation in the overall process.
- Flow sheets, showing all proposed equipment, piping, and general control methods, including:
  - Material and energy balances, projections showing all material flow rates, and energy requirements, such as heat generation and removal rates for each step of the process
  - Process monitoring and control, showing all proposed process monitoring instrumentation and describing the methods used to control the process
- A description and characterization of all process waste streams.
- A description of storage facilities for all feed materials and all wastes prior to the final disposition.
- A description of facilities for packaging and handling wastes prior to off-site shipping.
- Utility requirements, including process requirements for both fuel and electricity. Also include need for backup requirements to allow for emergency shutdown of the process and related pollution control systems.
- Feed materials requirements, including both quantities and qualities of all chemicals that are required, and the need for any special feed preparation.
- Equipment lists for all major pieces of equipment for the destruction process, secondary treatment systems, and pollution control systems.

- Any data generated from agent or simulant tests or data resulting from destruction of similar chemicals by the proposed processes.
- Equipment designs, including design sketches, sizing calculations and materials of construction for all major pieces of process equipment.
- Plant layout. The design should show the layout and working space for the major pieces of equipment, plot plans for the current storage facilities, and planned means for transport of agent containers from the storage area to the destruction facilities.

11. On written request, firms will be provided with information on: the baseline incineration system; the chemical stockpile disposal program schedule; and the current program for developing neutralization and neutralization followed by biodegradation. Firms may write or fax their requests to Dr. Francis W. Holm, Science Applications International Corporation, 9 Aberdeen Shopping Plaza, Aberdeen, MD 21001. Fax: (410) 273-1001.

12. The NRC will review those promising alternative technologies, if any, identified by the Army as well as neutralization and neutralization-biodegradation. Concurrently, proponents of technologies identified by the Army will be asked to furnish a notional program plan including: a rough, order of magnitude estimate of the projected cost and schedule and chemical agent destruction test data. Firms must perform testing to obtain actual chemical agent test data at an Army approved surety laboratory at the firm's expense. The test data must be available to the NRC for review by 31 May 1996.

13. As a note of caution, those considering participation should understand that chemical agents and munitions are significantly more toxic than many substances normally referred to as "hazardous and toxic material." Therefore, high standards of employee, public, and environmental protection are required.

14. This announcement is meant to offer industry the opportunity to make the Army aware of potential alternative technologies which can meet the needs of the chemical demilitarization program. The process outlined herein will not necessarily lead to any request for proposals (RFP) or contract awards. The government does not intend to reimburse firms for the cost of providing data originally submitted pursuant to this request.

15. Mr. Eric W. Braerman, Procurement Directorate, CBDCOM, is the point of contact for this announcement, (410) 671-4469.

ROBERT D. ORTON
Major General, U.S. Army Program Manager
for Chemical Demilitarization

Appendix B

Input from the Public

Chapter 9 discusses the rationale for public involvement in the panel's deliberations and describes interactions with the communities neighboring the Newport and Aberdeen sites and meetings with regulators. This appendix includes samples of the letters of invitation sent to individuals and organizations in Indiana and Maryland prior to the public forums and summarizes the categories of stakeholders contacted.

The Army Program Office, located in Maryland with established communications links to stakeholders interested in the chemical demilitarization process, assisted with the notification process within Maryland. Because of the independent notification by the Army, the panel sent fewer letters to Maryland stakeholders than to Indiana stakeholders.

Indiana

Letters to government officials (federal, state, and city) 85
Letters to Indiana Citizens Advisory Commission (CAC) members 2
Letters to other citizens, media, companies, etc. 254

Maryland

Letters to government officials (federal, state, and city) 7
Letters to Maryland Citizens Advisory Commission (CAC) members 9
Letters to other citizens, media, companies, etc. . . . 95

LETTER OF INVITATION TO THE CITIZENS OF NEWPORT, INDIANA

March 4, 1996

The National Research Council (NRC) has been asked by the Army to evaluate alternative technologies (alternatives to the Army's baseline incineration process) for the destruction of bulk chemical warfare agents stored at facilities near Aberdeen, Maryland and Newport, Indiana.

In August through October, 1995, the Army conducted an evaluation of chemical destruction processes that resulted in the selection of three technologies, plus the Army's two neutralization technologies, to be evaluated by the NRC. A Panel on Review and Evaluation of Alternative Chemical Disposal Technologies, called the AltTech Panel, was formed by the NRC. The AltTech panel will provide a report to the Army in August, 1996 that will make recommendations on whether any of the five technologies is suitable for pilot plant demonstration. In the Fall of 1996, the Army will present its recommendations to the Defense Acquisition Board (DAB) on which, if any, of the technologies should move forward to the pilot plant demonstration phase.

On Tuesday, March 12 at 7:00 p.m. representatives of the AltTech Panel will be present at North Vermillion High School, RR 1, Cayuga, Indiana, to solicit the public's views on these technologies. As Chairman of the AltTech Panel, I am writing to inform you of this information gathering meeting. In the past, public meetings like this have added greatly to the knowledge base of other NRC committees and have ensured views of all interested parties are heard and considered.

At the meeting you will be provided an opportunity to state your views about the five technologies. The AltTech panel has been informed that the Army provided information on the alternatives during its earlier meeting, and that you have also had the opportunity to review vendor-provided information and information placed by the Army in libraries. Therefore, the panel will not spend valuable time describing the technologies again at this meeting so that the time can be applied to the most important objective, hearing your input.

I will begin the public meeting by making a short presentation that describes the NRC panel schedule and data gathering methodology. After my presentation, you may make your statements. To enable as many as possible with an opportunity to speak, you will be asked to limit your remarks to five minutes or less. If you intend to speak, please ensure you have signed in prior to the meeting. You are also encouraged to submit your statements in written form at the meeting, whether you speak or not. If you cannot attend the meeting, and you wish the AltTech panel to consider your views, please provide a written statement to the National Research Council, 2101 Constitution Avenue, N.W., Washington, DC 20418, Attn.: Mr. Michael A. Clarke, HA258, by March 31, 1996

The sole purpose of this meeting is to provide the public an opportunity to state its insights, observations, concerns, and feelings about the various technologies under consideration. You should also know that the panel will not share its assessment of the technologies with you at this meeting. That would be premature and is reserved for the panel's final report in August. Therefore, it is very important that you state only your views when you address the panel. Please do not address questions to vendor or Army personnel present. This meeting is intended to be a dialog between the NRC and the public. Conversations with the Army or vendors present should take place in other locations than the formal meeting.

Your opinions on these important local and national issues are important to us. The panel members and I look forward to hearing from you.

Sincerely,

Richard S. Magee

Richard S. Magee, Chair
AltTech Panel

LETTER OF INVITATION TO THE CITIZENS OF ABERDEEN, MARYLAND

March 5, 1996

The National Research Council (NRC) has been asked by the Army to evaluate alternative technologies (alternatives to the Army's baseline incineration process) for the destruction of bulk chemical warfare agents stored at facilities near Aberdeen, Maryland and Newport, Indiana.

In August through October, 1995, the Army conducted an evaluation of chemical destruction processes that resulted in the selection of three technologies, plus the Army's two neutralization technologies, to be evaluated by the NRC. A Panel on Review and Evaluation of Alternative Chemical Disposal Technologies, called the AltTech Panel, was formed by the NRC. The AltTech panel will provide a report to the Army in August, 1996 that will make recommendations on whether any of the five technologies is suitable for pilot plant demonstration. In the Fall of 1996, the Army will present its recommendations to the Defense Acquisition Board (DAB) on which, if any, of the technologies should move forward to the pilot plant demonstration phase.

On Friday, March 15 at 8:00 p.m. representatives of the AltTech Panel will be present at the Kent County Courthouse, County Commissioner's Room, 103 Cross Street, Chestertown, MD and on Saturday, March 16, 1996 at 10:00 a.m. at Edgewood High School, Willoughby Beach Road, Edgewood, Maryland, to solicit the public's views on these technologies. As Chairman of the AltTech Panel, I am writing to inform you of this information gathering meeting. In the past, public meetings like this have added greatly to the knowledge base of other NRC committees and have ensured views of all interested parties are heard and considered.

At the meeting you will be provided an opportunity to state your views about the five technologies. The AltTech panel has been informed that the Army provided information on the alternatives during its earlier meeting, and that you have also had the opportunity to review vendor-provided information and information placed by the Army in libraries. Therefore, the panel will not spend valuable time describing the technologies again at this meeting so that the time can be applied to the most important objective, hearing your input.

I will begin the public meeting by making a short presentation that describes the NRC panel schedule and data gathering methodology. After my presentation, you may make your statements. To enable as many as possible with an opportunity to speak, you will be asked to limit your remarks to five minutes or less. If you intend to speak, please ensure you have signed in prior to the meeting. You are also encouraged to submit your statements in written form at the meeting, whether you speak or not. If you cannot attend the meeting, and you wish the AltTech panel to consider your views, please provide a written statement to the National Research Council, 2101 Constitution Avenue, N.W., Washington, DC 20418, Attn.: Mr. Michael A. Clarke, HA258, by March 31, 1996

The sole purpose of this meeting is to provide the public an opportunity to state its insights, observations, concerns, and feelings about the various technologies under consideration. You should also know that the panel will not share its assessment of the technologies with you at this meeting. That would be premature and is reserved for the panel's final report in August. Therefore, it is very important that you state only your views when you address the panel. Please do not address questions to vendor or Army personnel present. This meeting is intended to be a dialog between the NRC and the public. Conversations with the Army or vendors present should take place in other locations than the formal meeting.

Your opinions on these important local and national issues are important to us. The panel members and I look forward to hearing from you.

Sincerely,

Richard S. Magee

Richard S. Magee, Chair
AltTech Panel

Appendix C

Meetings and Site Visits

Panel Meeting: October 11–12, 1995
Washington, D.C.

Participants. Panel chair: Richard S. Magee. Panel members: Joan B. Berkowitz, Gene H. Dyer, Frederick T. Harper, Joseph A. Heintz, David A. Hoecke, David S. Kosson, Walter G. May, Alvin H. Mushkatel, Laurance Oden, George W. Parshall, L. David Pye, William Tumas; BAST liaison: Robert A. Beaudet; NRC staff members: Bruce Braun, Michael Clarke, Jacqueline Johnson, and Deborah Randall. Briefers.

Objectives. Welcome and introduce new members; complete administrative matters; complete composition and balance discussion; discuss and develop prototype criteria checklist; organize panel into subpanel teams; perform historical review for new members; receive status briefings from applicable Army officials on call for alternative technologies; and discuss November meeting requirements.

Panel Meeting: November 20–21, 1995
Washington, D.C.

Participants. Panel members, NRC staff, and briefers

Objectives. Welcome and introduce two new members; complete administrative matters; complete composition and balance discussion; discuss and develop prototype criteria checklist; develop and approve report concept; organize panel into technology assessment teams; receive briefings from applicable Army officials on alternative technology selection process; receive briefings from technology proponent company finalists; and discuss future meetings/vendor visits.

Site Visit: January 8–9, 1996
Fall River, Massachusetts

Participants. Panel members: Gene Dyer, and Laurance Oden. BAST liaison: Robert Beaudet.

Objectives. Receive presentations and information from technology proponent company for catalytic extraction process technology and perform site tour.

Site Visit: January 8–9, 1996, Ontario, Canada

Participants. Panel members: Walter May, Roger Staehle, and William Tumas

Objectives. Receive presentations and data from technology proponent company for gas-phase reduction technology and perform site tour.

Site Visit: January 14–16, 1996
Aberdeen/Dounreay, Scotland

Participants. Panel members: Roger Staehle, Joan Berkowitz, and Walter May. NRC staff member: Michael Clarke.

Objectives. Receive information from AEA and SubSea on the status of the electrochemical oxidation process.

Site Visit: January 18–19, 1996
Oak Ridge, Tennessee

Participants. Panel members: Gene Dyer and Laurance Oden. NRC staff member: James Zucchetto.

Objectives. Receive presentations and information from technology proponent company for catalytic extraction process technology and perform site tour.

Site Visit: January 18–19, 1996, Edgewood, Maryland

Participants. Panel members: George Parshall and David Kosson. NRC staff members: Bruce Braun and Donald Siebenaler.

Objectives. Receive presentations and data from Army and Army contractors on neutralization technologies.

Panel Meeting: February 1–2, 1996
Irvine, California

Participants. Panel. NRC staff members: Bruce Braun, Michael Clarke, Deborah Randall, and Shirel Smith.

Objectives. Welcome and introduce three new members; complete administrative matters; complete composition and balance discussion; receive Army briefing on AltTech program status; receive briefings from team leaders on vendor assessment visits; discuss plan for public meetings and meetings with state and federal agencies; and discuss report status and future activities.

Panel Meeting: March 14–15, 1996
Washington, D.C.

Participants. Panel chair: Richard S. Magee. Panel members: Joan B. Berkowitz, Gene H. Dyer, Frederick T. Harper, Joseph A. Heintz, David A. Hoecke, David S. Kosson, Walter G. May, Alvin H. Mushkatel, Laurance Oden, George W. Parshall. BAST liaison: Robert A. Beaudet. NRC staff members: Bruce Braun, Michael Clarke, and Deborah Randall.

Objectives. Complete composition and balance discussion for three new members; receive panel assessment team briefings on chapter draft status and data requirements; discuss comparison criteria for chapter 8; assemble first full message draft of report; discuss results of Newport public and regulator meetings and plan for Aberdeen meetings; discuss AltTech program status; and discuss future report activities and goals for April meeting.

Site Visit: March 16–17, 1996, Oak Ridge, Tennessee

Participants. Panel chair: Richard S. Magee. Panel members: Joseph Heintz, David Hoecke, and Laurance Oden.

Objectives. Receive follow-up information on status of catalytic extraction process technology.

Site Visit: April 2, 1996, Ontario, Canada

Participant. Panel member: Frederick Harper

Objectives. Receive presentations data from technology proponent company for gas-phase reduction technology and perform site tour.

Site Visit: April 8–9, 1996, Edgewood, Maryland

Participants. Panel members: David Kosson and George Parshall

Objectives. Receive follow-up information from Army and Army contractors on the status of neutralization technologies.

Writing Session: April 11–12, 1996
Washington, D.C.

Participants. Panel members: Alvin Mushkatel and Richard Magee. NRC staff member: Michael Clarke.

Objectives. Organize report and draft community and regulator chapter of report.

Panel Meeting: April 18–20, 1996, Washington, D.C.

Participants. Panel chair: Richard S. Magee. Panel members: Joan B. Berkowitz, Gene H. Dyer, Frederick T. Harper, Joseph A. Heintz, David A. Hoecke, David S. Kosson, Walter G. May, Alvin H. Mushkatel, George W. Parshall, L. David Pye, William Tumas. BAST liaison: Robert A. Beaudet. NRC staff members: Bruce Braun, Michael Clarke, and Deborah Randall.

Objectives. Assemble first full message draft of report; develop a strategy for the rapid development of a concurrence draft; discuss AltTech program status, including final data acquisition and surety testing; review preliminary hazard and operability report status; settle panel indemnity issue; and set goals for the May meeting.

Site Visit: April 26, 1996, Oak Ridge, Tennessee

Participants. Panel member: Frederick Harper

Objectives. Tour facilities and orient risk assessment panel member; discuss risk issues with the technology proponent company.

Site Visit: May 5–8, 1996, London, England

Participants. Panel members: Joan Berkowitz and Walter May

Objectives. Evaluate and assess the electrochemical reduction alternative technology as a candidate for pilot-plant demonstration by the U.S. Department of the Army for destruction of chemical agents; receive presentations from the technology proponent company on that technology and perform site tour.

Panel Meeting: May 15–17, 1996, Washington, D.C.

Participants. Panel chair: Richard S. Magee. Panel members: Joan B. Berkowitz, Gene H. Dyer, Frederick T. Harper, David A. Hoecke, David S. Kosson, Walter G. May, Alvin H. Mushkatel, Laurance Oden, William Tumas. BAST liaison: Robert A. Beaudet. NRC staff members: Bruce Braun, Michael Clarke, and Deborah Randall. Technical writer/consultant: Robert Katt.

Objectives. Assemble and sign off on concurrence draft of report; discuss AltTech program status and methodology for including surety testing data; settle panel indemnity issue; and discuss milestones leading to report review and publication.

Appendix D

Modification to Statement of Task

DEPARTMENT OF THE ARMY
OFFICE OF THE ASSISTANT SECRETARY
RESEARCH DEVELOPMENT AND ACQUISITION
103 ARMY PENTAGON
WASHINGTON DC 20310-0103

REPLY TO
ATTENTION OF

11 MAR 1996

National Research Council
Commission on Engineering
and Technical Systems
2101 Constitution Avenue
Washington, D.C. 20418

Dear Mr. Braun:

This letter confirms the conversation of March 7, 1996 that the National Research Council (NRC) may consider, as part of its study on the review and evaluation of alternative chemical disposal technologies, off-site shipping options of treated agent residuals to appropriately permitted disposal facilities. The Army intends to provide to the NRC in its final submission in April (the NRC's data cutoff date) the option of shipping agent-free hydrolysate from the neutralization process to off-site disposal facilities. We ask that you consider this option and include the result in your report.

Specifically, we request that you modify the Statement of Task as follows which adds an additional bullet after the fourth bullet that reads:

- o Consider the option of shipping treated effluents agent free) to off-site appropriately permitted disposal facilities; and

Sincerely,

John D. Gorrell
Colonel, GS
Director
Chemical Demilitarization

Appendix E

Electrochemical Oxidation

WATER BALANCE ISSUES

When HD is treated in the Silver II process, 75 kg•mols will be decomposed during the course of a campaign as the source of oxygen for agent oxidation. An additional amount of water will be lost by the parasitic reaction in which water is decomposed and O_2 gas evolves. Approximately 176 kg•mols of water will be carried from the anode compartment to the cathode compartment by electrical diffusion of hydrated hydrogen ions. The total of these water losses, more than 251 kg•mols, should be compared to the initial water content of the anode compartment of 2.5 m^3 or 139 kg•mols. Part of the loss, as yet unquantified, is made up by spontaneous osmotic diffusion from the cathode compartment back to the anode department, induced by the large difference in acid concentration between the two. (The anolyte is maintained at 8 molar in nitric acid, the catholyte at 4 molar.)

When VX is treated, the water losses are about 116 kg•mols from agent oxidation and 307 kg·mols from the transport of hydrated hydrogen ions. Total losses therefore exceed 423 kg•mols during the course of a VX campaign, compared with the initial water content in the anode compartment of 139 kg•mols. As in the HD case, there is an as yet undetermined osmotic flow of water back from the cathode compartment to the anode compartment.

MASS BALANCE DATA FOR SILVER II PROCESS

Tables E-1 and E-2 provide an elemental analysis of the mass balance data provided by the TPC for the Silver II process for treating VX and HD, respectively.

TABLE E-1 Elemental Breakdown of Mass Balances for VX Destruction

	Metric Tons							
Input	Total	C	H	S	N	P	O	Na
VX	2	0.99	0.19	0.24	0.1	0.23	0.24	
HNO_3	0.7		0.01		0.16		0.53	
H_20	0.3		0.03				0.27	
H_20_2	1.9		0.11				1.78	
H_20	3.6		0.4				3.2	
NaOH	1.8		0.05				0.72	1.04
H_20	0.1		0.01				0.09	
0_2	4.9						4.9	
N_2	0.5				0.5			
Total In	15.8	1.0	0.8	0.2	0.8	0.2	11.7	1.0
Output	Total	C	H	S	N	P	O	Na
CO_2	3.8	1.04					2.76	
0_2	0.1						0.1	
N_2	0.5				0.5			
NO_x	0.004							
HNO_3	1.12		0.02		0.25		0.85	
H_2O	3.9		0.43				3.47	
H_2O	3.6		0.42				3.2	
$NaNO_3$	0.6				0.1		0.34	0.16
Na_2SO_4	1.07			0.24			0.48	0.35
Na_3PO_4	1.23					0.23	0.48	0.52
Total Out	15.9	1.0	0.9	0.2	0.9	0.2	11.7	1.0

TABLE E-2 Elemental Breakdown of Mass Balances for HD Destruction

	Metric Tons							
Inputs	Total	C	H	S	Cl	N	O	Na
HD (mustard)	2	0.6	0.1	0.4	0.89			
HNO_3	0.4		0.01			0.09	0.3	
H_20	0.2		0.02				0.18	
H_20_2	1.1		0.06				1.04	
H_20	2		0.22				1.78	
NaOH	2		0.05				0.8	1.15
H_20	0.1		0.01				0.09	
0_2	2.8						2.8	
N_2	0.3					0.3		
Total In	10.9	0.6	0.5	0.4	0.9	0.4	7.0	1.2
Outputs	Total	C	H	S	Cl	N	O	Na
CO_2	2.2	0.6					1.6	
0_2	0.1						0.1	
N_2	0.3					0.3		
NO_x	0.002							
HNO_3	0.63		0.01			0.14	0.48	
H_2O	2.2		0.24				1.96	
Na_2SO_4	1.8			0.4			0.81	0.58
NaCl	1.5				0.9			0.59
H_20	2.2		0.24				1.96	
Total Out	10.9	0.6	0.5	0.4	0.9	0.4	6.9	1.2

Appendix F

Gas-Phase Reduction

THERMODYNAMIC CALCULATIONS

The ECO LOGIC process is described as a gas-phase chemical reduction process in which waste materials react with hydrogen and steam at high temperature. The reaction conditions are very different from the reaction conditions in industrial hydrogenation processes, which are usually carried out at much higher hydrogen pressure and lower temperature than the ECO LOGIC conditions and require a catalyst because of the lower temperature. In this appendix, the AltTech Panel has used thermodynamic data to examine likely chemical reactions and reaction products that will result from processing agent.

Data on free energy of formation were used for these calculations. The data were taken primarily from the JANAF Thermochemical Tables (JANAF, 1985); a few of the data are from Perry's *Chemical Engineers Handbook* (Perry et al., 1984). Data at 1100 K were used as representative of reactor conditions; data at 298 K were used as representative of quenched reactor products.

Feed material in the main reactor is at a high enough temperature for cracking (breakup of the carbon chain into smaller fragments) to occur rapidly. Molecular fragments can then react with the hydrogen and steam in the reactor environment. The end products indicated by thermodynamic considerations are discussed below for carbon and for each of the heteroatoms.

Carbon

Methane is the only hydrocarbon with significant thermodynamic stability at 1100 K in the presence of hydrogen as illustrated by the following possible product reactions:

		Equilibrium Constant
Ethylene	$C_2H_4 + 2H_2 = 2CH_4$	$K_{1100\,K} = 1417$
Acetylene	$C_2H_2 + 3H_2 = 2CH_4$	$K_{1100\,K} = 4.2 \times 10^{12}$
Benzene	$C_6H_6 + 9H_2 = 6CH_4$	$K_{1123\,K} = 2 \times 10^{17}$

The panel concludes that at this, reaction temperature and with this hydrogen content in the main reactor, these hydrocarbons would react almost completely to form methane. Methane itself, however, is not expected at high concentration; reaction with hydrogen should result in solid carbon and only low concentration of methane.

$$C + 2H_2 = CH_4 \qquad K_{1100\,K} = 0.0356$$

With a hydrogen content of 70 percent in the product gas, the equilibrium methane concentration is calculated to be only 1.7 percent. This does not conform to experimental observation, however. The observed methane content, which is reported to be as high as 15 percent, probably represents a nonequilibrium, rate-controlled product. A possible alternative explanation is that carbon formed in high temperature reactions sometimes has a higher free energy than graphite (so-called "Dent" carbon). At 500°C, the free energy of this carbon form may be 15 KJ above the free energy of graphite. This difference would lead to a larger equilibrium constant for the reaction ($K_{1100\,K} = 0.1859$) and a possible equilibrium methane content of 9 percent.

This calculation suggests that high methane content is probably a result of the reaction sequence during the decomposition process. It also suggests that solid carbon should be expected as a product. The TPC assumes that 10 percent of the carbon in the feed will show up as solid elemental carbon in the reactor effluent gas. Precursors to the solid carbon, such as polycyclic aromatics, would then also be expected.

Steam can also react with carbon (and methane), and in fact thermodynamic equilibrium would result in complete conversion.

$$C + H_2O = CO + H_2 \qquad K_{1100\,K} = 11.16$$

However, reaction rates with industrial carbon at 1100 K are very slow and are inhibited by the presence of hydrogen (Gadsby et al., 1946; May et al., 1958).

Reaction with carbon from a decomposing hydrocarbon could be faster, however, and significant CO would be expected.

Much of the CO produced would react via the water gas shift reaction to form CO_2.

$$CO + H_2O = CO_2 + H_2 \qquad K_{1100\,K} = 0.988$$

The rate for this reaction is high enough to approach equilibrium.

Heteroatoms

Chlorine

Chlorine (in mustard) should react almost completely to HCl.

$$H_2 + Cl_2 = 2HCl \qquad K_{1100\,K} = 66000$$

Sulfur

Sulfur should go primarily to H_2S in the reactor.

$$H_2 + S = H_2S \qquad K_{1100\,K} = 51.9$$

A very small amount of H_2S could react with steam.

$$H_2S + H_2O = 3H_2 + SO_2 \qquad K_{1100\,K} = 7.6 \times 10^{-7}$$

A small amount of sulfur should be expected in the quench of the product stream—probably in the HCl product solution.

Phosphorus

Phosphine and oxides of P(II) and P(IV) do not appear to be very stable relative to elemental phosphorus at either reactor (1100 K) or quench (298 K) conditions; only very small concentrations would therefore be expected. For example:

Phosphine: $P + 1.5H_2 = PH_3$

$$K_{1100\,K} = 5.2 \times 10^{-4}$$
$$K_{298\,K} = 3.8 \times 10^{-6}$$

P-Oxides: $P + H_2O = PO + H_2$

$$K_{1100\,K} = 1 \times 10^{-4}$$
$$K_{298\,K} = 1.4 \times 10^{-32}$$

$$P + 2H_2O = PO_2 + 2H_2$$
$$K_{1100\,K} = 0.0016$$
$$K_{298\,K} = 2 \times 10^{-25}$$

Higher-valence oxides, P_4O_6 in particular, are much more stable under both reactor and quench conditions.

$$4P + 6H_2O = P_4O_6 + 6H_2$$
$$K_{1100\,K} = 2.8 \times 10^{28}$$
$$K_{298\,K} > 10^{100}$$

The trivalent oxide appears to be the most stable oxide under the reducing conditions of the process. It is considerably more stable than the divalent and tetravalent oxides, as well as the pentavalent oxide, at both reactor and quench conditions.

$$P_4O_6 + 4H_2O = P_4O_{10} + 4H_2$$
$$K_{1100\,K} = 6.7 \times 10^{-22}$$
$$K_{298\,K} = 1.5 \times 10^{-58}$$

Under oxidizing conditions, P_4O_{10} would be the stable species.

$$P_4O_6 + 2O_2 = P_4O_{10}$$
$$K_{298\,K} > 10^{100}$$

It appears likely that the phosphorus species produced in the reactor will be the oxide of trivalent phosphorus, P_4O_6.

A number of phosphorus acids might form in solution when the reactor vapor is quenched. The stable one appears to be the orthophosphorous acid (Moeller, 1952).

$$P_4O_6 + 6H_2O = 4H_3PO_3$$

This form is unstable in an oxidizing atmosphere and would presumably convert to the pentavalent orthophosphoric acid, H_3PO_4. The rate of conversion to the higher oxide is not known.

Nitrogen

Nitrogen would be expected in the form of molecular nitrogen, ammonia, and possibly some N-oxide species. An interesting possibility that will need further examination is the potential to produce hydrogen cyanide (HCN). At reactor conditions, this material would be expected at parts-per-million concentration, though at virtually zero concentration at room temperature.

$$1/2H_2 + 1/2N_2 + C = HCN$$
$$K1100\ K = 2.2 \times 10^{-5}$$
$$K298\ K = 14 \times 10^{-22}$$

The rates of reaction are unknown. Because nitrogen is associated with carbon (in VX), HCN would probably be formed in the reactor. Whether it will be at its equilibrium level and whether it will persist (at above equilibrium level) during the quench are questions that will need evaluation.

MATERIAL BALANCES

Two sets of material and energy balances were submitted by the TPC, the first on January 30, 1996, the second on April 4, 1996. The panel has examined the balances for HD. The two balances differed in the feed rate of HD: 5.0 liter/min. for the first, 2.736 liter/min. for the second. The numbers that follow are taken from the second balance unless otherwise stated. The feed rate of 2.736 liter/min. corresponds to a destruction rate of 5 metric tons per day (5.5 English tons per day).

The feed to the reactor consists of four streams: feed HD; gas from the steam reformer; waste steam; and waste water. A stream from the SBV (sequencing batch vaporizer) would also go to the reactor when the SBV is operating. Some product gas from the product gas blower might also be recycled directly back to the reactor, bypassing the catalytic reformer.

The largest gas stream is the reformer gas, which constitutes approximately 85 percent of the total gram-mols of feed. This gas is at a high temperature (775°C) and has a large H_2 content.

Reformer Gas	g-mols/min.	vol%
H_2	755	74.0
CH_4	15.3	1.5
CO	35.3	3.5
CO_2	55.3	5.4
H_2O	159.8	15.7

Table F-1 shows flow rates and compositions into and out of the reactor. The hydrogen content of the product gas is kept high, above 55 percent (wet basis) in this case. There may also be trace quantities (parts per million) of other materials not shown in the product gas analysis above are possible (SO_2, for example).

The TPC assumes that 10 percent of the carbon in the HD feed will be solid carbon in the product. Most of the TPC's experience has been with aromatic feed stocks, such as PCBs (polychlorinated biphenyls), which would presumably yield relatively large carbon residues. The carbon residue from HD (or VX) might be lower than 10 percent.

The methane content of the product gas is well above the thermodynamic equilibrium value. It may simply represent a nonequilibrium product limited by the reaction rate. The methane presumably forms from CH_2 radicals (see section on Thermodynamic Calculations).

Most of the gas feed to the reactor is at high temperature; the reformer gas, which is 85 percent of the total, is at 775°C, and the direct recirculation gas is heated to

TABLE F-1 Material Flows to and from GPCR Reactor

	Feeds to Reactor		Products from Reactor	
Material Species	g-mols/min.	vol%	g-mols/min.	vol%
H_2O[a]	270.2	23.4	239.0	20.9
H_2	755	65.4	654.0	57.2
CH_4	15.3	1.3	77.8	6.8
Hydrocarbon	0.8	.07	–	–
HD (mustard)	21.84	1.89	–	–
CO	35.3	3.06	43.5	3.8
CO_2	55.3	4.79	62.9	5.5
H_2S	–	–	21.8	1.9
HCl	–	–	43.6	3.8
Solid Carbon	–	–	8.7	–

[a]The hydrogen and oxygen in feed and product do not balance exactly.

TABLE F-2 Material Balance for HD in the ECO LOGIC Process

Material In[a]		Material Out	
Material Species	g-atoms/min.	Material Species	g-atoms/min.
Carbon from HD	87.35	Solid carbon	8.7
		CO_2 in HCl solution	5.5
		CO_2 in MEA offgas	51.2
		CH_4/CO/ CO_2 in gas to burner	22.0
		Total carbon out	87.4
Hydrogen		**Hydrogen**	
from HD	174.7	Gas to Burner	
from H_2O		H_2	235.6
Reactor	62.4	H in CH_4	53.7
Catalytic Reformer	~130	HCl solution	43.45
		H_2S from MEA	43.45
Total H_2 in	367.1	Total H_2 out	366.2
Sulfur from HD	21.84	H_2S in MEA offgas	21.84
Chlorine in HD	43.68	HCl in quench solution	43.68

[a] Based on HD feed rate of 21.84 g-mols/min.

600°C. The electric heaters in the reactor then supply energy to raise the gas mixture to between 850 and 900°C. The reaction itself is a combination of hydrocracking (to produce methane), which is exothermic, and steam reforming (to produce CO), which is endothermic. Overall the reaction appears to be slightly exothermic (about 1,400 kJ/kg of HD processed, equivalent to less than 10 percent of the heat of combustion of HD).

The product gas from the reactor is quenched with water to produce an HCl solution of moderate concentration together with suspended carbon. The quench will also dissolve some of the H_2S (and the possible low concentration [ppm range] of SO_2), as well as some CO_2. The suspended carbon must be filtered out before disposal of the HCl solution. The TPC has estimated that the quench will remove 43.8 g-mols/min. of HCl and 5.5 g-mols/min. of CO_2. (On the assumption that the HCl solution will be fairly concentrated, perhaps 30 percent, the CO_2 removal rate appears too high.)

The H_2S and most of the CO_2 will be recovered in the methanolamine scrubber. The product gas from this scrubber will be:

H_2S	21.8 g-mols/min.; 29.9 vol%
CO_2	51.2 g-mols/min.; 70.1 vol%

The scrubbed gas will have the following composition (dry basis).

	Composition	
Gas	g-mols/min	vol%
H_2	654.0	83.7
CH_4	77.8	10.0
CO	43.5	5.6
CO_2	6.2	0.8

In the material balances submitted by the TPC, no aromatic hydrocarbons are shown for the product gas, and the submitted design makes no provision for hydrocarbon scrubbing. The TPC does recognize that some high-molecular-weight hydrocarbon may be present (a precursor to solid carbon) and that a scrubber for removal may be necessary.

Part of the scrubbed product gas is recycled (mostly via the catalytic reformer); part is burned to supply steam. Overall, the material balance indicates that approximately 17 percent of the scrubbed gas will be burned. The products of HD destruction then show up in the streams shown in Table F-2. The scrubbed product gas, which consists mainly of hydrogen (83 percent) and methane (10 percent), should burn cleanly, that is, with negligible products of incomplete combustion.

REFERENCES

Gadsby, J., C.N. Hinshelwood, and K.W. Sykes. 1946. Kinetics of the Reactions of the Steam-Carbon System. Proceedings of the Royal Society, London, Series A. 187: 129–151.

JANAF. 1985. JANAF Thermochemical Tables. 3rd Ed. Journal of Physical and Chemical Reference Data 14, Suppl.1.

May, W.G., R.H. Mueller, and S.B. Sweetser. 1958, Carbon steam reaction kinetics from pilot plant data. Industrial and Engineering Chemistry 50: 1289–1296.

Moeller, T. 1952. Inorganic Chemistry. New York: John Wiley and Sons, Inc.

Perry; R. H., D.W. Green, and J.O. Maloney (eds). 1984. Perry's Chemical Engineers Handbook. 6th Ed. New York: McGraw-Hill.

Appendix G

Mass Balances for HD Neutralization

This appendix contains mass balance matrices for the four HD neutralization configurations. For each configuration, there is a matrix for process inputs and one for process outputs. The stream numbers in the column headings are keyed to the numbered input and output streams shown in the process diagrams preceding each set of matrices. Each process diagram consists of two sheets: sheet 1 is the left side of a full diagram, sheet 2 is the right side. Input streams are numbered from 1; output streams are numbered from 100.

The diagrams and the mass balance data are derived either from the April 4, 1996, design package submitted by the Army Alternative Technology Program or from more recent data.

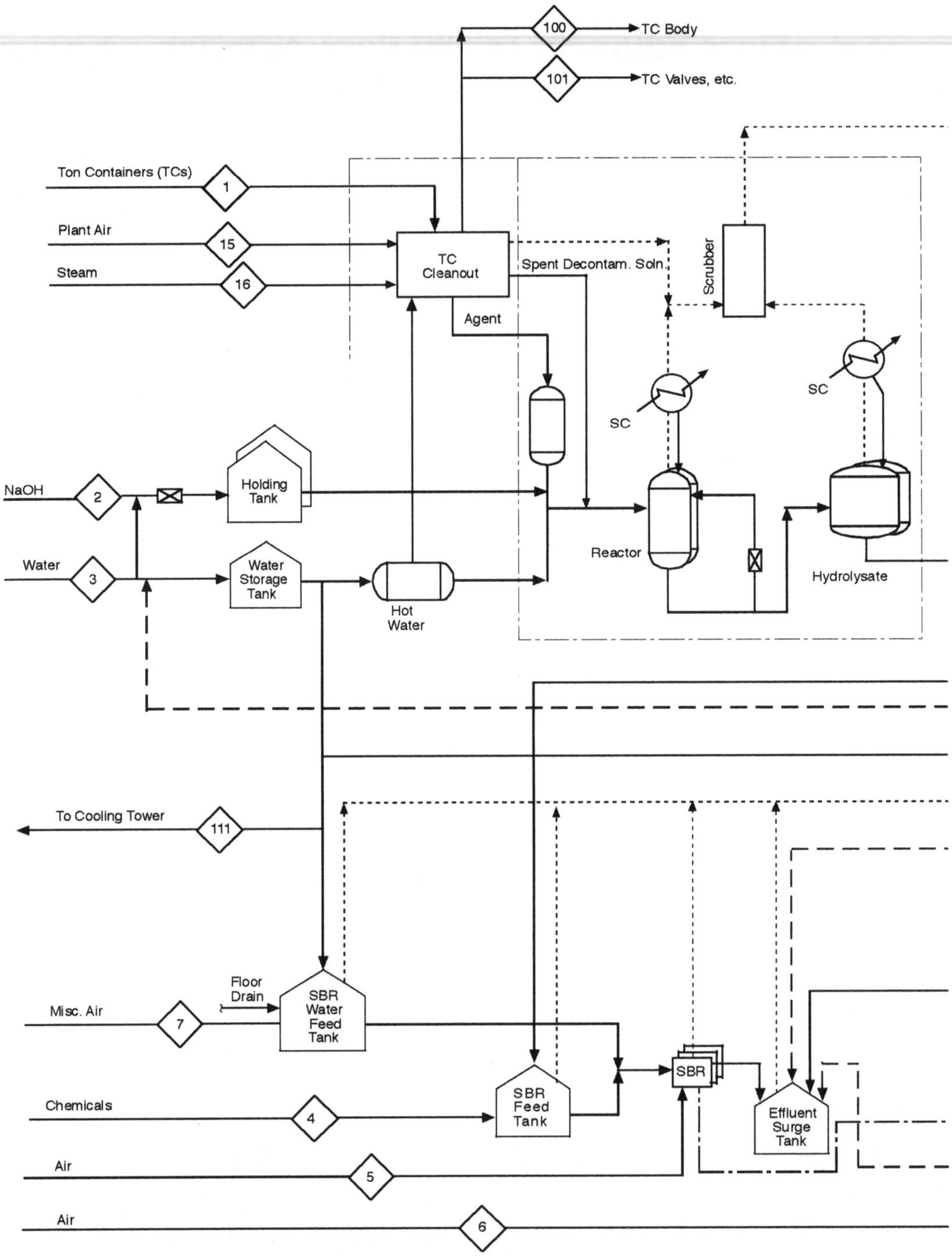

FIGURE G-1 HD neutralization, configuration 1. Neutralization followed by on-site biodegradation, including water recycling and photochemical oxidation of VOCs.

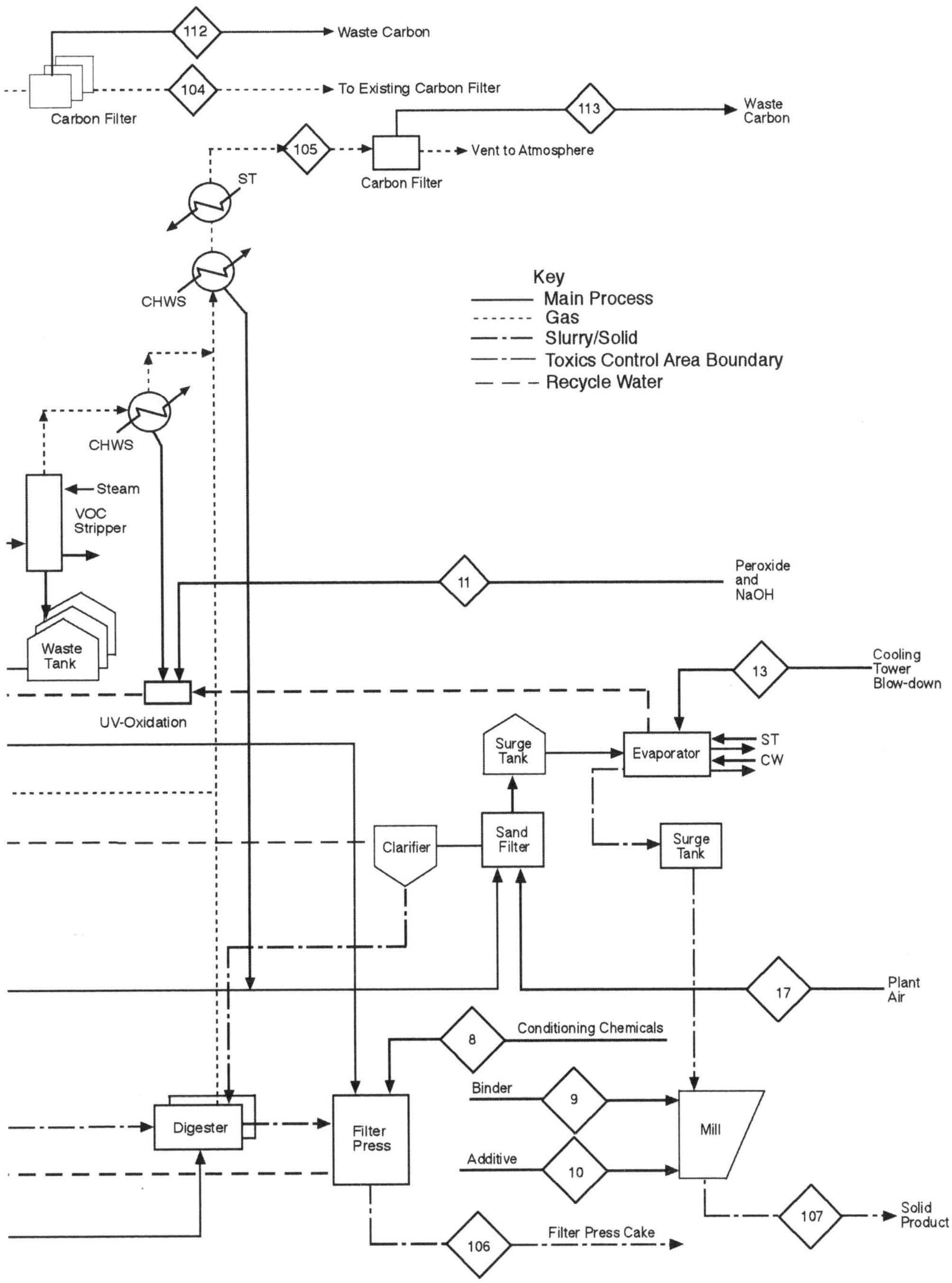
112
Waste Carbon
104
To Existing Carbon Filter
Carbon Filter
113
Waste Carbon
105
Vent to Atmosphere
Carbon Filter
ST
CHWS
Key
Main Process
Gas
Slurry/Solid
Toxics Control Area Boundary
Recycle Water
CHWS
Steam
VOC Stripper
11
Peroxide and NaOH
Waste Tank
13
Cooling Tower Blow-down
UV-Oxidation
Surge Tank
Evaporator
ST
CW
Clarifier
Sand Filter
Surge Tank
17
Plant Air
8
Conditioning Chemicals
Binder
9
Mill
Digester
Filter Press
Additive
10
107
Solid Product
106
Filter Press Cake

TABLE G-1 Process Inputs for HD Neutralization, Configuration 1

	Neutralization			Biodegradation					Solidify & Stabilize		Photochem. Oxidation	Ton Container Cleanout		Water Recycle		
Stream Number (see figures)	1	2	3	4	5	6	7	8	9	10	11	15	16	13	17	
Description	TCs with Agent	NaOH (aq.)	Water	Nutrients and Buffer	Air	Air	Air	Conditioning Chemicals	Binder	Additives		Air	Steam	Cooling Tower	Sand Filter Air	Total Inputs
Process Conditions																
Total flow, kg/1,000 kg	1,020	1,002	0	1,262	62,296	6,671	2,409	62	2,736	547	171	1,648	227	34,166	31	114,248
Pressure, psig		15	60	0	0	0	0	0	0	0	30	80	125	0	25	
Temperature, °F	70	70	47	70	70	70	70	70	70	70	70	100	351	75	110	
Physical state, solid (S), liquid (L), or gas (G)	S, L & G	L	L	S & L	G	G	G	L	S	S	L	G	G	L	G	
Feeds and Components																
HD ($C_4H_8Cl_2S$), kg/1,000 kg	904															904
Sodium hydroxide (NaOH), kg/1,000 kg		501									11					512
Water(H_2O), kg/1,000 kg		501	0	174	921	105	36	40			106		227	34,166		36,276
Sulfur-containing impurities, kg/1,000 kg	82															82
Chlorinated aliphatic hydrocarbons, kg/1,000 kg	7															7
Process Chemicals, Other																
$NaHCO_3$, kg/1,000 kg				1,008												1,008
KNO_3, kg/1,000 kg																0
KCl, kg/1,000 kg				15												15
Na_2SO_4, kg/1,000 kg																0
NH_3, kg/1,000 kg				37												37
Na_3PO_4, kg/1,000 kg																0
Wolin salts, kg/1,000 kg				8												8
H_3PO_4, kg/1,000 kg				20												20
H_2O_2, kg/1,000 kg											54					54
Organics, kg/1,000 kg																0

Polymer, kg/1,000 kg					4					4
Fe^{+2}, Fe^{+3}, kg/1,000 kg	10									10
Cl^{-}, kg/1,000 kg	17									17
Activated carbon (estim.), kg/1,000 kg										0
Biosolids, kg/1,000 kg										0
$Fe(OH)_3$, kg/1,000 kg					18					18
Binder compound (TBD), kg/1,000 kg						2,736				2,736
Cement additive (TBD), kg/1,000 kg							547			547
Gases										
O_2, kg/1,000 kg		14,208	1,520	549				380	7	16,664
N_2, kg/1,000 kg		44,239	4,733	1,711				1,189	23	51,895
CO_2, kg/1,000 kg		2,928	313	113				78	2	3,434
TC Shells	1.23									1.23
TC Valves	2.49									2.49
TC Plugs	7.45									7.45
TC Cuttings (3 lb/TC, estim.), kg/1,000 kg	1.69									1.69

TABLE G-2 Process Outputs for HD Neutralization, Configuration 1

	Ton Container Clean-Out and Neutralization				Biodegradation Process			Solidify & Stabilize	Water Recycle	
Stream Number (see figures) Description	100 Ton Container Bodies	101 Valves Plugs, etc.	104 Vent Gas	112 Activated Carbon	105 Vent Gas	106 Biomass (from filter)	113 Activated Carbon	107 Solid Product	111 To Cooling Tower	Total Outputs
Process Conditions										
Total flow, kg/1,000 kg			1,651	2	71,339	972	4	8,754	31,478	114,200
Pressure, psig			-5	0	0	0	0		5	
Temperature, °F	70	70	110		100	90		70	125	
Physical state, solid (S), liquid (L), or gas (G)	S	S	G	S	G		S	S	L	
Feeds and Components										
HD ($C_4H_8Cl_2S$), kg/1,000 kg										0
Sodium hydroxide (NaOH), kg/1,000 kg										0
Water (H_2O), kg/1,000 kg					811				31,478	32,289
Sulfur-containing impurities, kg/1,000 kg										0
Chlorinated aliphatic hydrocarbons, kg/1,000 kg										0
Process Residuals, Other										
$NaHCO_3$, kg/1,000 kg										0
NaCl, kg/1,000 kg						7		740		747
KNO_3, kg/1,000 kg								2		2
KCl, kg/1,000 kg								7		7
Na_2SO_4, kg/1,000 kg						7		792		799
NH_3, kg/1,000 kg								1		1
$NaNO_3$, kg/1,000 kg						1		28		29
$NaNO_2$, kg/1,000 kg								8		8
Na_3PO_4, kg/1,000 kg								7		7
Wolin salts, kg/1,000 kg								8		8
H_3PO_4, kg/1,000 kg										0

H_2O_2, kg/1,000 kg									0
Organics, kg/1,000 kg						1		66	67
Polymer, kg/1,000 kg						4			4
Fe^{+2}, Fe^{+3}, kg/1,000 kg									0
Cl^-, kg/1,000 kg									0
Activated carbon (estim.), kg/1,000 kg				2			4		6
Biosolids, kg/1,000 kg						163			163
$Fe(OH)_3$, kg/1,000 kg						32			32
Binder compound (TBD), kg/1,000 kg								2,736	2,736
Cement additive (TBD), kg/1,000 kg								547	547
Gases									
O_2, kg/1,000 kg			378		50,720				51,098
N_2, kg/1,000 kg			1,273		15,246				16,519
CO_2, kg/1,000 kg					4,563				4,563
TC Shells	1.23								1.23
TC Valves		2.49							2.49
TC Plugs		7.45							7.45
TC Cuttings (3 lb/TC, estim.), kg/1,000kg		1.69							1.69

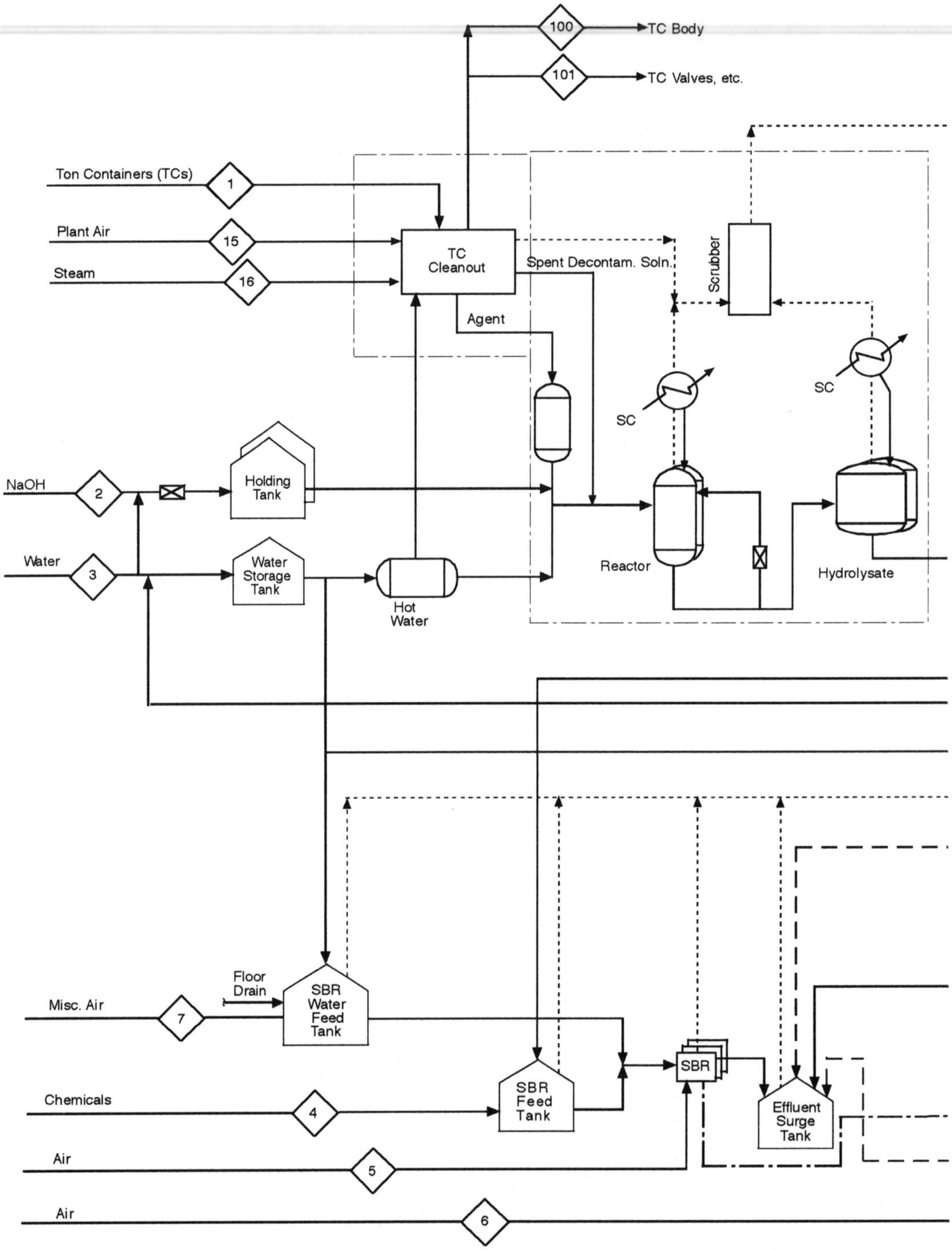

FIGURE G-2 HD neutralization, configuration 2. Neutralization followed by on-site biodegradation. VOCs are treated by photochemical oxidation. Biodegradation process effluent is discharged to a FOTW.

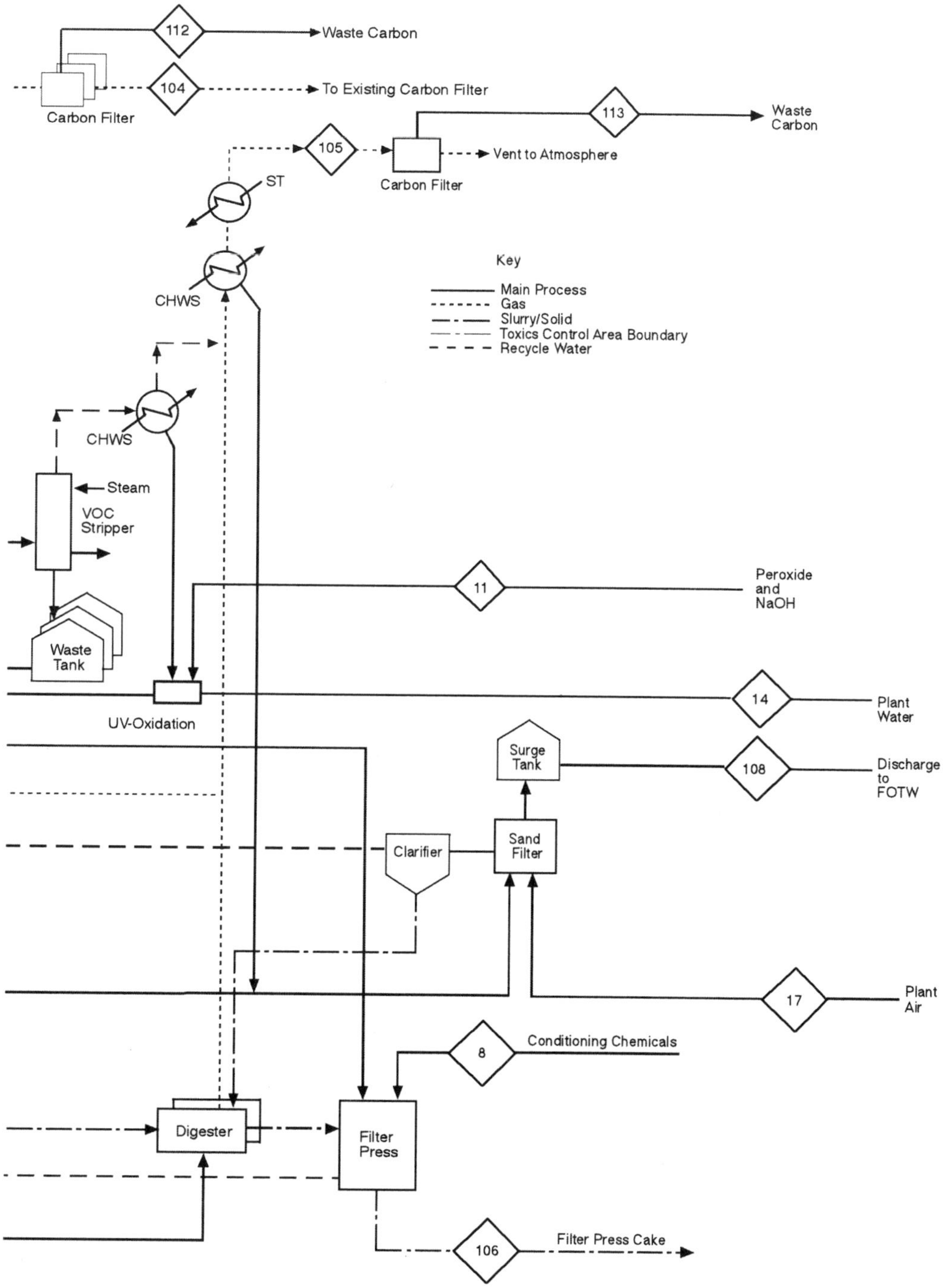
112
Waste Carbon
104
To Existing Carbon Filter
Carbon Filter
113
Waste Carbon
105
Vent to Atmosphere
Carbon Filter
ST
Key
Main Process
Gas
Slurry/Solid
Toxics Control Area Boundary
Recycle Water
CHWS
CHWS
Steam
VOC Stripper
11
Peroxide and NaOH
Waste Tank
14
Plant Water
UV-Oxidation
Surge Tank
108
Discharge to FOTW
Clarifier
Sand Filter
17
Plant Air
8
Conditioning Chemicals
Digester
Filter Press
106
Filter Press Cake

TABLE G-3 Process Inputs for HD Neutralization, Configuration 2

	Neutralization			Biodegradation					Photochemical Oxidation		Ton Container Cleanout		Sand Filtration	
Stream Number (see figures) Description	1 TCs with Agent	2 NaOH (aq.)	3 Water	4 Nutrients and Buffer	5 Air	6 Air	7 Air	8 Conditioning Chemicals	11 Oxidation Chemicals	14 Dilution Water	15 Air	16 Steam	17 Air	Total Inputs
Process Conditions														
Total flow, kg/1,000 kg	1,020	1,002	6,727	1,262	62,296	6,671	2,409	62	34	78,571	1,648	227	31	161,960
Pressure, psig		15	60	0	0	0	0	0	30	60	80	125	25	
Temperature, °F	70	70	47	70	70	70	70	70	70	47	100	351	110	
Physical state, solid (S), liquid (L), or gas (G)	S, L & G	L	L	S & L	G	G	G	L	L		G	G	G	
Feeds and Components														
HD ($C_4H_8Cl_2S$), kg/1,000 kg	904													904
Sodium hydroxide (NaOH), kg/1,000 kg		501							4					505
Water (H_2O), kg/1,000 kg		501	6,727	174	921	105	36	40	24	78,571		227		87,326
Sulfur-containing impurities, kg/1,000 kg	82													82
Chlorinated aliphatic hydrocarbons, kg/1,000 kg	7													7
Process Chemicals, Other														
$NaHCO_3$, kg/1,000 kg				1,008										1,008
KNO_3, kg/1,000 kg														0
KCl, kg/1,000 kg				15										15
Na_2SO_4, kg/1,000 kg														0
NH_3, kg/1,000 kg				37										37
Na_3PO_4, kg/1,000 kg														0
Wolin salts, kg/1,000 kg				8										8
H_3PO_4, kg/1,000 kg				20										20
H_2O_2, kg/1,000 kg									6					6
Organics, kg/1,000 kg														0
Polymer, kg/1,000 kg								4						4

Fe^{+2}, Fe^{+3}, kg/1,000 kg	10							10
Cl^{-}, kg/1,000 kg	17							17
Activated carbon (estim.), kg/1,000 kg								0
Biosolids, kg/1,000 kg								0
$Fe(OH)_3$, kg/1,000 kg					18			18
Gases								
O_2, kg/1,000 kg		14,208	1,520	549		380	7	16,664
N_2, kg/1,000 kg		44,239	4,733	1,711		1,189	23	51,895
CO_2, kg/1,000 kg		2,928	313	113		78	2	3,434
TC Shells	1.23							
TC Valves	2.49							
TC Plugs	7.45							
TC Cuttings (3 lb/TC, estim.), kg/1,000 kg								

TABLE G-4 Process Outputs for HD Neutralization, Configuration 2

	Ton Container Cleanout and Neutralization				Biodegradation Process			Sand Filtration	
Stream Number (see figures) Description	100 Ton Container Bodies	101 Valves, Plugs, etc.	104 Vent Gas	112 Activated Carbon	105 Vent Gas	106 Biomass (from filter)	113 Activated Carbon	108 Effluent (to FOTW)	Total Outputs
Process Conditions									
Total flow, kg/1,000 kg			1,651	2	71,339	972	4	87,947	161,915
Pressure, psig			-5	0	0	0	0	0	
Temperature, °F	70	70	110		100	90		94	
Physical state, solid (S), liquid (L), or gas (G)	S	S	G	S	G		S	L	
Feeds and Components									
HD ($C_4H_8Cl_2S$), kg/1,000 kg									0
Sodium hydroxide (NaOH), kg/1,000 kg									0
Water (H_2O), kg/1,000 kg					811	757		86,278	87,846
Sulfur-containing impurities, kg/1,000 kg									0
Chlorinated aliphatic hydrocarbons, kg/1,000 kg									0
Process Residuals, Other									
$NaHCO_3$, kg/1,000 kg									0
NaCl, kg/1,000 kg						7		740	747
KNO_3, kg/1,000 kg								2	2
KCl, kg/1,000 kg								7	7
Na_2SO_4, kg/1,000 kg						7		792	799
NH_3, kg/1,000 kg									0
$NaNO_3$, kg/1,000 kg						1		28	29
$NaNO_2$, kg/1,000 kg								8	8
Na_3PO_4, kg/1,000 kg								7	7
Wolin salts, kg/1,000 kg								8	8
H_3PO_4, kg/1,000 kg									0
H_2O_2, kg/1,000 kg									0
Organics, kg/1,000 kg						1		76	77
Polymer, kg/1,000 kg						4			4
Fe^{+2}, Fe^{+3}, kg/1,000 kg									0
Cl^-, kg/1,000 kg									0
Activated carbon (estim.), kg/1,000 kg				2			4		6

TABLE G-4 *Continued*

	Ton Container Cleanout and Neutralization				Biodegradation Process			Sand Filtration	
Stream Number (see figures) Description	100 Ton Container Bodies	101 Valves, Plugs, etc.	104 Vent Gas	112 Activated Carbon	105 Vent Gas	106 Biomass (from filter)	113 Activated Carbon	108 Effluent (to FOTW)	Total Outputs
Biosolids, kg/1,000 kg						163			163
$Fe(OH)_3$, kg/1,000 kg						32			32
Binder compound (TBD), kg/1,000 kg									
Cement additive (TBD), kg/1,000 kg									
Gases									
O_2, kg/1,000 kg			378		50,720				51,098
N_2, kg/1,000 kg			1,273		15,246				16,519
CO_2, kg/1,000 kg					4,563				4,563
TC Shells	1.23								1.23
TC Valves		2.49							2.49
TC Plugs		7.45							7.45
TC Cuttings (3 lb/TC, estim.), kg/1,000 kg		1.69							1.69

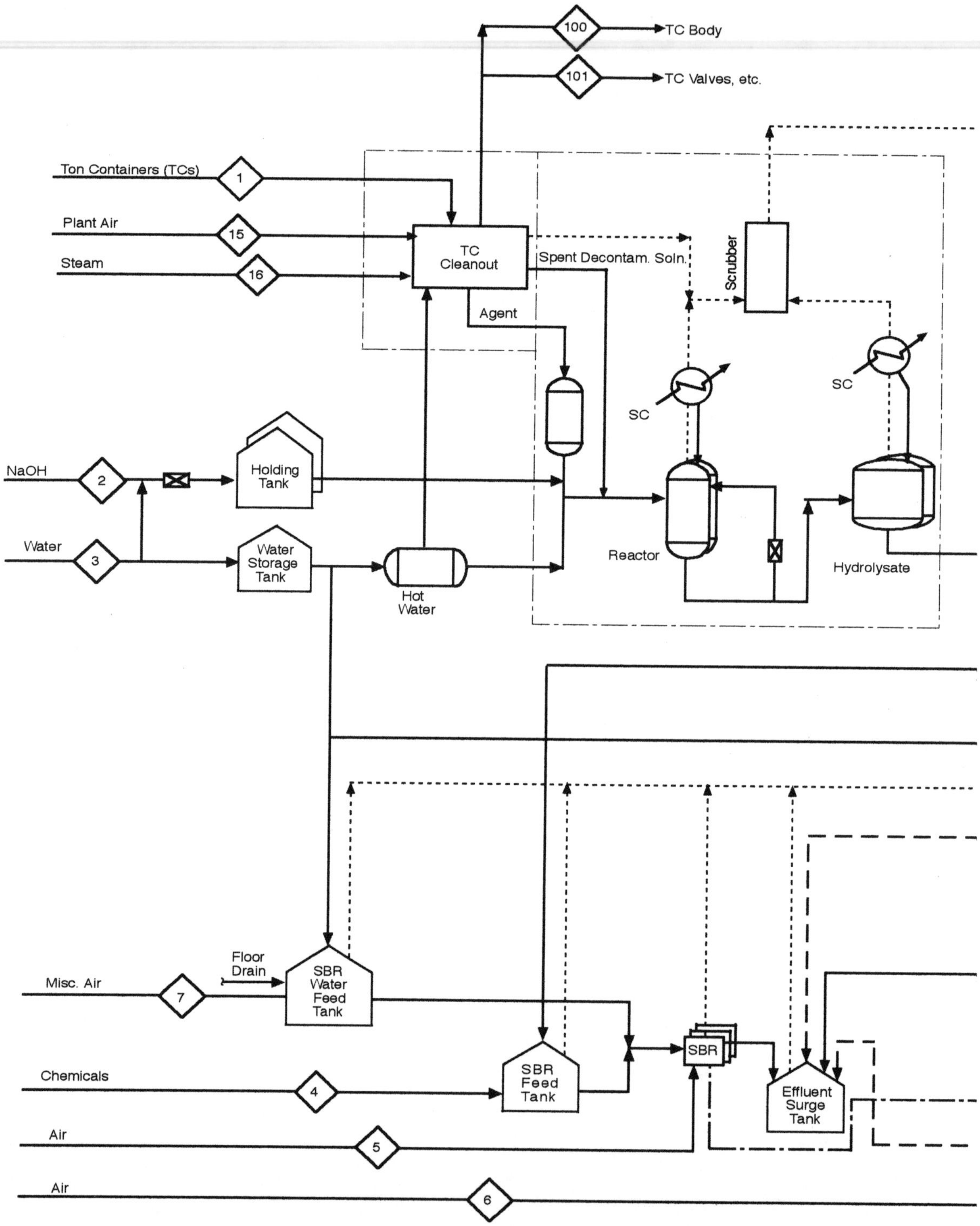

FIGURE G-3 HD neutralization, configuration 3. Neutralization followed by on-site biodegradation. VOCs are shipped to an off-site TSDF. Biodegradation process effluent is discharged to a FOTW.

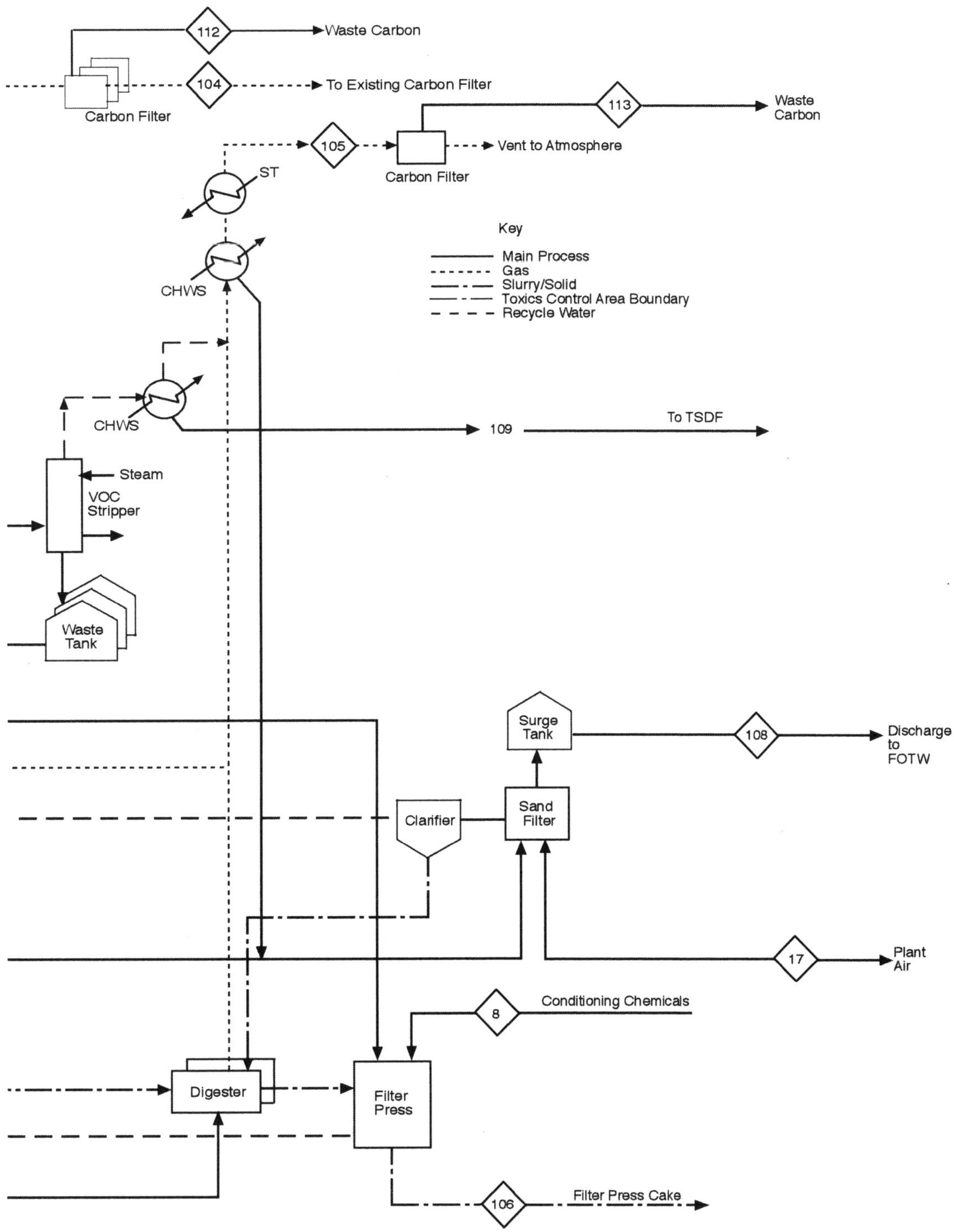
112
Waste Carbon
104
To Existing Carbon Filter
Carbon Filter
113
Waste Carbon
105
Vent to Atmosphere
Carbon Filter
ST
Key
Main Process
Gas
Slurry/Solid
Toxics Control Area Boundary
Recycle Water
CHWS
CHWS
109
To TSDF
Steam
VOC Stripper
Waste Tank
Surge Tank
108
Discharge to FOTW
Clarifier
Sand Filter
17
Plant Air
8
Conditioning Chemicals
Digester
Filter Press
106
Filter Press Cake

TABLE G-5 Process Inputs for HD Neutralization, Configuration 3

	Neutralization			Biodegradation					Ton Container Cleanout		Sand Filtration	
Stream Number (see figures) Description	1 TCs with Agent	2 NaOH (aq.)	3 Water	4 Nutrients and Buffer	5 Air	6 Air	7 Air	8 Conditioning Chemicals	15 Air	16 Steam	17 Air	Total Inputs
Process Conditions												
Total flow, kg/1,000 kg	1,020	1,002	6,727	1,262	62,296	6,671	2,409	62	1,648	227	31	161,960
Pressure, psig		15	60	0	0	0	0	0	80	125	25	
Temperature, °F	70	70	47	70	70	70	70	70	100	351	110	
Physical state, solid (S), liquid (L), or gas (G)	S, L & G	L	L	S & L	G	G	G	L	G	G	G	
Feeds and Components												
HD ($C_4H_8Cl_2S$), kg/1,000 kg	904											904
Sodium hydroxide (NaOH), kg/1,000 kg		501										501
Water (H_2O), kg/1,000 kg		501	87,691	174	921	105	36	40		227		87,326
Sulfur-containing impurities, kg/1,000 kg	82											82
Chlorinated aliphatic hydrocarbons, kg/1,000 kg	7											7
Process Chemicals, Other												
$NaHCO_3$, kg/1,000 kg				1,008								1,008
KNO_3, kg/1,000 kg												0
KCl, kg/1,000 kg				15								15
Na_2SO_4, kg/1,000 kg												0
NH_3, kg/1,000 kg				37								37
Na_3PO_4, kg/1,000 kg												0
Wolin salts, kg/1,000 kg				8								8
H_3PO_4, kg/1,000 kg				20								20
Organics, kg/1,000 kg												0
Polymer, kg/1,000 kg								4				4
Fe^{+2}, Fe^{+3}, kg/1,000 kg	10											10

Cl^-, kg/1,000 kg	17							17
Activated carbon (estim.), kg/1,000 kg								0
Biosolids, kg/1,000 kg								0
$Fe(OH)_3$, kg/1,000 kg					18			18
Gases								
O_2, kg/1,000 kg		14,208	1,520	549		380	7	16,664
N_2, kg/1,000 kg		44,239	4,733	1,711		1,189	23	51,895
CO_2, kg/1,000 kg		2,928	313	113		78	2	3,434
TC Shells	1.23							
TC Valves	2.49							
TC Plugs	7.45							
TC Cuttings (3 lb/TC, estim.), kg/1,000 kg								

TABLE G-6 Process Outputs for HD Neutralization, Configuration 3

	Ton Container Cleanout and Neutralization				Biodegradation Process			Sand Filtration	Stripped VOCs	
Stream Number (see figures) Description	100 Ton container bodies	101 Valves, plugs, etc.	104 Vent gas	112 Activated carbon	105 Vent gas	106 Biomass (from filter)	113 Activated carbon	108 Effluent (to FOTW)	109 VOCs to TSDF	Total Outputs
Process Conditions										
Total flow, kg/1,000 kg			1,651	2	71,339	972	4	87,947	2,352	164,267
Pressure, psig			-5	0	0	0	0	0	0	
Temperature, °F	70	70	110		100	90		94	35	
Physical state, solid (S), liquid (L), or gas (G)	S	S	G	S	G	S	S	L	L	
Feeds and Components										
HD ($C_4H_8Cl_2S$), kg/1,000 kg										0
Sodium hydroxide (NaOH), kg/1,000 kg										0
Water (H_2O), kg/1,000 kg					811			86,278	2,347	89,436
Sulfur-containing impurities, kg/1,000 kg										0
Chlorinated aliphatic hydrocarbons, kg/1,000 kg									5	5
Process Residuals, Other										
$NaHCO_3$, kg/1,000 kg										
NaCl, kg/1,000 kg						7		740		747
KNO_3, kg/1,000 kg								2		2
KCl, kg/1,000 kg								7		7
Na_2SO_4, kg/1,000 kg						7		792		799
NH_3, kg/1,000 kg										0
$NaNO_3$, kg/1,000 kg						1		28		29
$NaNO_2$, kg/1,000 kg								8		8
Na_3PO_4, kg/1,000 kg								7		7
Wolin salts, kg/1,000 kg								8		8
H_3PO_4, kg/1,000 kg										

H_2O_2, kg/1,000 kg									
Organics, kg/1,000 kg						1		76	77
Polymer, kg/1,000 kg						4			4
Fe^{+2}, Fe^{+3}, kg/1,000 kg									
Cl^-, kg/1,000 kg									0
Activated carbon (estim.), kg/1,000 kg				2			4		6
Biosolids, kg/1,000 kg						163			163
$Fe(OH)_3$, kg/1,000 kg						32			32
Gases									
O_2, kg/1,000 kg			378		50,720				51,098
N_2, kg/1,000 kg			1,273		15,246				16,519
CO_2, kg/1,000 kg					4,563				4,563
TC Shells	1.23								1.23
TC Valves		2.49							2.49
TC Plugs		7.45							7.45
TC Cuttings (3 lb/TC, estim.), kg/1,000 kg		1.69							1.69

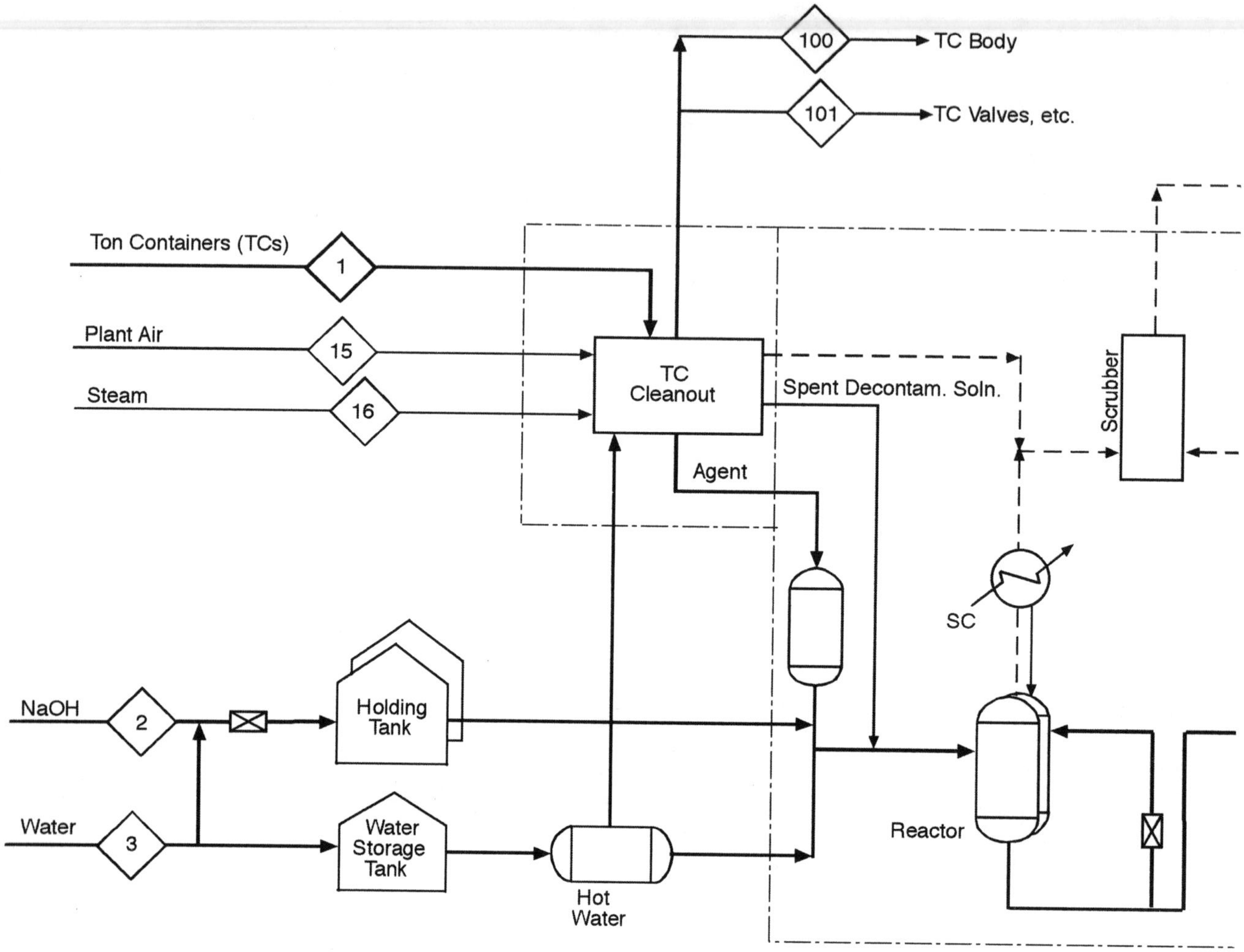

FIGURE G-4 HD neutralization, configuration 4. Neutralization followed by off-site biodegradation of hydrolysate at a TSDF. VOCs remain in the hydrolysate.

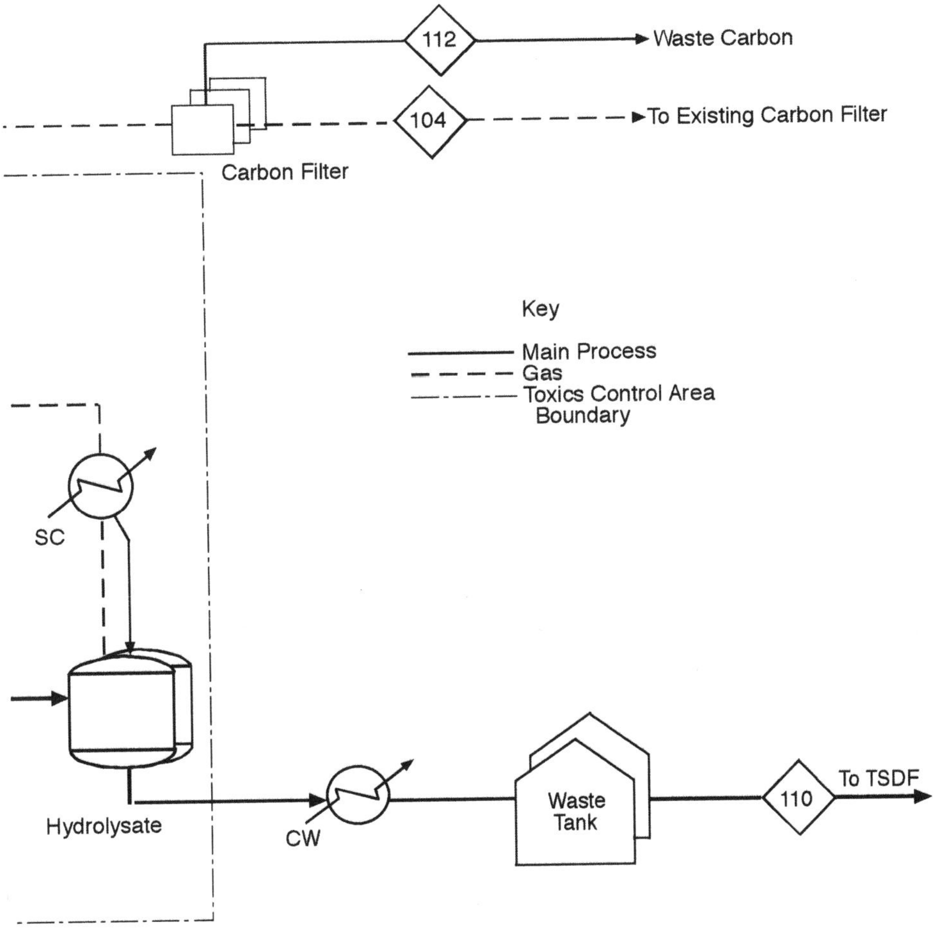
112
Waste Carbon
104
To Existing Carbon Filter
Carbon Filter
Key
Main Process
Gas
Toxics Control Area
Boundary
SC
Hydrolysate
CW
Waste
Tank
110
To TSDF

TABLE G-7 Process Inputs for HD Neutralization, Configuration 4

	Neutralization			Ton Container Clean-Out		
Stream Number (see figures)	1	2	3	15	16	
Description	TCs with Agent	NaOH (aq.)	Water	Air	Steam	Total Inputs
Process Conditions						
Total flow, kg/1,000 kg	1,020	1,002	27,304	1,648	227	31,201
Pressure, psig		15	60	80	125	
Temperature, °F	70	70	47	100	351	
Physical state, solid (S), liquid (L), or gas (G)	S, L & G	L	L	G	G	
Feeds and Components						
HD ($C_4H_8Cl_2S$), kg/1,000 kg	904					904
Sodium hydroxide (NaOH), kg/1,000 kg		501				501
Water (H_2O), kg/1,000 kg		501	27,304		227	28,032
Sulfur-containing impurities, kg/1,000 kg	82					82
Chlorinated aliphatic hydrocarbons, kg/1,000 kg	7					7
Process Chemicals, Other						
Fe^{+2}, Fe^{+3}, kg/1,000 kg	10					10
Cl^-, kg/1,000 kg	17					17
Activated carbon (estim.), kg/1,000 kg						0
Gases						
O_2, kg/1,000 kg				380		380
N_2, kg/1,000 kg				1,189		1,189
CO_2, kg/1,000 kg				78		78
TC Shells	1.23					1.23
TC Valves	2.49					2.49
TC Plugs	7.45					7.45
TC Cuttings (3 lb/TC, estim.), kg/1,000 kg						

TABLE G-8 Process Outputs for HD Neutralization, Configuration 4

	Ton Container Clean-Out and Neutralization					
Stream Number (see figures) Total Inputs	100 Ton Container Bodies	101 Valves, Plugs, etc.	104 Vent Gas	110 Hydrolysate to TSDF	112 Activated Carbon	Total Outputs
Process Conditions						
Total flow, kg/1,000 kg			1,651	29,468	2	31,121
Pressure, psig			-5	20	0	
Temperature, °F	70	70	110	120		
Physical state, solid (S), liquid (L), or gas (G)	S	S	G	L	S	
Feeds and Components						
HD ($C_4H_8Cl_2S$), kg/1,000 kg						0
Sodium hydroxide (NaOH), kg/1,000 kg				12		12
Water (H_2O), kg/1,000 kg				27,949		27,949
Sulfur-containing impurities, kg/1,000 kg				17		17
Chlorinated aliphatic hydrocarbons, kg/1,000 kg				7		7
Process Residuals, Other						
Thiodiglycol, kg/1,000 kg				624		624
Other hydrolysis products, kg/1,000 kg				125		125
NaCl, kg/1,000 kg				715		715
Fe^{+2}, Fe^{+3}, kg/1,000 kg						0
Cl^-, kg/1,000 kg						0
Activated carbon (estim.), kg/1,000 kg					2	2
Gases						
O_2, kg/1,000 kg			378			378
N_2, kg/1,000 kg			1,273			1,273
CO_2, kg/1,000 kg						0
TC Shells	1.23					1.23
TC Valves		2.49				2.49
TC Plugs		7.45				7.45
TC Cuttings (3 lb/TC, estim.), kg/1,000 kg		1.69				1.69

Appendix H

Mass Balances for VX Neutralization

This appendix contains mass balance matrices for neutralization of VX followed by off-site treatment of oxidized hydrolysate, as described in Chapter 8. There is one matrix for process inputs (Table H-1) and one for process outputs (Table H-2). The stream numbers in the column headings are keyed to the numbered input and output streams shown in the process diagram (Figure H-1). Input streams are numbered from 1; output streams are numbered from 100.

The process diagram and the mass balance data are derived either from the April 4, 1996, design package submitted by the Army Alternative Technology Program or from more recent data.

TABLE H-1 Process Inputs for VX Neutralization

	Neutralization				Ton Container Cleanout			
Stream Number (see figures) Description	1 Ton Containers with Agent	2 NaOH (aq.)	3 Water	4 NaOCl (aq)	5 Air	6 Steam	7 Decontam. Fluid	Total Inputs
Process Conditions								
Total flow, kg/1,000 kg	1,000	1,028	2,660	4,958	1,728	174	510	12,058
Pressure, psig		15	65	15	80	125	15	
Temperature, °F	70	70	47	70	70	351	70	
Physical State, solid (S), liquid (L), or gas (G)	S,L	L	L	L	G	G	L	
Major Feed Components								
VX, $C_{11}H_{26}NO_2PS$, kg/1,000 kg	937							937
Water, kg/1,000 kg		617	2,660	4,214		174	495	8,160
NaOH, kg/1,000 kg		411						411
NaOCl, kg/1,000 kg				744			15	759
NaCl, kg/1,000 kg								0
Air, kg/1,000 kg					1,728			1,728
Agent Impurities								
Diisopropylamine, kg/1,000 kg	1							1
Diisopropylcarbodiimide (stabilizer), kg/1,000 kg	17							17
O-Ethyl methylethylphosphinate, kg/1,000 kg	2							2
Diethyl methylphosphonate, kg/1,000 kg	1							1
2-(Diisopropylamino)ethanethiol, kg/1,000 kg	9							9
O,O-Diethyl methylphosphonothioate, kg/1,000 kg	2							2
O,S-Diethyl methylphosphonothioate, kg/1,000 kg	1							1
2-(Diisopropylamino)ethyl ethyl sulfide, kg/1,000 kg	1							1
Diethyl dimethylpyrophosphonate ("Pyro"), kg/1,000 kg	10							10
O,O-Diethyl dimethylpyrophosphonothioate, kg/1,000 kg	2							2
O-(2-Diisopropylaminoethyl) O-ethylmethylphosphonate, kg/1,000 kg	3							3
1,2-bis(ethyl methylphosphonothiolo)ethane, kg/1,000 kg	6							6
Unknowns, kg/1,000 kg	7							7
Ton Containers, no./1000 kg	1.52							1.52
TC Valves, no./1000 kg	3.06							3.06
TC Plugs, no./1000 kg	9.19							9.19
TC Cuttings (3 lb/TC estimated), kg/1,000 kg								0
Activated Carbon (estimated), kg/1,000 kg								0

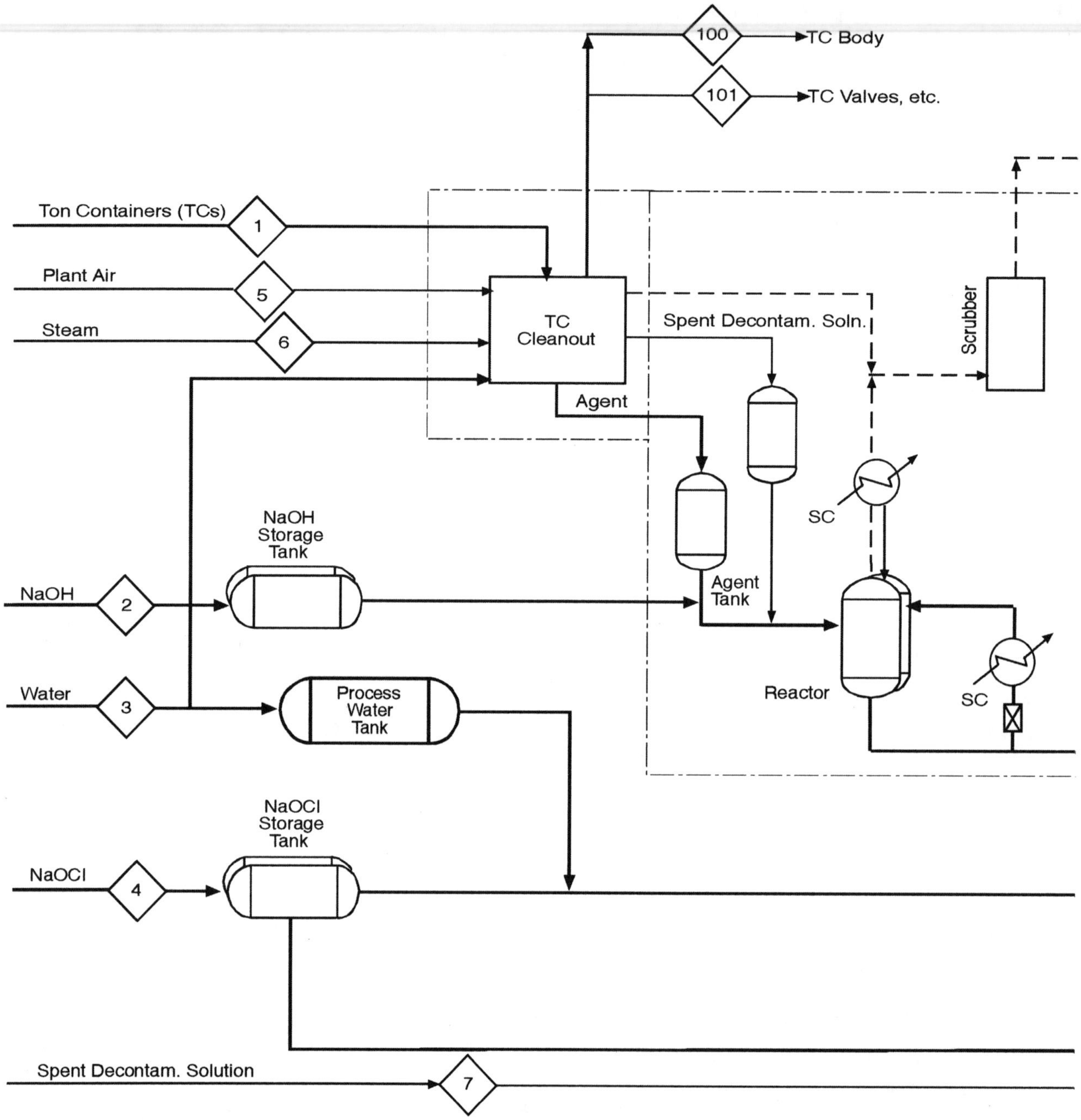

FIGURE H-1 VX neutralization and treatment with oxidizing agent, followed by off-site treatment of oxidized hydrolysate. (Off-site treatment method not shown).

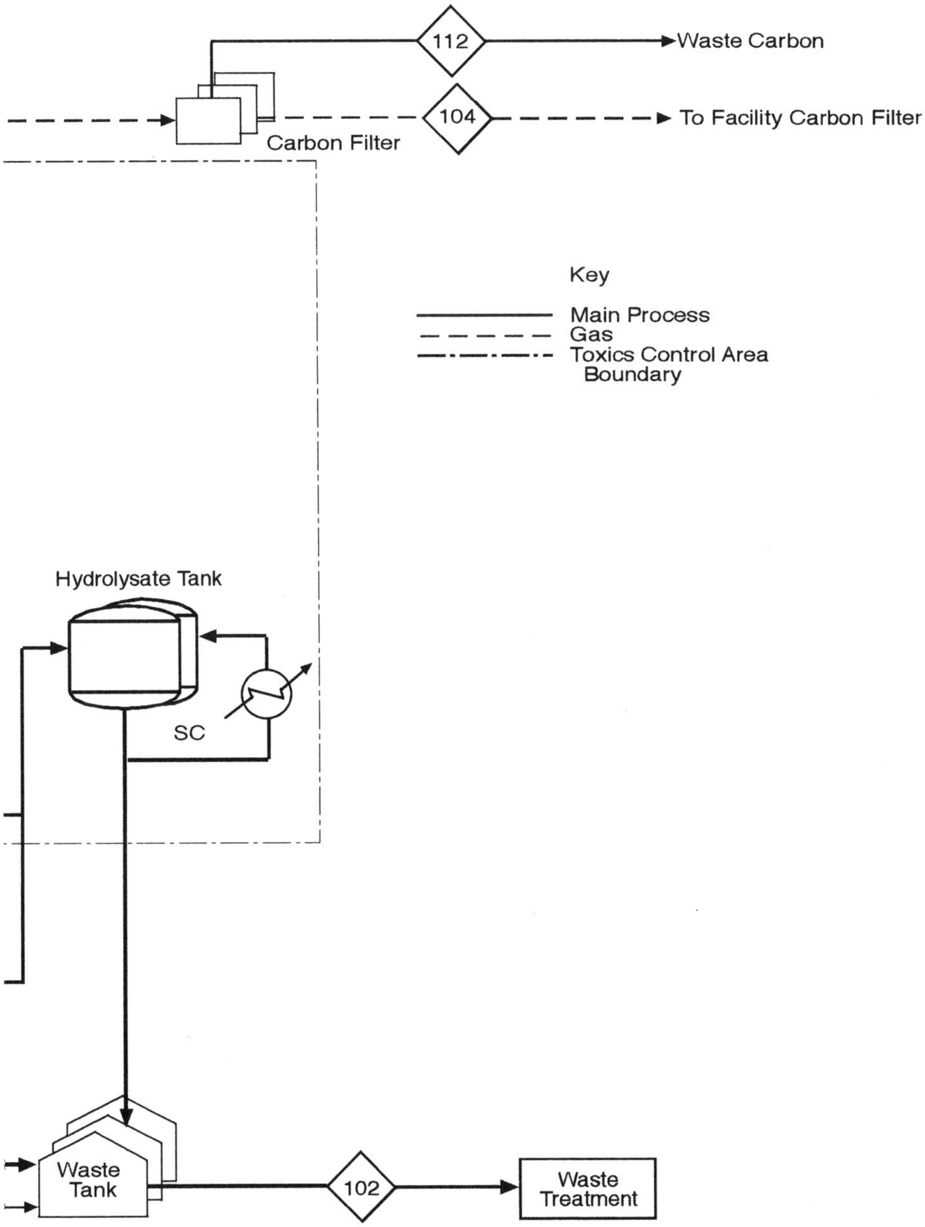
112
Waste Carbon
104
To Facility Carbon Filter
Carbon Filter
Key
Main Process
Gas
Toxics Control Area
Boundary
Hydrolysate Tank
SC
Waste
Tank
102
Waste
Treatment

TABLE H-2 Process Outputs for VX Neutralization

	Ton Container Cleanout and Neutralization					
Stream Number (see figures) Description	100 Ton Container Bodies	101 Valves, Plugs, etc.	102 Hydrolysate to TSDF	103 Vent Gas	104 Activated Carbon	Total Outputs
Process Conditions						
Total Flow, kg/1,000 kg	see below	see below	10,330	1,736	2	12,068
Pressure, psig			10	-1		
Temperature, °F	70	70	70	110	110	
Physical State, solid (S), liquid (L), or gas (G)	S	S	L	G	S	
Major Feed Components						
VX, $C_{11}H_{26}NO_2PS$, kg/1,000 kg						0
Water, kg/1,000 kg			7,575			7,575
NaOH, kg/1,000 kg			120			120
NaOCl, kg/1,000 kg						0
NaCl, kg/1,000 kg			1,192			1,192
Air, kg/1,000 kg				1,736		1,736
Agent Impurities						
Diisopropylamine, kg/1,000 kg			1			1
O-Ethyl methylethylphosphinate, kg/1,000 kg			2			2
O,O-Diethyl methylphosphonothioate, kg/1,000 kg			2			2
O,S-Diethyl methylphosphonothioate, kg/1,000 kg			1			1
2-(Diisopropylamino)ethyl ethyl sulfide, kg/1,000 kg			1			1
O,O-Diethyl dimethylpyrophosphonothioate, kg/1,000 kg			2			2
1,2-bis(ethyl methylphosphonothiolo)ethane, kg/1,000 kg			6			6
Unknowns, kg/1,000 kg			7			7
Process Residuals, Other						
EMPA-Na, kg/1,000 kg			463			463
MPA-2Na, kg/1,000 kg			49			49
EA-2192, Na salt, kg/1,000 kg			2			2
Chloroform, kg/1,000 kg			43			43
Chloroamine, kg/1,000 kg			2			2
Diisopropylamino ethylsulfonic acid, kg/1,000 kg			251			251
Diisopropylamino ethylsulfinic acid, kg/1,000 kg			534			534
Dicyclohexylurea (DCHU) or Diisopropylurea, kg/1,000 kg			14			14
Sodium methylphosphinate, kg/1,000 kg			1			1
EMPSA, Na salt, kg/1,000 kg			5			5
Methyl phosphonothioates, as salts, kg/1,000 kg			10			10
Other methylphosphonates, as salts, kg/1,000 kg			2			2
Disulfide, kg/1,000 kg			11			11
Trisulfide, kg/1,000 kg			16			16
Other sulfides and amines, kg/1,000 kg			18			18
TC, no./1000 kg	1.52					1.52
TC Valves, no./1000 kg		3.06				3.06
TC Plugs, no./1000 kg		9.19				9.19
TC Cuttings (3 lb/TC estimated, kg/1,000 kg		2.08				2.08
Activated Carbon (estimated), kg/1,000 kg					2	2.00

Appendix I

Biographical Sketches of Panel Members

Richard S. Magee, *chair,* is a professor in the Department of Mechanical Engineering and the Department of Chemical Engineering, Chemistry, and Environmental Science and is executive director of the Center for Environmental Engineering and Science at New Jersey Institute of Technology (NJIT). He also directs the U.S. Environmental Protection Agency Northeast Hazardous Substance Research Center as well as the Hazardous Substance Management Research Center, which is jointly sponsored by the National Science Foundation and the New Jersey Commission on Science and Technology, both headquartered at NJIT. He is a fellow of the ASME (American Society of Mechanical Engineers) and a diplomate of the American Academy of Environmental Engineers. Dr. Magee's research expertise is in combustion, with a major interest in the incineration of municipal and industrial wastes. He has served as vice chairman of the ASME Research Committee on Industrial and Municipal Wastes and as a member of the United Nations Special Commission (under Security Council Resolution 687) Advisory Panel on the Destruction of Iraq's Chemical Weapons Capabilities. He is presently a member of the North Atlantic Treaty Organization Science Committee Priority Area Panel on Disarmament Technologies. Dr. Magee is also the current chair of the NRC Committee on Review and Evaluation of the Army Chemical Stockpile Disposal Program (Stockpile Committee).

Joan B. Berkowitz graduated from the University of Illinois with a Ph.D. in physical chemistry and from the Sloan School Senior Executive Program at M.I.T. Dr. Berkowitz is currently the managing director of Farkas Berkowitz and Company. She has extensive experience in the area of environmental and hazardous waste management, a knowledge of available technologies for the cleanup of contaminated soils and groundwater, and a background in physical and electrochemistry. She has contributed to several EPA studies, been a consultant on remediation techniques, and assessed various destruction technologies. Dr. Berkowitz has written numerous publications on hazard waste treatment and environmental subjects.

Gene H. Dyer graduated with a bachelor of science degree in chemistry, mathematics, and physics from the University of Nebraska. Over a 12-year period, he worked for General Electric as a process engineer, the U.S. Navy as a research and development project engineer, and the U.S. Atomic Energy Commission as a project engineer. In 1963, he began a more than 20-year career with the Bechtel Corporation, first as a consultant on advanced nuclear power plants and later as a program supervisor for nuclear facilities. From 1969 to 1983, he was manager of the Process and Environmental Department, which provided engineering services related to research and development projects, including technology probes, environmental assessment, air pollution control, water pollution control, process development, nuclear fuel process development, and regional planning. As a senior staff consultant for several years, he was responsible for identifying and evaluating new technologies and managing further development and testing for practical applications. Mr. Dyer is a member of the American Institute of Chemical Engineers and a registered professional engineer. He recently served as a member of the NRC Committee on Alternative Chemical Demilitarization Technologies and is currently a member of the NRC Committee on Review and Evaluation of the Army Chemical Stockpile Disposal Program (Stockpile Committee).

Frederick T. Harper is the manager of the Accident Analysis and Consequence Assessment Department at the Sandia National Laboratory, Albuquerque, New Mexico. His areas of expertise are the probabilistic assessment of accident progression, including the physical response of systems to accident conditions and the transport of toxicological and radiological contaminants; assessment of the release of contaminants; and the structural and thermal response of systems to fire and explosion. Dr. Harper has served on an international committee in the area of consequence uncertainty and has been a prime developer of computer codes for assessing toxicological consequences and accident progression. Dr. Harper earned a bachelor's degree from Yale University in physics, a master's degree from the University of Virginia in nuclear engineering, and a doctorate, also in nuclear engineering, from the University of New Mexico. He is a member of Tau Beta Pi, the American Physical Society, and the American Nuclear Society.

Joseph A. Heintz recently retired from the Atlantic Richfield Oil Corporation where he was engineering manager for many years. Mr. Heintz attended the University of Illinois and Purdue University where he received degrees in electrical engineering. He is an expert in mechanical design, plant layout, process configuration, and process monitoring. He has supervised the designing of pressure vessels, overseen stress analysis studies, coordinated engineering standards and instrumentation groups responsible for developing process control strategies, prepared detailing piping and instrumentation diagrams, identified control system components, and prepared control system functional specifications. In addition, he has participated in the selection of control system vendors. His is a member of the Instrument Society of America.

David A. Hoecke, president and CEO of Enercon Systems, Inc., is an expert in the fields of waste combustion, pyrolysis, heat transfer, and gas cleaning. He graduated with a B.S.M.E from the Cooper Union in 1960 and rose from project engineer to R&D manager to chief engineer for incineration at Midland-Ross Corporation and later founded his own company. Mr. Hoecke has been responsible for the design and construction of numerous combustion systems, including solid waste incinerators, thermal oxidizers, heat recovery systems, gas-to-air heat exchangers, and high velocity drying ovens. This hands-on experience gives him the expertise needed to participate in the assessment of alternative destruction technologies for chemical agents. He has served as the co-chair of the ASME Subcommittee on Vitrification Systems. He also recently served on the ASME Board on Research and Technology Development.

David S. Kosson graduated with a bachelor of science degree in chemical engineering, a master's degree in chemical and biochemical engineering, and a doctorate in chemical and biochemical engineering from Rutgers-The State University of New Jersey. He joined the faculty at Rutgers in 1986 as an associate professor, with tenure in 1990. He became a full professor in 1996. Dr. Kosson teaches graduate and undergraduate chemical engineering courses and conducts research for the Department of Chemical and Biochemical Engineering, where considerable work is under way in developing microbial, chemical, and physical treatment methods for hazardous waste. He is responsible for project planning and coordination, from basic research through full-scale design and implementation. He has published extensively in the fields of chemical engineering, waste management and treatment, and contaminant fate and transport in soils and groundwater. Dr. Kosson is a participant in several Environmental Protection Agency advisory panels involved in waste research and is the director of the Physical Treatment Division of the Hazardous Substances Management Research Center in New Jersey. He is a member of the American Institute of Chemical Engineers. He recently served as a member of the NRC Committee on Alternative Chemical Demilitarization Technologies and is currently a member of the NRC Committee on Review and Evaluation of the Army Chemical Stockpile Disposal Program (Stockpile Committee).

Walter G. May graduated with a bachelor of science degree in chemical engineering and master of science degree in chemistry from the University of Saskatchewan and a doctor of science degree in chemical engineering from the Massachusetts Institute of Technology. He joined the faculty of the University of Saskatchewan as a professor of chemical engineering in 1943. In 1948, he began a distinguished career with Exxon Research and Engineering Company, where he was a senior science advisor from 1976 to 1983. He was professor of chemical engineering at the University of Illinois from 1983 until his retirement in 1991. There he taught courses in process design, thermodynamics, chemical reactor design, separation processes, and industrial chemistry and stoichiometry. Dr. May has published extensively, served on the editorial boards of *Chemical Engineering Reviews* and *Chemical Engineering Progress*, and has obtained numerous patents in his field. He is a member of the National Academy of Engineering and a fellow of the American Institute of Chemical Engineers, and he has received special awards from the American Institute of Chemical Engineers and ASME. Dr. May's particular interest is in separations research. He is a registered professional engineer in the state of

Illinois and recently served as a member of the NRC Committee on Alternative Chemical Demilitarization Technologies. He is currently a member of the NRC Committee on Review and Evaluation of the Army Chemical Stockpile Disposal Program (Stockpile Committee).

Alvin H. Mushkatel, professor and director of the Office of Hazards Studies, and Professor of the School of Planning and Landscape Architecture, Arizona State University, is an expert in emergency response and risk perceptions. His research interests include emergency management, natural and technological hazards policy, and environmental policy. He has been a member of the NRC Committee on Earthquake Engineering and the Committee on the Decontamination and Decomissioning of Uranium Enrichment Facilities. His most recent research focuses on conflicts in intergovernmental policy involving high-level nuclear waste disposal and the role of citizens in technological policy decision-making. He has published extensively on issues relating to siting controversies. Dr. Mushkatel is currently a member of the NRC Committee on Review and Evaluation of the Army Chemical Stockpile Disposal Program (Stockpile Committee).

Laurance Oden is a retired senior researcher in the Pyrometallurgy Subdivision of the Process Metallurgy Division of the Albany Research Center, U.S. Bureau of Mines, Albany, Oregon. Dr. Oden's expertise is in the fields of high-temperature phase equilibria, superconductivity, the corrosion chemistry of metals and nonmetals, the thermochemistry of high temperature reactions, and the processing of metals and slags. He has written or co-written 94 publications and formal presentations and is the holds 15 patents. Dr. Oden received his bachelor's degree in chemistry from Oregon State University and his Ph.D. from Oregon State in mathematics and metallurgy.

George W. Parshall is a member of the National Academy of Sciences and a retired member of the Central Research Department of E.I. du Pont de Nemours & Company where he served for nearly 40 years, including 13 years as director of chemical science. Dr. Parshall is an expert in conducting and supervising chemical research, particularly in the area of catalysis and inorganic chemistry. He is a past member of the NRC Board on Chemical Science and Technology and has played an active role in NRC and National Science Foundation activities. He is currently a member of the NRC Committee on Review and Evaluation of the Army Chemical Stockpile Disposal Program (Stockpile Committee).

L. David Pye is currently dean of the College of Ceramics at Alfred University. Having received his undergraduate degree at Alfred, Dr. Pye started as a research engineer in the Melting and Forming Laboratory of PPG Industries, followed by Army service and a stint at Bausch and Lomb. After completing graduate studies at the University of Rochester and Alfred, he embarked on a long and distinguished career at Alfred University. In the course of his rise from assistant professor to dean, Dr. Pye has published more than 70 technical articles, presented more than 100 lectures and papers, established numerous international symposia, and set up the first Ph.D program in glass science in the United States. Dr. Pye is a fellow of the American Ceramic Society and the American Institute of Chemists and many other professional societies. He received the Dominick Labino Award from the Glass Art Society in 1995 and numerous other awards.

Roger W. Staehle is currently an industrial consultant and adjunct professor of chemical engineering and materials science at the University of Minnesota. He is a member of the National Academy of Engineering and has received the Whitney Award from the National Association of Corrosion Engineers (NACE) for outstanding work in corrosion science. He was a dean of the Institute of Technology and professor of chemical engineering and materials science at the University of Minnesota. Before that, Dr. Staehle was a professor at Ohio State University. Dr. Staehle has organized the two largest centers of corrosion science in the United States, one at Ohio State, called the Fontana Corrosion Center, and the other at the University of Minnesota. He was appointed first chair in corrosion science and technology at Ohio State when he received the International Nickel Chair. He was an editor of *Corrosion Journal* and *Advances in Corrosion Science and Technology*, has edited 23 books, and has written 160 papers. He is a fellow of NACE and the American Society for Metals. He has been a reactor engineer with the nuclear submarine program and a consultant on the subject of corrosion and degradation for industries in all major fields in the United States and many foreign countries.

William Tumas is currently the group leader for the Waste Treatment and Minimization Science and

Technology Group at Los Alamos National Laboratory. He is a senior chemist known primarily for his science and engineering research on waste treatment and minimization. His work has included research and development technology, industrial waste applications, and environmental restoration for DOE. At Los Alamos he has studied supercritical fluids, oxidation, and organic transformations. Dr. Tumas has written numerous papers and is a member of several professional organizations.

Appendix J

Questionnaires Sent to Technology Proponent Companies and Environmental Regulators

The AltTech Panel developed a questionnaire to guide panel members as they gathered information during visits and subsequent interactions with the three TPCs (technology proponent companies) and the Army's Alternative Technology Program, which was treated as the proponent for the neutralization technologies. This appendix includes samples of the cover memo sent to the TPCs and the memo sent to the Army, as well as the questionnaire.

December 15, 1995

MEMORANDUM

TO: Technology Firms

FROM: Mike Clarke, AltTech Panel, Study Director

As currently planned, representatives of the NRC's AltTech Panel will be visiting each of you during the month of January. These visits will necessarily be brief and to the point, as the assessment team's time is limited. Thank you all for the support you have already provided.

The list of questions that follows is provided to each of you to facilitate discussion and to ensure that you have the opportunity to plan for the requisite company representation at the meetings. I make no assertion that the list is all-inclusive, that there are no redundancies, or that some of the information is included in your submissions; only that these represent the body of data sought. The assessment teams are free to range over a wide spectrum of pertinent subjects, but, clearly, if they receive clear and concise answers to this list, they will have achieved most of their data gathering goals. In preparing an agenda for this visit, please allow adequate time for this purpose, even if it is at the expense of other important activities such as tours or company briefings. Thanks in advance for your help.

Recognizing that the holidays are rapidly approaching, and I'd like to take this opportunity to wish you "Happy Holidays" and a safe and prosperous New Year, if you choose to answer some or all of these questions in writing either in advance or for delivery at the meetings, that would be very much appreciated. It might help you with your responses and reduce the amount of note taking the assessment teams will have to do.

There are, of course, other areas that will be investigated that do not involve the companies, including meetings with the Army, state and federal regulators, and the interested public. This process should be completed by March.

Attachment: Questionnaire

December 15, 1995

MEMORANDUM

TO: LTC Steve Landry, Chief
Applied Technology Branch

FROM: Mike Clarke, AltTech Panel, Study Director

As currently planned, representatives of the NRC's AltTech Panel will be visiting you during the month of January. For review of the Army's neutralization technologies, this visit is scheduled for 18 and 19 January at Aberdeen. The visit will necessarily be brief and to the point, as the assessment team's time is limited. Thank you all for the support you have already provided.

The list of questions that follows is provided to each technology proponent to facilitate discussion and to ensure that you have the opportunity to plan for the requisite representation at the meeting. I make no assertion that the list is all-inclusive, that there are no redundancies, or that some of the information is included in your submissions; only that these represent the body of data sought. The assessment team is free to range over a wide spectrum of pertinent subjects, but, clearly, if it receives clear and concise answers to this list, it will have achieved most of its data gathering goals. In preparing an agenda for this visit, please allow adequate time for this purpose, even if it is at the expense of other important activities such as tours or technology briefings. Thanks in advance for your help.

Recognizing that the holidays are rapidly approaching, and I'd like to take this opportunity to wish you "Happy Holidays" and a safe and prosperous New Year, if you choose to answer some or all of these questions in writing either in advance or for delivery at the meetings, that would be very much appreciated. It might help you with your responses and reduce the amount of note taking the assessment team will have to do.

Attachment: Questionnaire

QUESTIONNAIRE FOR TECHNOLOGY ASSESSMENT

1. Operational Requirements and Considerations

1.1 Feed Streams

- Has waste handling received attention so that one can be confident that there will be no surprises?
- What equipment is necessary for waste feeding and handling? At what scale has it been demonstrated?
- Is any pretreatment required? How are gels, solids, and other inhomogeneities handled and fed?
- How are the ton containers handled and what are the feed requirements to clean them?

1.2 Process Operation

For agent detoxification:

- What is the maximum residual concentration of agent in each process effluent?
- Materials and Energy Balance: What is the quantity (per unit of agent), physical state (gas, liquid, solid, slurry) and chemical composition (major components, unreacted reactants, organic reaction products, inorganic reaction products) for each process effluent? Specify for each agent type to be processed. What are the analytical detection limits for each species in each phase?
- Are any of the process reactions reversible to the extent that agent can be reformed?
- What type of toxicity evaluation, if any, has been carried out on process residuals?

For ton container cleanout:

- What is the proposed method for removal and detoxification of residual agent in bulk containers?
- How is detoxification/cleaning of ton containers ensured to 3X? 5X? What analytical methods will be necessary?
- How will the ton containers be managed (recycled, landfilled, etc.) after clean out?
- Materials and Energy Balance:
- What is the quantity (per unit of agent), physical state (gas, liquid, solid, slurry) and chemical composition (major components, unreacted reactants, organic reaction products, inorganic reaction products) for each process effluent (e.g. decontamination fluid)? Specify for each agent type to be processed. What are the analytical detection limits for each species in each phase?

1.3 Process Effluent Streams

For bulk agent and ton container cleanout:

- What is total amount of solid, liquid, aqueous, slurry and gaseous waste products produced from treatment?
- What is the proposed management scenario (e.g. aqueous discharge to wastewater treatment facility, solidification/stabilization, landfill, atmospheric emission, recycling) for each process effluent?
- What additional treatment will be required to achieve disposal requirements under the proposed management scenario? What testing has been carried out for these treatment requirements and at what scale and on what wastes?
- What commercial facilities have been identified as potential recipients for each effluent waste stream? What are the permit requirements for the proposed management option?

For non-process wastes:

- How will non-process wastes (e.g. entry suits, dunnage, facility decontamination fluids) be managed?
- What additional treatment will be required to achieve disposal requirements under the proposed management scenario? What testing has been carried out for these treatment requirements? What commercial facilities have been identified as potential recipients for each non-process waste type?

1.4 Process Instrumentation and Controls

- What are the process monitoring requirements, e.g. detection limits for the feed and product streams? For process control? For effluents?

- How stringent are the process monitoring and control requirements?
- Does proven monitoring technology exist to meet process control and effluent discharge requirements? What is the operational experience with these monitoring systems?
- If new monitoring technology is required, what is the status of its development?

2. Materials of Construction

2.1 System and Materials

- What is the overall system diagram of piping and components?
- What are the materials of construction of the piping and components? Alloys, specifications.
- Where are the welds and what is the state of their stress relief?
- What kinds of inspections of welds and joints are being made?

2.2 Environmental Chemistry

- What are the nominal chemical environments, temperatures, pressures, residual stresses, and flow rates in each of the pipes and components?
- What are the exterior environments for the piping and components, i.e., the environments on the side opposite the process side? E.g., insulation, relative humidity, atmospheric contamination, and leached chemicals?
- What is the major environment and its nominal composition?
- What impurities are in the environment?
- What kinds of crevices are there in the piping and components in terms of gaskets, tight geometries, thermal sleeves, weld under penetrations, surface deposits, and/or bottom deposits?
- Where are heat transfer surfaces? What are the heat fluxes? Is there any heat flux in crevice geometries such as at tube supports?
- What are the startup and shutdown procedures?
- What are the procedures for deoxygenating or similar steps on startup? What is the temperature change rate on startup?
- What is the design life of the system and materials?

2.3 Qualification of Materials in the Application

- What work has been done to qualify materials of construction for the design life in the way of corrosion and mechanical testing?
- If no laboratory work has been done, what literature references support the application of the materials?

2.4 Failure Definition

- What modes of failure have you considered for the various materials and components in your total system?
- What are the bases for considering the various failure modes?

2.5 Monitoring and Inspection

- What factors are you planning to monitor in the operating system? E.g., chemistry (what species?), temperature, pressure? Is the monitoring continuous or batch?
- How frequently will the system be inspected, and what locations are inspected for what observations?

2.6 Previous Experience

- What similar engineering or field experience is available on this or similar systems? What failures have occurred? What have been the results of inspections?
- What prototype facilities or laboratory systems have been operated using your system? What is their experience, operation time, failures, inspections?

3. Process Stability, Reliability and Robustness

3.1 Stability

- Can deviation from "normal" operation lead to an out-of-control situation where the system will find another operating regime that is quite different from the one desired?
- Are there process mechanisms, e.g. uncontrolled reactions, that could lead to a catastrophic facility failure? What are the safeguards against such events?

- How does the system respond/adjust to modest reaction condition changes, e.g. will a temperature rise lead to uncontrolled temperature increases.
- What is the total amount of stored energy in the system at any one time?

3.2 Reliability

- Does the mechanical equipment have a good record of performance?
- Are there backup systems to rescue the operation in case of failure of a component?
- How quickly will the backup system respond?

3.3 Robustness

- Will the process operate satisfactorily over a wide and varying range of operating conditions: temperature, pressure, energy input (mechanical, electrical, thermal) and composition of feed. How does the system respond to upsets in feed, reaction conditions or energy input?
- What control mechanisms are necessary to ensure operation with varying conditions and feeds?
- Will operation be continuous (days?, weeks?, months?) or intermittent? Which is better? Can other modes be employed?

4. Operations and Maintenance

4.1 Operations:

- What are the staffing requirements for normal operation? for normal shutdown/restart? for emergency shutdown/restart?
- What are the training requirements for staff (e.g. Ph.D electrical engineering vs. chemical plant operator vs. municipal sewage treatment operator)?
- What is the operational experience (documented) of the technology? On what kinds of wastes has the operational experience been obtained?
- What operational safeguards are built into the system?
- What control systems are necessary? What control systems have previously been demonstrated/employed? What does the control room look like?
- What experience is available on downtime vs. operational time? on what types of waste streams and at what scale?

4.2 8 hour versus 24 hour operation:

- Can the operation be reasonably run 8 hours a day? continuously for 24 hours a day?
- Does the system work better continuously or in an 8 hour operation shift?
- What are the requirements for shutdown/ready mode?

4.3 Startup/Shutdown:

- What is the procedure and can the system be shut-down and restarted with minimal upsets during normal operations?
- What are the procedures for emergency shutdown?
- What is the procedure for restarting after emergency shutdown?

4.4 Maintenance:

- What routine maintenance is required for normal operation?
- What documented record of performance is available concerning operation and maintenance of equipment? How much down-time is typical for normal operation? What is the operation/maintenance history of the technology?
- Are maintenance manuals and documented procedures available?
- What is the lifetime of equipment and what are the main consumables? What is the documented record of performance of equipment? How is the equipment replaced or maintained?
- What measures are taken to assure worker safety/exposure during routine maintenance?
- What staffing is required for normal maintenance?

5. Utility requirements

- What are the electrical, water and fuel requirements for the process?

6. Scale-Up Requirements

- What is the state of development of the process? How novel is the process?
- What scale of operation has been demonstrated and on what types of waste? To what extent will the plant be modified from the largest operation demonstrated?
- To what extent is the process, or parts of it, demonstrated commercial technology?
- Has the process been demonstrated with agent, i.e. feedstock and range of feedstock anticipated for the plant?
- To what extent have processes that would be used for ton container cleanout been demonstrated and on what types of waste?
- Very high conversions, e.g. 6 nines, will be required. How does the reactor design allow for this, e.g. batch reaction, staged reactors, etc. Will it change with scale-up? How does the system scale with mass, volume? What are the economic scaling factors?
- How much is understood about mass and energy transfer and will there be differences in "mixing" and "heat transfer" between small and large scale equipment (e.g. impellers and vessel size/shape, flow Reynolds number, Froude number (2-phase))?
- Has any catalysis been adequately demonstrated over reasonable time of operation with the range of feeds and possible poisons that will be encountered? How is system regenerated after poisoning?
- How well are the reaction mechanisms and intermediates understood for the destruction process? Frequently a reaction requires an "intermediate" that is built up during the reaction itself; the reaction may exhibit an "induction period" as a consequence. Is the reaction mechanism understood well enough to anticipate this?
- How many unit operations are involved in the entire treatment process, including treatment of secondary wastes?

7. Facility Decommissioning

- How will the disposal facility be decommissioned?
- What wastes (type and quantity) will be generated from facility decommissioning and how will they be managed?

8. Process Safety

8.1 Plant Safety and Health Risks

Risk of catastrophic failure and agent release:

- What are the possible modes of failure in feed systems, equipment, process operations, and monitoring systems that could give rise to a sudden release of agent?
- What influence could external factors have on the possibility of agent release (e.g. earthquake, vibration, ambient temperature, humidity, electrostatic discharge)?
- What measures can be taken to prevent the sudden release of agent and/or processing products? What is the proposal for secondary containment?
- What measures can be taken to mitigate the effects of an agent release on base personnel and the surrounding population if an agent release does occur?

Risk of exposing plant workers to agent:

- What are the possible modes of exposure of workers to agent over the duration of the disposal program?
- What is the expected level and duration of exposure for each of the identified modes?
- What are the known human and health effects of such exposures?
- What can be done to prevent worker exposure?
- Risk of plant worker exposure to other hazardous chemicals:
- What other hazardous chemicals could workers be exposed to?
- What are the associated human health effects at the possible levels and durations of exposure?
- What can be done to prevent worker exposure, and to mitigate the effects of exposure if it does occur?

8.2 Community Safety, Health and Environmental Risks

Risks of agent release and exposure due to normal operations:

- What are the possible sources and duration of agent release during normal operations?
- Following release, what are the pathways of agent migration outside of plant boundaries?
- What are the possible routes, levels and duration of exposure?
- What are the health effects that might result?
- What damages might result to natural resources and man-made structures?

Other risks due to normal operations:

- What other hazardous chemicals could be released during normal operations?
- What are the possible sources and duration of such releases?
- What are the pathways of migration outside of plant boundaries?
- What are the possible routes, levels and duration of exposure?
- What are the health effects that might result?
- What damages might result to natural resources and man-made structures?

Risks due to abnormal events:

- What is the largest possible release of agent?
- What is the largest possible release of other hazardous chemicals?
- How large an area would be affected?
- What are the possible adverse effects on human health and the environment?
- What emergency preparedness and emergency response measures can be taken to mitigate adverse effects?

Accident risk assessment:

- Has an accident risk assessment been done? If so, what were the results?

Health and environmental risk assessment:

- Has a health and environmental risk assessment been done? If so, what were the results?

Liability insurance:

- What type of liability insurance, if any, covers the use of the proposed technology?
- Has a risk assessment been conducted in support of an application for insurance? If so, what were the results?

9. Schedule

- What is the schedule for pilot-scale design and construction?
- What is the schedule for pilot-scale testing and evaluation?
- What is the effect on facility construction of scale-up requirements from pilot-plant to commercial operations?
- What is the time required for facility construction?
- What is the time required for facility systemization?
- What is the effect on facility construction imposed by regulatory requirements? Permitting requirements?
- What is the effect of public acceptance on technology implementation?
- What is the expected duration of operations?
- What is the schedule for facility closure and site remediation?

REGULATORY REVIEW AND PERMITTING IMPACTS

The following questions were solicited from the AltTech panel members as those they would most like to discuss in meetings with state environmental officials in Maryland and Indiana. There are redundancies, but to avoid the omission of any subtleties, the questions are included as written. The general intent is to determine the extent to which regulatory and permitting impacts may affect the eventual use of the five alternative technologies under consideration by the Army and being evaluated by the NRC. There is no prioritization to the questions.

1. What are the primary restrictions on quantity, composition and toxicity for aqueous waste disposal?

2. What are the primary restrictions on quantity, composition and toxicity for solid waste disposal from a chemical agent destruction facility at a land disposal facility? What testing is required to verify attainment of requirements?

3. What are the primary restrictions on quantity, composition and toxicity for atmospheric emissions from a chemical agent destruction process (combustion and non-combustion emissions)? What testing is required to verify attainment of requirements?

4. What information will the regulatory agencies require to approve the use of these technologies?

5. What are the permitting requirements and schedule for treatment technology systems? Have these systems been permitted on other wastes and at what scale?

6. How can the AltTech Panel obtain a copy of the state or federal regulations governing the management of hazardous wastes, water effluents, and air emissions?

7. Has your state been granted authority to administer: the RCRA program? NPDES permits? Air permits? What role does the EPA play?

8. Are mustard and/or VX listed as hazardous under state regulations?

9. If not, what would each state need to know to determine whether they are characteristically hazardous under state regulations?

10. Would a state RCRA permit be required to treat mustard (Maryland) or VX (Indiana) for each of the following processes:

    - neutralization
    - high temperature/high pressure hydrogen reduction
    - low temperature/ambient pressure electrochemical oxidation
    - molten metal bath agent destruction with recoverable byproducts

11. What specific regulatory subtitles would apply to each of the above treatment processes?

12. If any of the treatment processes would have to be permitted under the state equivalent of RCRA Subtitle X, what experience has the state had in Subtitle X permitting?

13. What are the steps involved in applying for a RCRA permit? How long does the process typically take from submission of an application to final approval?

14. Are there different regulatory requirements for full-scale treatment and for bench or pilot-scale treatment for purposes of R&D? If so, what are the limitations on throughput under an R&D permit?

15. What are the steps involved in applying for an air, NPDES or SPDES permit? How long does the process typically take from application submission to final approval?

16. What are the steps involved in applying for a construction permit?

17. What additional permits or approvals would be required prior to startup of operations?

18. M4 Environmental asserts that its technology is a recycling process. Do the states of Maryland and

Indiana concur with this characterization, or is a RCRA permit required? What about the other technologies?

19. What air permits are required for a gas turbine/power generator fired with syngas (from agent)?

20. Would it allay concerns if the agents were neutralized before being treated by the alternative technologies?

21. Are there reasons to believe that any of these technologies would be prohibited in Maryland or Indiana?

22. Would your state permit shipment of hydrolysate produced by agent neutralization to a toxic waste treatment facility?

23. How does the Clean Water Act provision restricting disposal of agent-derived waste into navigable waters affect the disposal of agent hydrolysate in your state?

24. How will the combustion of off-gases from the Eco Logic process be regulated?

What restrictions will be placed on the NO_x emissions from the AEA process?